1986

Communication
Graphics

Communication Graphics

WENDELL C. CROW

California State University, Fullerton

Prentice-Hall, Englewood Cliffs, New Jersey 07632

Library of Congress Cataloging-in-Publication Data

CROW, WENDELL C. (DATE)
  Communication graphics.

  Bibliography: p.
  Includes index.
    1. Communication—Graphic methods.   2. Editing.
3. Printing, Practical.   4. Graphic arts.   I. Title.
P93.5.C76 1986   001.55'2'022   85-12104
ISBN 0-13-153792-X

Editorial/production supervision: Hilda Tauber
Cover concept by Wendell C. Crow
   Color photography by Ron Cable;
   cover props: Pace Graphics;
   color separations: Orbus Graphics,
   Fullerton, CA
Manufacturing buyer: Harry P. Baisley
Art production: Gail Cocker, Peggy Finnerty

Printed in the United States of America

10 9 8 7 6 5 4 3 2 1

ISBN 0-13-153792-X   01

PRENTICE-HALL INTERNATIONAL (UK) LIMITED, *London*
PRENTICE-HALL OF AUSTRALIA PTY. LIMITED, *Sydney*
PRENTICE-HALL OF CANADA INC., *Toronto*
PRENTICE-HALL HISPANOAMERICA, S.A., *Mexico*
PRENTICE-HALL OF INDIA PRIVATE LIMITED, *New Delhi*
PRENTICE-HALL OF JAPAN, INC., *Tokyo*
PRENTICE-HALL OF SOUTHEAST ASIA PTE. LTD., *Singapore*
EDITORA PRENTICE-HALL DO BRASIL, LTDA., *Rio de Janeiro*
WHITEHALL BOOKS LIMITED, *Wellington, New Zealand*

Contents

TEXT PREPARATION: Editing, Marking, and Copyfitting 57

GRAPHIC IMAGES: Powerful Partners of Words in Print 75

TEXT AND IMAGE PROCESSING: Capturing Words and Pictures Through Electronics 105

8 MODERN DESIGN PROBLEMS: From Identity Symbols to Effective Page Structures 136

9 PUBLICATIONS DESIGN: Planning a Creative, Cohesive Visual Package 154

10 PASTEUP: Assembling Text and Graphic Images for Reproduction 186

Preface

This book is written for everyone interested in the fundamentals of communicating through graphics in print. Those studying, teaching, supervising, or working professionally in all areas of graphic design, publications editing, or printing should find useful information here.

Emphasis has been placed on providing current, practical information for the reader. Each chapter or subject area focuses on the key concepts, explains the terminology, and discusses in detail the tools, machines, and materials used. Most chapters also explain professional applications of the content, and include practical examples, problems, and solutions from the working world.

In our fast moving, media-intensive society, nearly everyone today has an ingrained, if subconscious, understanding of modern graphic communication. Advertisements—perhaps the most sophisticated applications of mass-audience graphics in television, magazines, newspapers, and other media—have become such a pervasive force that hundreds, perhaps thousands, enter our homes weekly. During the past two decades we have witnessed a great popular surge of creative photography, contemporary graphic design, electronically generated imagery, and other phenomena, all helping to produce a populace for which few visual surprises remain.

Today, satisfactory application of a person's natural appreciation of and aptitude for communications graphics probably is limited more by a lack of technical expertise and knowledge than by deficiencies in creative thought. By stressing a breadth of practical graphics preparation and experience, this book seeks to help readers discover and develop whatever talents they possess.

No claim is made that everyone can become a professional graphics specialist, nor that a single book can make this possible. Serious students of communications graphics will consult many of the books that inspired this one, and will examine critically the works and philosophies of successful designers, artists, and editors. Those wishing to become professionals should be prepared to spend several years in concentrated academic study or professional practice in the field. This book is intended to provide either the beginning of, or a useful supplement to, serious study of the subject.

Most people, while not wanting to become professional designers, do need a working familiarity with graphics fundamentals. Thousands of publications editors, public relations practitioners, photographers, business executives, students, and other communicators face graphics problems regularly, sometimes daily. Nearly all of us, too, find ourselves confronted with planning and executing some form of printed communication, and a trip to a printing plant can be a frustrating, if not frightening, experience for the uninitiated. An appreciation of the mutually dependent relationship between graphics and printing will help make the experience of communicating design and printing needs to professionals much more satisfying. Certainly, it will make the final printed piece more successful.

Early artists and designers often became their own printers by necessity, priding themselves on learning the finest technical secrets of their craft. They knew a beautiful work of art or creative design could be no better for a mass audience than its lowest common denominator, the printed reproduction.

Today, a massive printing and publishing industry—the tenth largest in the United States—provides the highly sophisticated reproductions still demanded by artists, photographers, designers, editors, advertisers, and consumers. This means that today, more than ever, those who prepare materials for print must learn to speak the language of the printer. Rich in the colorful heritage of humankind's evolving love affair with the printed word, this language is at once specific, confusing, even charming. It is also *necessary*, as anyone with a storage closet piled high with less-than-satisfactory printed matter can verify.

A disappointing printed piece is the operational definition of "garbage in, garbage out," computer slang for "poor communications means poor printing." Should readers of this book retain nothing but a useful, accurate graphics language, they will be much more confident in their relations with graphics professionals.

Because an understanding of printing is so vital for successful work in graphics, this book includes a special section (Chapter 15) on a subject rarely discussed in general graphics books—the procedures, practices, and attitudes of printers and the professionals with whom they work. The reader will find a full discussion of trade customs, information on estimates, quotations, and contracts, even an overview of pricing economics and how the computer has revolutionized this and most other aspects of the industry.

In short, this book seeks to address the subject of graphics communications from the logical, practical point of view of the working professional.

ACKNOWLEDGMENTS

Many people gave their time and support to this endeavor. I regret that the list is far too long for me to include everyone, but I must thank the following individuals for contributions of critical importance to the success of the book:

• Steve Dalphin and Hilda Tauber, my editors at Prentice-Hall, who guided me through a seemingly endless series of hurdles in search of a quality book.

• Teachers of journalism and communication who reviewed the manuscript and made helpful suggestions include: George Brown, Southern Illinois University at Carbondale; William Korbus, University of Texas at Austin; Bryce T. McIntyre, Stanford University, California; and Dr. Paul W. Sullivan, Temple University, Philadelphia.

• My colleagues at Cal State Fullerton—particularly Ed Trotter for persuading me to do it in the first place, Russ Romain for constant encouragement (and reminding me of Michener's thoughts on the value of learning the proper names of tools), and both Pat O'Donnell and Dave DeVries for their excellent photographic work.

• Local professionals Gregg Green of Smith Printing, Tustin; Dana Courdrey of Premier Printing, Fullerton; and Mike and Jack Patten of Creative Press of Orange County, Anaheim, for technical assistance and advice.

• Douglas Covert, former colleague and now photojournalist/professor at Indiana University, Indianapolis, for his major contribution to the writing of Chapter 6 on graphic images.

• Dr. Mark Guldin, dean of the College of Graphic Arts & Photography, Rochester Institute of Technology, for making available the paintings which illustrate the history of graphic communication in Chapter 1.

• International Typeface Corporation, for use of the ITC alphabets in Chapter 3 and Appendix A, and Bert De Pamphilis of World Typeface Center, for helpful information on typographic design.

• The American Paper Institute, Inc. and Stuart J. McCampbell, manager, for suggestions regarding paper technology; also, the Graphic Arts Technical Foundation, especially Ray Blair, editor-in-chief, for permission to use images and technical descriptions from *What the Printer Should Know about Paper* by William Bureau in Chapter 13. Also to William H. Frohlich of the National Paper Trade Association, Inc., for many helpful sources on paper.

• Rose Marie Kenny of Hammermill Paper Group, Hammermill Paper Company, for photographs depicting paper manufacturing stages; and Westvaco Corporation's William Henchey for providing an excellent papermaking flow diagram.

• James E. Renson of the National Association of Printing Ink Manufacturers, Inc., and the Sun Chemical Corporation, Printing Ink Division, for furnishing important images and information in Chapter 13.

• Elinor Selame, president of Selame Design, for her encouragement and permission to use the unqiue corporate identity system depicted in *Developing a Corporate Identity: How to Stand Out in the Crowd.*

• W. G. Fredrickson, vice-president, technology, Harris Corporation, for information about his research on the future of communication graphics.

• Stephen Dunbar of Penta Systems International Inc., Barbara S. Yagerman of Compugraphic Corporation, Cliff Kolovson of Atex, Inc., Tim Malett of Autologic Corporation, and Janis Friesen of Mergenthaler Linotype, for extensive material on typesetting systems.

• Mario Garcia of The Poynter Institute for Media Studies, for permission to use diagrams from his book, *Contemporary Newspaper Design.* Also, Roger Fidler, former editor of The Society for Newspaper Design

Notebook, for his design checklist; and to Louis Silverstein of *The New York Times*, Francis Pike of the Columbia (Missouri) *Daily Tribune*, and Christian Anderson of the *Orange County Register*, for permission to reproduce several page designs from their papers.

• Christopher Mayhall of Byers Corporation, William C. Bollinger of Bernal Rotary Systems, Inc., Burton B. Ritchey of H. Whitney Cutler Advertising, Roger R. Reinke of Screen Printing Systems, Inc., Christine Skrak of Miller Printing Equipment, H. M. Williams of Brandtjen & Kluge, Inc., G. J. Meyers of Consolidated International Corp., Bruce Mabel of Heidelberg West, Inc., John Jacobsen of Allied Gear and Machine Co., Vincent Smart of Dow Chemical U.S.A., Jenny Wade of Harris Corporation, Robert W. Layton of The Gerber Scientific Instrument Company, Janice Wilcox of BASF Systems Corporation, and John Klustner of *The Denver Post* for substantial contributions of material to the chapter on printing processes.

• Hammermill Paper Company, Joyce Schwartz of Pantone, Inc., Frances J. Sullivan of Eastman Kodak Company, Joan Monahan of HCM Graphic Systems, Inc., and Y. Tsuji of Dainippon Screen Manufacturing Company Limited for furnishing superb color originals, films, and technical information regarding color reproduction.

• Fred B. Dempsey of S. D. Warren Paper Company for permission to use several charts and diagrams based on information from their fine booklet, *The Warren Paper Estimating Guide*.

• Andy Gritzuk of Stewart Systems Corporation, and Elman Snow of Porte Publishing Company for material on estimating printing costs.

Thanks are also due to the many companies and organizations which allowed the reproduction of their advertisements, photographs, page designs, and other materials that enhance this book.

Finally, I am grateful to my students, who make such a massive undertaking bearable. May we continue learning from one another.

This book is dedicated to my wife, Jan, who unfailingly believed in this project and sacrificed much to help me see it through; to my children, Chris, Amy, Brook, and Jason, who understood far beyond their years when "the book" beckoned; and to my parents, Wendell and Betty Crow, from whose example I learned the value of work and the worth of my profession.

Graphic Communication

Where We've Been
Where We Are
Where We're Going

Graphic communication is as old as recorded history. The story of its evolution from simple pictographic symbols to elegant alphabets and our painfully slow progress toward practical reproduction of graphic images is truly a chronicle of civilization.

Historic Highlights

The following time line is a brief overview of highlights in the fascinating struggle to record human communication. Perhaps it will help us remember that most of the wondrous communication tools we take for granted stem from centuries-old human ingenuity and imagination.

First Use of Symbols—35,000–15,000 B.C.

Early peoples in what is now France and Spain lined the walls of caves with simple *pictographs* of animals, hunters, and natural phenomena. Hunters may have believed that the symbols gave them special powers over the animals depicted. The images were probably used to record successful kills and other events in daily life.[1]

First Exact Reproduction of Images—4500–3500 B.C.

Prior to the invention of writing, *stamp seals*, made from stone or metal, were carved in relief and used to impress ownership marks into moist clay. About 3500 B.C., *cylinder seals* containing duplicate relief symbols introduced what eventually became the principle of the modern rotary press. When held between the fingers and rolled across soft clay, these cylindrical stones produced a permanent, repetitive image pattern when the clay hardened.[2]

Earliest Writing—3300 B.C.

Ancient Sumerian scribes used a wedge-shaped stylus to press *cuneiform* images into clay. Simple pictographs were combined to form different expressions, syllables, and sounds. Later, more complex ideograms expressing concepts such as "to conquer" were introduced.[3]

About the same time, the Egyptians began to develop pre-alphabetic symbols called *hieroglyphics*, or "sacred engravings." These symbols were used as concept ideograms, as phonograms to indicate word sounds, or to explain other symbols.[4] About 3200 B.C. the Egyptians also introduced *papyrus*, a writing substrate somewhat similar to rough paper. Made from cross-hatched strips of papyrus reed, it accepted almost any kind of ink. Papyrus was made in both royal and popular varieties, and its use lasted about 42 centuries.[5]

Movable Relief Images for Inkless Printing—1700 B.C.

In southern Asia Minor, use of typelike relief symbols impressed into a soft clay was introduced. The

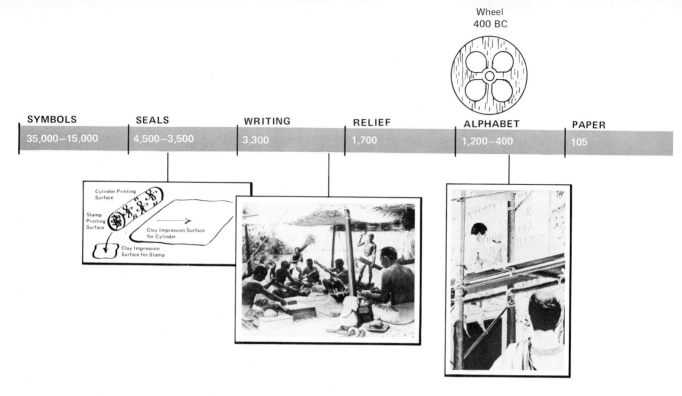

Wheel
400 BC

SYMBOLS	SEALS	WRITING	RELIEF	ALPHABET	PAPER
35,000–15,000	4,500–3,500	3,300	1,700	1,200–400	105

3500 B.C. Stamp/cylinder seals.
Reprinted with permission from Karlen Mooradian, "The Dawn of Printing," Journalism Monographs No. 23, May 1972. Copyright the Association for Education in Journalism and Mass Communication.

3200 B.C. Making papyrus.

500 B.C. The Latin alphabet.

circular pattern of syllables found on the *Phaestos disk* on the island of Crete suggests an alphabetic structure and the first known use of reusable images similar to relief types.[6]

Origins of the Latin Alphabet—1200–400 B.C.

A Semitic-Egyptian race known as the Phoenicians developed an alphabet of some 22 symbols about 1200 B.C.[7] Around 900 B.C., the Greeks adopted the system and added vowels. By 500 B.C., the symbols had fallen into the hands of the Romans who refined it into a form nearly identical to today's *Latin alphabet*.[8]

The Invention of Papermaking—105 A.D.

Ts'ai Lun, a Chinese court official, discovered how to make *paper* from a pulp mixture of mulberry bark, hemp, linen rags, and water spread on a mat and sun-dried. Muslims carried the art to Europe in 751. About 20 centuries elapsed between the invention of paper and its modern manufacture, primarily from wood fiber.[9]

The First Printing Ink—200

The earliest writing inks—made from charcoal and glue—probably originated with the Egyptians who required a substance for use on papyrus.[10] Between 200 and 400 the Chinese refined this formula into one closer to today's *printing ink*, a combination of lampblack pigment and an oil vehicle.[11]

The First Mass Printings—764–868

While historians credit the Chinese with wood block printing at least as early as 400 A.D., the Japanese Empress Shotoku commissioned the first *mass publication*—a million copies of wood block-printed sheets containing Buddhist prayers—between 764 and 777. The Chinese, however, produced the first block-printed book, *The Diamond Sutra*, in 868.[12]

First Movable Type for Printing in Ink—1043–1403

Centuries before Gutenberg, *movable types* were used in printing. A Chinese alchemist, Pi Sheng, invented types made from baked clay about 1043. A font of 60,000 wooden types was produced for the Korean ruler Wang Chen in 1313, and a bronze type foundry existed in Korea by the year 1403. Success of these type fonts was greatly limited because of the large number and complexity of pieces required for printing oriental languages.[13]

2

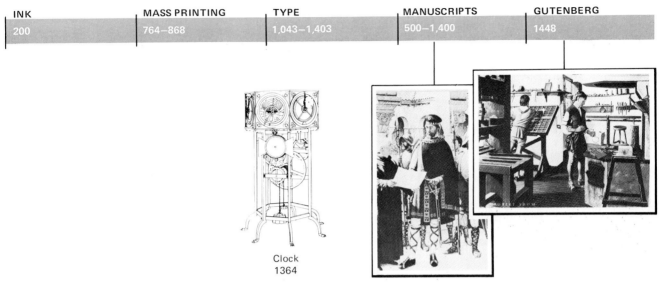

INK	MASS PRINTING	TYPE	MANUSCRIPTS	GUTENBERG
200	764—868	1,043—1,403	500—1,400	1448

Clock
1364

Ninth century A.D. *Charlemagne and scribe.*

1448. Gutenberg.

The Preservation Role of Manuscripts— 500–1400

Following the decline of the Roman Empire, civilization entered a period of stagnation known as the Dark Ages. Monks in St. Benedict's sixth century monastic orders were set to work hand-lettering *manuscripts* which helped preserve knowledge. Illuminated initial letters and decorations in red, blue, and gold were added to enhance the meticulous calligraphy of the scribes.[14] Charlemagne, the ninth century King of the Franks, is credited with having his scribes similarly record for all humanity the culture of ancient Greece and Rome.[15] A single volume of a twelfth-century German manuscript, *The Gospel of Henry the Lion*, recently brought $12 million at auction.

The First Practical Movable Type and Printing System—1448

Johann Gutenberg, a German stone cutter, is generally given credit for inventing the first really practical method of casting individual pieces of *movable type*. He not only devised a method to produce precise type molds and punches, but also invented a formula for molten metal casting, converted a wine press for printing, and even manufactured his own ink and paper. His famous 42-line Bible is dated 1448, but he is thought to have printed the poem *Judgement Day* in 1444.[16] The art of typography is also attributed to Gutenberg, who painstakingly designed his types to simulate the hand-lettered text of his time.

Gutenberg is remembered today for bringing together and refining the elements necessary for true printing.

Intaglio Printing—1446

At about the same time Gutenberg was perfecting his inventions for relief printing, a process employing exactly the opposite principle—that of printing from an engraved or depressed surface—was being developed. Labelled *intaglio* (pronounced inTALyo), this process was used for copper engraving and was performed as early as 1446. By carefully inscribing (scratching) depressions into the soft surface, metalsmiths produced intricate designs in armor and shields. Ink placed on the finished engraving and wiped off the top surface would remain in the incised depressions. This ink could then be transferred to paper pressed against the image.[17]

Two modern printing methods, steel engraving and rotogravure, use the intaglio principle.

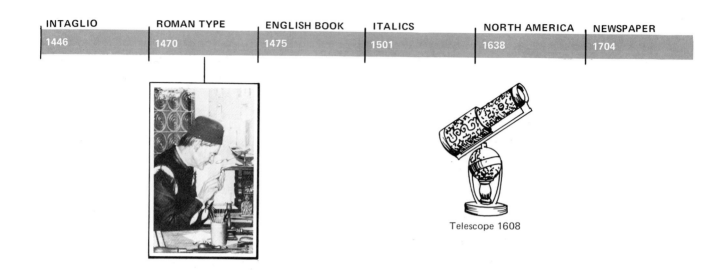

INTAGLIO	ROMAN TYPE	ENGLISH BOOK	ITALICS	NORTH AMERICA	NEWSPAPER
1446	1470	1475	1501	1638	1704

Telescope 1608

1470. Jenson—first Roman type.

First Roman Type—1470

While Gutenberg's type resembled hand-lettered styles of the period, Nicholaus Jenson, a French mint master, designed the first *roman typeface* with thick and thin strokes and serifs.[18] His alphabets, based on simplicity and beautiful stroke and space proportions, still influence type design five hundred years later.

First English Language Book—1475

William Caxton, a retired merchant, brought printing to England in 1476, one year after he had printed the *first book in English* at a printing establishment in the Netherlands. Caxton is noted for his printed translations of classic literature.[19]

First Printing in Italic Type—1501

Aldus Manutius, an Italian printer and student of literature, established a printing house in Venice about 1489. There, he printed the works of Homer, Plato, Aesop, and others. In 1501, he printed the first book containing *italics*, a right-leaning variation of roman type more reminiscent of cursive handwriting.[20]

Printing Comes to North America—1638

Sailing from England with his employer, a clergyman named Jose Glover who died on the journey, Steven

Daye was faced with the task of establishing the printing operation Glover wanted. He set up Glover's press in Cambridge near Boston. His *Freeman's Oath*, produced on a single sheet, is thought to be the first printing in North America.[21]

The First Newspaper in America—1704

The Boston News-Letter, published in 1704, is usually considered the first true *American newspaper*, although Benjamin Harris had printed a single issue of a one-page *Publick Occurrences* in 1690. Established by John Campbell and printed in the shop of Bartholomew Green, the *News-Letter* lasted until 1776.[22]

Giants of Early Type Design—Eighteenth Century

Three great names in early type design were William Caslon and John Baskerville of England, and Giambattista Bodoni of Italy. Each developed widely acclaimed *type designs* that bear their last names and became the forerunners of major divisions of roman type. Caslon, produced about 1734, is a perennially popular old style design. Bodoni, 1788, is a modern style, with heavy stroke contrast and thin serifs. Baskerville, 1757, represents a transitional stage between the two.[23]

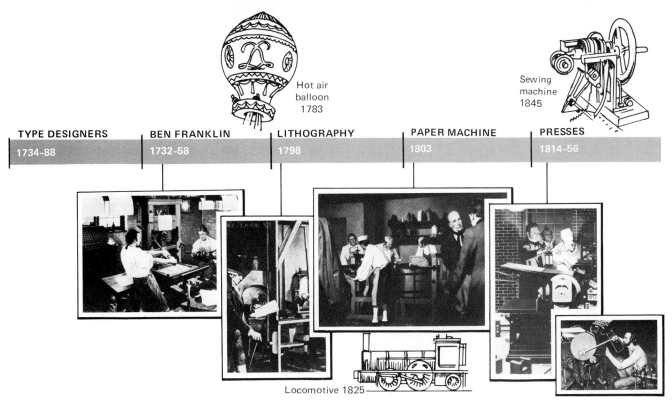

Hot air balloon 1783

Sewing machine 1845

TYPE DESIGNERS	BEN FRANKLIN	LITHOGRAPHY	PAPER MACHINE	PRESSES
1734–88	1732–58	1798	1803	1814–56

Locomotive 1825

1732. Ben Franklin, colonial printer.

1798. Alois Senefelder invents lithography.

1803. The Fourdrinier papermaking machine.

1814. Koenig and his cylinder press.

1856. Gordon's platen press.

Benjamin Franklin, Colonial Printer— 1732–1758

Most reknowned of early American printers, Benjamin Franklin was also a famous patriot, scientist, inventor, writer, and statesman. Franklin's *Pennsylvania Gazette*, begun in Philadelphia in 1732, and *Poor Richard's Almanac* ensured his prominence in colonial society long after he left the printing trade.[24]

Lithography Is Born—1798

Credit for today's most common printing process belongs largely to Alois Senefelder, a German playwright. He discovered that images sketched with a greasy pencil on a porous stone would repel water and accept ink that could be transferred to paper. Called *lithography* (stone writing), the principle is often labelled planography, since a flat surface is used. In 1905, while working with metal plates in common use by then, Ira Rubel discovered that lithographic images work better when printed first on a rubber blanket, then transferred to paper. This was the beginning of *offset* printing.[25]

The First Papermaking Machine—1803

Nicholas Louis Robert, a clerk in a paper making mill in Essenay, France, devised the first machine to produce a continuous roll of paper instead of handmade sheets in 1798. This hand-cranked

papermaking machine was later perfected and manufactured by the Fourdrinier brothers of London, after whom the device is named.[26]

Press Improvements—1814–1856

Mass production of printed material became easier in the nineteenth century through a series of improvements in *press designs*. Frederich Koenig sold the first power-driven cylinder press to *The London Times* in 1814. It greatly speeded printing by carrying paper in a circular fashion around a cylinder to contact a flat form of type below.[27] In 1846, R. Hoe & Co. produced a type-revolving press, the first to use a rotary (cylinder-to-cylinder) principle. Its design required that forms containing loose pieces of type be wrapped around one cylinder. This design was replaced by Hoe in 1871 with a true rotary press which used single-piece stereotype plates.[28]

George P. Gordon devised a treadle-operated version of today's platen presses in 1850. In 1856, William Bullock produced the first web-fed, perfecting press— capable of printing on both sides of a continuous roll of paper.[29]

Important Platemaking Discoveries— 1815–1861

A young goldsmith and coin maker for the State of Massachusetts discovered in 1815 the method by

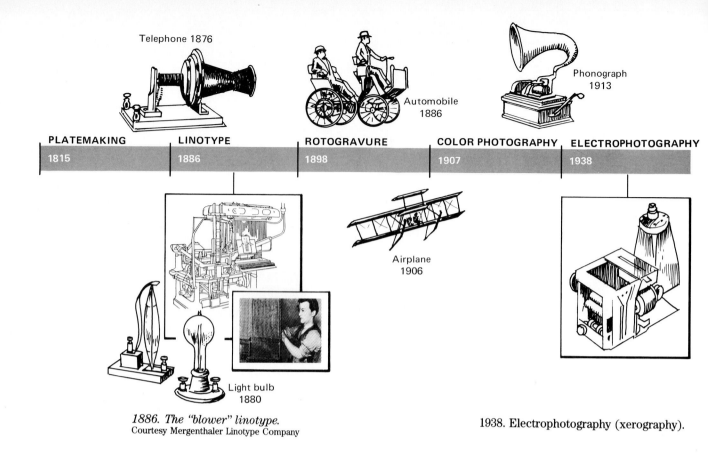

Telephone 1876

Automobile 1886

Phonograph 1913

Airplane 1906

PLATEMAKING	LINOTYPE	ROTOGRAVURE	COLOR PHOTOGRAPHY	ELECTROPHOTOGRAPHY
1815	1886	1898	1907	1938

Light bulb 1880

1886. The "blower" linotype.
Courtesy Mergenthaler Linotype Company

1938. Electrophotography (xerography).

which money, stamps, and fine printing is now produced—*steel plate engraving*. Jacob Perkins found a way to harden and soften steel plates so that when a design engraved in hardened steel was pressed into soft steel, duplicate plates could be made that could themselves be hardened.[30] *Photoengraving*, the use of photographic methods to etch a printing plate, was first done by Joseph Niépce about 1824.[31] *Halftone photography*, the method used to break pictures into tiny dots for printing, was discovered by William Talbot in 1852.[32] It was not until 1878 that Frederic E. Ives used a ruled glass screen to make a relief halftone plate and print photographs on paper.[33]

The Linotype Machine—1886

Truly one of the most significant inventions of all time was the automatic, hot metal *linecasting machine*, the Linotype, which cast full lines of type instead of individual pieces. Perfected by Ottmar Mergenthaler after significant previous work by many others, this "miracle of cams" completely revolutionized publishing, allowing far more pages to be printed in each issue of newspapers and magazines. Despite its unparalleled significance when invented, the Linotype is almost completely obsolete a century later.

The Appearance of Rotogravure—1898

Printing from an etched copper cylinder using the intaglio principle was first done by Karl Klic (Klisch). This process is known as *rotogravure*, and utilizes thousands of tiny cells engraved to different depths, depending upon how much ink is needed for the density of desired images. It is one of four major printing methods used today.[34]

Color Photography—1907

The first practical color process—Autochrome—was invented by Louis and Auguste Lumiere in 1907. Their "additive" process required that the exposure be made to a glass plate coated with a screen pattern of red, green, and blue dots. The technique owed much to discoveries made earlier by James C. Maxwell.

Kodachrome®, which uses the popular "subtractive" color process, was developed in 1936.[35]

Electrophotography—1938

The process that helped make possible a copier—and soon, perhaps, a printing plant—in offices everywhere

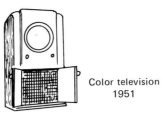

Color television
1951

COMPUTER	PHOTOTYPESETTER	LASER	COLOR SCANNER
1946	1946	1953	1955

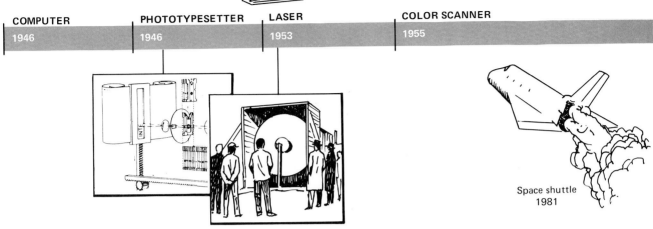

Space shuttle
1981

1946. The Fotosetter.
Reprinted with permission from *The Printing Industry* by Victor Strauss © 1967 Printing Industries of America, Inc.

1953–60. Masers/lasers.

was invented in 1938 by Chester F. Carlson. Xerography, or "dry writing," was accomplished by reflecting the image of an original document from a mirror onto an electrically charged selenium drum. A copy was created as special dry-ink powder, attracted to the dark image on the drum, was transferred and fused to paper.

Electrophotography has become increasingly important in "demand" printing as electronic imaging and sophisticated bindery and finishing attachments have made the machines competitive with small duplicating presses and traditional binding methods.[36]

The Modern Computer—1946

In 1941 a German engineer, Konrad Zuse, produced the first successful "universal adding machine" with electromagnetic relays. His original model was destroyed in an air raid. A forerunner of the first real computer is credited to Harvard University professor Howard H. Aiken, in 1944, with assistance from International Business Machines Corporation (IBM). In 1946 J. G. Brainerd, J. Presper Eckert, and John W. Mauchley supervised the project at the University of Pennsylvania, Philadelphia, that helped usher in the computer age with their Electronic Numeral Integrator and Calculator (ENIAC).

Innovations such as integrated circuits, transistors, and microprocessing chips have so dramatically affected the size, speed, and capacity of computers that they now have evolved into tools of everyday life. They are also essential to high-speed typesetting and many other functions in communications graphics.[37]

The First Commercial Phototypesetter—1946

The first practical typesetter that required no hot metal to produce images was the Fotosetter, tested in the United States Government Printing Office in 1946 by Intertype Corporation. While this machine's outward appearance resembled the hot metal linecasters of the day, its imaging system was quite different. The small brass matrices (mats) which circulated inside the machine contained not tiny letter molds but negatives, each with an *image* of a letter or number. As a matrix passed a specified point in its travel through the machine, a beam of light flashed through the negative in its side, exposing the image onto photographic paper.

The Fotosetter, which enjoyed a brief success, was soon displaced by photo-optic systems of simpler design.[38]

Masers and Lasers—1953–1960

An American physicist and engineer, Charles H. Townes, developed a device in 1953 that was capable of concentrating and amplifying microwaves, thus making possible highly accurate transmission of communications signals over long distances. His first *maser* (*m*icrowave *a*mplification by *s*timulated *e*mission of *r*adiation) operated by electronically "exciting" ammonia molecules as they passed through a cylinder made of charged metal rods. The result was powerful microwave radiation.

An optical maser (laser) was developed in 1960 by Theodore Maiman. It consisted of a synthetic ruby cylinder into which was coiled a flashtube. The device amplified the light waves into a thin, highly intensified beam.

Uses for lasers in today's communications graphics include color separation, typesetting, plate and cylinder exposure and engraving, and die cutting.[39]

Color Separation Scanning—1955

Printing Developments, Inc., a subsidiary of Time-Life, Inc., produced the first commercially available electronic color scanner in 1955. Although films for four-color printing had been produced on cameras and enlargers for some time, the scanner greatly speeded the process. The device scanned each original slide (transparency) with a beam of light, converting each tiny area of the image into color signals. These signals—separate records of the amount of yellow, red, blue, or black needed to reproduce each area of the original image—were used to expose four pieces of film. These films had to be exposed through a halftone screen to produce printable dots for each color. Many of today's devices, using laser technology, produce dots directly on film during scanning.[40]

Graphic Communication Today

Little did early inventors and visual arts enthusiasts imagine how expansive and important the graphic communications professions and related industries would become. Consider the scope of activity in the United States alone:[41]

The Industry as a Whole

Printing, publishing, and related industries combined are counted among the top ten manufacturing activities in the United States with the largest number of plants—more than 50,000. Shipments of goods worth about $100 billion annually help its 1.3 million emloyees earn a $20 billion payroll. And, unlike many of today's largest manufacturing segments, the industry has continued to grow, even in difficult times.

Newspapers

In a society showered with electronic alternatives, newspapers still lead all media in advertising revenue—more than $17 billion. The nation's 1,760 dailies and 8,000 weeklies claim about 100 million in combined circulation. It takes nearly half a million workers and 12 million tons of newsprint to satisfy newspaper readers each year.

Magazines

Nearly 11,000 periodicals with a shipment value of more than $10 billion are produced annually in 3,000 plants. Of the roughly 200 consumer, business, and trade magazines that begin publication every year, about half survive. Magazines enjoy nearly $4 billion in advertising revenues each year, but average 45 percent of revenues from circulation to readers.

Book Printing and Publishing

With nearly 50,000 new titles offered each year, book publishers and their printers produce goods worth about $10 billion for a variety of consumer, education, and trade markets. This activity is made possible by 50,000 employees working at 1,750 sites across the nation.

Commercial Printing

Another half million workers make up the five segments of the commercial printing industry, which include publications, advertising, general, financial/legal, and package printing. Nearly $24 billion in shipped goods, about 70 percent produced by offset, were shipped in 1982. Another $5 billion worth were shipped by business forms printers.

Reprographic Printing

A familiar fixture in shopping and commercial centers is the so-called *quick printer* which provides a variety of copying and duplicating services in addition to low-cost printing. Some 15,000 such printers, many of them under franchise, account for $2 to $3 billion in sales each year. Between 60 and 100,000 *in-plant* printing centers operated by businesses, government, and large organizations produce considerably more reprographic work—$12 billion worth—in addition to increasingly sophisticated printing, binding, and mailing.

Related Professions

In addition to those directly producing printed communications, thousands of advertising and public relations agencies, companies, government agencies, and commercial studios employ graphics specialists. These art directors, designers, illustrators, and production artists create and prepare material for print, and their efforts often support the work of countless freelancers.

Industries in Transition

So quickly does change occur today in graphics industries that it is hard to distinguish present from future. In this electronic age, microprocessors and lasers have begun to simplify almost every task associated with producing printed images. Breakthroughs are no assurance that familiar, trusted methods will disappear overnight, however. In today's uncertain economic times the enormous cost of buying or developing new equipment, modifying facilities, training personnel, and testing new systems is being met with increasing caution. Yet the changes already in place are remarkable, and many printers, publishers, editors, and designers do have at their disposal tools and systems that foreshadow what eventually all may have.

New Rules for the Graphics Game

Regardless of subject or format, most graphic communications traditionally have required all or most of the following:

- a plan or design
- a rough layout or dummy
- written copy for headlines, text, and captions
- type set from the copy
- photographs and other illustrations
- correct positioning or assembly of images

- plates
- printing
- binding

All of the above are still necessary, but how these tasks are accomplished today may be far from traditional. Consider these developments:

Copy completely researched, written, edited, and approved in electronic form without need for paper.

Typesetting done by laser directly from copy transmitted over phone lines by a client who may be working from a home computer.

Illustrations that originate from or are modified through computer manipulations, edited and stored in digital form.

Image assembly accomplished electronically from stored text and graphics instead of on pasted pages or film.

Plate preparation that requires no film or other intermediate material, instead done directly from computer signals.

Printing in high quality color by methods considered inferior only a few years ago—some requiring no printing plate at all.

Bindery integrated with printing so completely that blank rolls of paper are converted into finished, bound books by a single machine in minutes.

It would be a mistake for those entering the graphics professions to assume that the innovations suggested here and detailed in Chapter 7 have made most of the more familiar jobs unnecessary. The new machines may operate at the speed of light, but human beings still have a comforting habit of evaluating, analyzing, planning—and worrying—before they leap headlong into new ventures. People today enter a graphics job market where the familiar and the dazzling exist side by side. And most graphics employment studies suggest a continuing shortage of highly skilled personnel.

Anthony Smith, prominent media futurist and the author of *Goodbye, Gutenberg,* has said that "in the information business all techniques and technologies are either experimental or obsolescent."[42] We would do well to accept a working world where the only constant is change and be prepared to play an important role in it.

Notes

1 Edmund C. Arnold, *Ink on Paper 2* (New York: Harper & Row, Publishers, Inc., 1972), p. 11.

2 Karen Mooradian, "The Dawn of Printing," *Journalism Monographs* 23 (May 1972), pp. 1–24.

3 J. A. Toussaint, ed., *Highlights of Printing History* (Government Printing Office, 1941), p. 5.

4 "Hieroglyphic," The New Columbia Encyclopaedia (1975), Vol. 4, p. 1241.

5 Information booklet on papyrus, untitled (Pyramids, Egypt: Pyramids Papyrus Institute, 1980), pp. 7–8.

6 Mooradian, "Dawn," p. 25.

7 Toussaint, *Highlights*, p. 6.

8 Arnold, *Ink on Paper 2*, p. 15.

9 "How Paper Came To America," folder published by the American Paper Institute, 1978.

10 Arnold, *Ink on Paper 2*, p. 238.

11 *Printing Ink Handbook*, 4th ed. (New York: National Association of Printing Ink Manufacturers, Inc., 1980), p. 7.

12 Paul Lunde, "A Missing Link," *Aramco World Magazine* 32, No. 2 (March–April, 1981), pp. 26–30.

13 Lunde, "A Missing Link," p. 26.

14 Toussaint, *Highlights*, p. 12.

15 *Graphic Communications Through the Ages* (Neenah, Wisc.: Kimberly-Clark Corporation, 1971), p. 4.

16 Toussaint, *Highlights*, p. 14.

17 Arnold, *Ink on Paper 2*, p. 199.

18 *Through the Ages*, p. 8.

19 Toussaint, *Highlights*, p. 21.

20 Toussaint, p. 18.

21 Toussaint, p. 25.

22 R. Randolph Karch, *Graphic Arts Procedures* (Chicago: American Technical Society, 1957), p. 21.

23 Carl Dair, *A Typographic Quest*, No. 1 (West Virginia Pulp and Paper Company, 1964), pp. 14–19; and Toussaint, *Highlights*, pp. 19–22.

24 *Through the Ages*, p. 13.

25 *Through the Ages*, pp. 14, 21.

26 *Paper and Paper Manufacturing* (New York: American Paper Institute, 1981), p. 3.

27 *Through the Ages*, p. 17.

28 Toussaint, *Highlights*, p. 45.

29 Toussaint, p. 44.

30 Toussaint, p. 37.

31 *Pocket Pal: A Graphics Arts Production Handbook* (New York: International Paper Company, 1974), p. 15.

32 *Pocket Pal*, p. 15.

33 *Through the Ages*, p. 19.

34 Toussaint, *Highlights*, p. 39.

35 Patrick Harpur, ed., *The Timetable of Technology* (New York: Hearst Books, 1982), p. 58.

36 Anthony Feldman and Peter Ford, *Scientists and Inventors* (New York: Facts on File Publications, 1979), pp. 299–300.

37 Sigvard Strandth, *A History of the Machine* (New York: A & W Publishers, Inc., 1979), p. 192.

38 Victor Strauss, *The Printing Industry* (New York: R. R. Bowker Company; and Washington, D.C.: Printing Industries of America, Inc., 1967), pp. 105–108.

39 Feldman and Ford, *Scientists*, p. 312.

40 Strauss, *The Printing Industry*, p. 163.

41 The figures given are approximate, based on data from the following sources:

Paul Gallanda, "Quick Printing Part 3: Franchises—Ten Times As Many as a Decade Ago," *Printing Impressions* 23, No. 12 (May 1981), pp. 8–10, 73.

Tom Johnson, "The Future of the Newspaper Industry," speech presented to the Audit Bureau of Circulations annual conference, Chicago, 1981.

Robert J. Haiman, "Separating the Winners From the Losers," *Presstime* 4, No. 8 (August 1982), pp. 24–25.

William C. Marcil, "Health of the Newspaper Industry," speech presented to the South Dakota Press Association, Sioux Falls, April 17, 1982.

Paul Gallanda, "Quick Printing Part 4: Independents—An 80% Share of the QP Market," *Printing Impressions* 24, No. 1 (June 1981), pp. 22–24, 45.

Roger Ynostroza, "Printing and Publishing: The Size of the Industry/Part 1," *Graphic Arts Monthly* 55, No. 1 (January 1983), 52–53.

1982 Technology Forecast (Pittsburgh: Graphic Arts Technical Foundation, 1982), pp. 1–8.

Printing and Publishing: Quarterly Industry Report, U.S. Government Printing Office, Washington, D.C., Spring 1982.

Facts About Newspapers, American Newspaper Publishers Assn., Washington, D.C., 1982.

42 Anthony Smith, "From a Bright Past to an Uncertain Future," *ASNE Bulletin*, No. 647 (Dec./Jan. 1982), p. 10.

2 The Design Process

Planning for Visual Impact

Creating truly effective graphic communications isn't easy, especially for those just learning how. Before you begin, remember to stop and think carefully about who you are trying to reach, what you are trying to say, and how graphic images and designs can help. It's likely that your design solutions will be about as good as the planning you put into them.

Introduction

In our search for effective ways to communicate through graphics, certain distinctions between types of visual images may help. We can begin by noting the difference between graphic art and graphic communication through design.

Few of us would attempt a precise definition of art. We sense that it is both process and product, some wonderfully intangible mixture of human imagination, technical mastery, and the desire for creative fulfillment through self-expression. We usually think of works of art as personal statements with intrinsic value. Viewed by others, such statements may evoke emotion, stir both the imagination and the intellect, and elicit feelings ranging from ecstasy to disgust. The capacity to trigger a response, *not the nature of that response*, is often considered an important measure of creative success.

Graphic designs represent our interest in predictable, *focused* responses to visual images. These designs may appear to be works of art. Indeed, they often are. But the works of professional graphic designers are nearly always produced with certain objectives in mind. These objectives, established before the work begins, help guide the designer and provide a clear basis for judging effectiveness.

All forms of design, by definition, involve *planning*. They are means to ends, ways of reaching specific goals.

We sometimes describe any patterned or structured image, however simple or complex, as a design. *Graphic design is the process of planning and creating visual presentations for achieving predetermined objectives relating to known or predicted audiences.*

Sometimes the result of the design process is a personal statement of the designer alone. More often, the resulting statement is a collective synthesis of designer talents, client wishes, and audience needs.

What Is Graphic Communication?

Graphic communication is the dissemination of graphic design to a larger audience, usually by means of a mass medium. Some common printed mass media include newspapers, magazines, books, direct mail, posters, billboards, and point-of-purchase displays. Sometimes signage of all kinds is considered a mass medium, as well. In the context of this book, we will consider graphic communications to be all purposive, print-oriented graphics designed for audiences, however limited in scope.

One of the most respected theorists in the field of mass communication, Wilbur Schramm, has defined the process of communication as "the sharing of an orientation toward a set of informational signs."[1]

Graphic communication involves the use of iconic (visual) codes or symbols to convey information. Usually we find that language (verbal) symbols containing semantic codes are also necessary to make our messages fully effective.

How well our combinations of visual and verbal symbols convey information depends upon how well readers understand the meanings of our symbols. For example, tourists in Saudi Arabia might buy and frame a beautiful color poster promoting a local soft drink. Later, if those same tourists recognize the product label in a restaurant, they might try a bottle. We could then say that at least one objective of the graphic image is being met: product recognition.

Meanwhile, however, if the local populace is busy carting home cases of the same soft drink because the poster had announced—in Arabic—a half-price sale on the drink, it's obvious that the tourists are at a disadvantage. Since they couldn't interpret all the symbols in the message, they miss out.

The degree, then, to which a graphic design achieves its purposes depends first upon how accurately readers can interpret symbols used by the communicator. Simply put, do the parties involved share a similar *frame of reference* toward the symbols? The designer's task is to see that they do.

Test yourself on the meanings of the symbols in Figure 2–1. If a graphic designer had used these symbols, would you have understood the message?

Graphic communicators employ well-known design principles to convey information efficiently. They know that uncluttered layout, exciting color, and readable type go a long way toward attracting and holding the attention of their image-consumers.

FIG. 2–1

a. The standard symbol used in electronic data processing to indicate a document.

b. A stylized symbol representing a leaf-type shutter or iris diaphragm in a camera.

c. Readers of Omni magazine will recognize this as—what else?—a road sign of the future for an android service station.

To summarize, the graphic communications discussed in this text:

• are designed to achieve predetermined objectives;

• are intended to reach target audiences, usually through mass media;

• attempt to provide information that will be understood similarly by all members of those audiences;

• combine principles of visual design and aesthetics in ways that enhance understanding of and appreciation for messages and their originators.

Beginning the Design Process

To plan effective graphic communications, we begin by considering:

• the nature of the audience we wish to address;

• the best way to construct messages to reach that audience;

• the medium of communication we wish to use; and,

• the elements which might either interfere with or enhance our messages. We are especially concerned with the first two of these in this chapter and in Chapter 8.

Identifying Audiences

Warren Paper Company, in the first of a series of books on how to plan printing to promote business,[2] explains just how diverse the audiences of an organization such as a corporation may be. They identify several potential internal and external audiences which might be addressed by management, including:

Known audiences—those who may be reached directly and personally because their names and locations are known. These include *employees, investors, dealers, customers,* and *company intermediaries.*

Unknown audiences—those who may be reached only by impersonal, broad-based methods. These include: the *volunteer audience*—those who seek out a product or service; the *classified audience*—those who share technical and professional interests with the company; the *segment audience*—those who share special but nonprofessional interests with the company; and *geographical audiences*—all persons who share a common location in relation to the company; that is, local, regional, national, and international audiences.

From the descriptions above, we can see that people may be members of more than one kind of audience at the same time.

Every organization or enterprise has its own list of audiences which may be addressed in different ways for different reasons. A first step is to identify them so that each may be reached effectively.

Determining Appropriate Visual Structure

Visual message structure is what this book is about. Of course, images with both visual and verbal content make up most printed communications. Structure not only includes the images, but also the format in which they are presented.

The links between *perception*, the process of our becoming aware of information through the senses, and *cognition*, whereby we begin to organize and make sense of that information, is not yet fully understood. However, we have begun to identify types of visual presentations and understand how well they convey information to different audiences.

In his thought-provoking book, *A Primer of Visual Literacy*, Donis A. Dondis maintains that we send and receive visual messages on three levels: representational, abstract, and symbolic.[3]

Representation is what we see and recognize from our environment and experience. A highly detailed drawing or photograph is an example.

Abstraction is the kinesthetic quality of a visual event reduced to its basic visual components. Cartoon characters are examples of abstraction of the human form. Alphabets and words are highly developed forms of language abstraction.

Symbolism is the system of codes which man has created arbitrarily and to which he has attached meaning. The skull and crossbones on a bottle of poison is a common example.

These three levels of visual communication interact with each other constantly in our daily lives, says Dondis. However, we can do much to tailor appropriate visual levels to each audience.

Young children who do not yet know the complex symbols of their culture respond well to realistic pictures or simple, abstract images. Eventually, they

will learn to understand symbols as they are taught the meaning of each. For example, a heart symbol will begin to represent love to a child who is exposed to stories and songs about love and who is taught to send valentines each year.

Even though adult readers are able to interact with visual information on highly symbolic levels, they regularly require other levels of visual presentation. When they go car shopping, they expect to pick up showroom brochures that contain representational color photographs of the car models, abstract drawings of the interiors and engines, and symbolic wild animal or space-age logos designed to convey the unique qualities of each car.

Special audiences have special needs. Younger readers, as a group, require symbols—alphabets and drawings—that are larger and less complex. Older readers may understand complex symbols, but also need them to be larger for easier reading.

Better educated readers can understand concepts through abstract language symbols without the need for excessive visual detail. Still, an audience of doctors, highly adept at absorbing large amounts of verbal abstraction, still would demand that their reference books show detailed photographs of major surgical procedures.

Some Tips from the Theorists

When we know an audience's characteristics, we can structure a design to reach its members at appropriate levels of visual processing. When we don't know much about an audience, we can at least use visual stimuli that produce similar and fairly predictable behavior in most humans. Research in fields as diverse as perceptual, cognitive, and Gestalt psychology, information processing, neurophysiology, and consumer behavior provides some useful guidelines.

Here are a few major design implications suggested from this research:

1. Human beings prefer images to be in a state of equilibrium or balance. When one half of a design is the mirror image of the other, we call this formal or *symmetrical* balance. *Asymmetrical* or informal balance may be achieved if images varying in size, weight, tone, or other characteristics are positioned in a way that gives a design stability without symmetry.

2. Contrast between visual elements is necessary for readers to process designs effectively. Contrast may relate to the size, color, brightness, or density of an image in relation to its background. This is often referred to as figure-field contrast. Contrast also relates to relative sizes, forms, and positions of type, illustrations, and other elements.

3. A visual design should be seen as an integrated whole rather than a collection of pieces.

4. Cognitive judgment about perceived messages is improved if we simplify a design. We also improve cognition by repeating elements of messages visually or verbally. This is known as redundancy.

5. Three simple concepts of design—proximity, alignment, and consistency of style—provide effective ways to organize the elements of a design. Glenn Hanson of the University of Illinois, a longtime student of the design process, labelled these concepts "layout integration." They are derived from the Gestaltist principles of attraction, continuity, and similarity. As shown in Figure 2–2, the three principles are:

  a. If design elements are in *proximity* to one another (that is, close together), they will seem more related than elements not grouped.

  b. If design elements are *aligned* vertically,

horizontally, obliquely, or in any continuous line, they will seem more related than nonaligned elements.

  c. If design elements are *consistent* or similar in style, shape, size, or other characteristics, they will appear to be more related than if they are not consistent.

Full design integration is most likely when all three elements—*proximity, alignment, consistency of style*—are present.

6. Many attempts have been made to analyze the flow of the eye through a page, but few solid rules have been established. Some general findings:

  a. People in Western cultures who read from left to right, top to bottom, tend to scan pages in roughly this same manner.

  b. Researchers disagree on whether the lower left, lower right, or lower center is the most valid terminal point for the eye—that is, where the eye initially comes to rest. Since the eye continues to scan constantly anyway, the importance of an initial resting point may be exaggerated.

FIG. 2–2
Elements of layout integration.

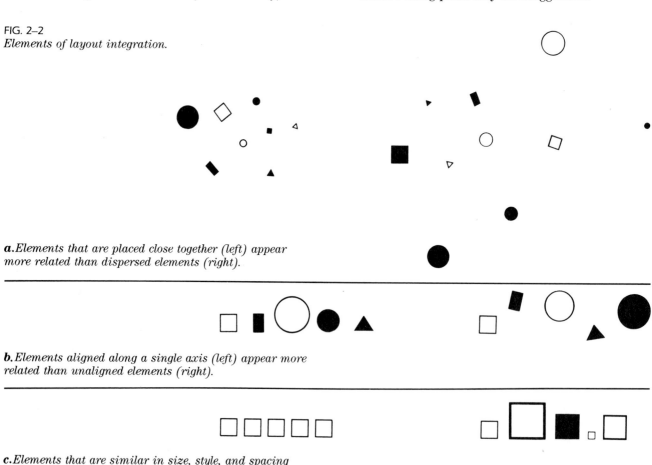

a.Elements that are placed close together (left) appear more related than dispersed elements (right).

b.Elements aligned along a single axis (left) appear more related than unaligned elements (right).

c.Elements that are similar in size, style, and spacing (left) appear more related than dissimilar elements (right).

c. Studies have identified some typical patterns of eye movement in a selected number of designs. However, since there may be hundreds, even thousands of variations possible for most page designs, the application of the research findings is limited. Designers would do well to concentrate on *guiding* eye movement in individual designs by selecting attractive elements and arranging them in ways that best fit the subject matter rather than depending too heavily upon innate eye flow patterns.

d. Eye movement appears to be influenced by the specific content of an illustration and its interaction with other elements on a page. Click and Baird state that *visual syntax* is etablished when we are sensitive to the interplay of images and their logical sequencing in the design.[4] Little is known about how the eye moves within the content of a *single* illustration appearing with others on a page, or how significantly the time allotted to each image affects comprehension or behavior.

e. Images or groups of images that appear to point to other elements are widely regarded as strong determinants of eye flow. Images usually are made to point into rather than off a page, and toward, not away from, one another.

In Chapter 8 we will discuss alternative forms of visual presentation and how the principles and ideas introduced in the present chapter are applied.

The Path to a Design Decision

Gregg Berryman, in *Notes on Graphic Design and Visual Communication,* identifies six stages through which visual designs may pass:[5]

1. *Problem identification*—determining what needs to be resolved through design and why. Setting the limits and criteria to use in finding solutions.

2. *Preliminary search for solutions*—the production of small sketches (thumbnails), and the generation of concepts, ideas, and alternative approaches.

3. *Refinement*—the narrowing down of solution possibilities to a more realistic and manageable number. More detailed layouts (roughs) may be produced showing the best ideas.

4. *Analysis*—a hard look at the best possibilities for a solution. Have all good alternatives been considered?

5. *Decision making*—selection of the best solution from the field of good possibilities. Highly realistic layouts (comprehensives) may be produced to aid in evaluating the top contenders.

6. *Implementation*—executing the design in final form and reproducing it.

Berryman suggests that the process of decision making in design may not follow these steps in a neat fashion. Often, continuous feedback and evaluation take place at all stages in the process. Figure 2–3 diagrams a linear and a cyclic design path.

Graphic design is but a part of the many design decisions that must be made for most printed materials. An annual report for a corporation, for example, passes through many stages: *concept formation, research, writing, design and layout, illustration (art and photography), financial data generation, prepress production, printing, and distribution.*

Many of these stages involve designing the overall structure and organization, not the visual elements,

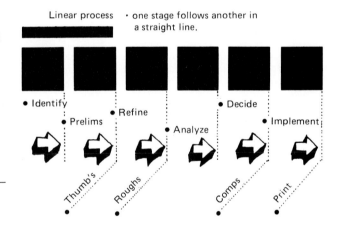

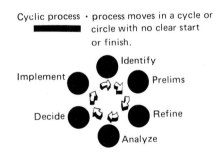

FIG. 2–3
Design path. Reproduced with permission from Notes on Graphic Design and Visual Communication *by Gregg Berryman.*

	Concept	Writing & Research	Design	Illustration & Photography	Financial Data	Typography & Key Line	Printing	Distribution
1st Period	Establish Theme							
2nd Period	Research	Rough Draft	Rough Layout		Financial Data Outline			
3rd Period			Storyboard Layout					
4th Period		Approval	Approval	Photography				
5th Period			Comprehensive Layout					
6th Period		Final Copy	Layout Approved / Secure Printing Bids		Financial Data Copy			
7th Period		Copy Approved	Select Printer			To Typesetter		
8th Period				Charts, Graphs	Final Financial Figures	Reader Proofs / Corrected Copy to Typesetter	Color Separations	
9th Period			Paste-up			Final Typography	Paper Stock & Envelopes	Mailing Lists
10th Period			Deliver to Printer					
11th Period							Brownline Print/Color Proofs	Addressing Envelopes
12th Period							On Press	
13th Period							Bindery	Mailing
14th Period								

FIG. 2–4
A flow chart for design and production of an annual report.
Courtesy Simpson Paper Company.

but it is design nonetheless. Figure 2–4 illustrates how other stages of the production of an annual report might affect, directly or indirectly, the decisions about graphic design.

Notes
1 Wilbur Schramm and Donald F. Roberts, eds., *The Process and Effects of Mass Communication* (Urbana, University of Illinois Press, 1971), p. 13.

2 *Business Management and Printing* (Boston, Mass.: S. D. Warren Paper Company, n.d.), pp. 19–21.

3 Donis A. Dondis, *A Primer of Visual Literacy* (Cambridge, Mass.: The MIT Press, 1973), pp. 67–84.

4 J. W. Click and Russell N. Baird, *Magazine Editing and Production,* 2nd ed. (Dubuque, Iowa: Wm. C. Brown Company, Publishers, 1979), p. 195.

5 Gregg Berryman, *Notes on Graphic Design and Visual Communication* (Los Altos, Calif.: William Kaufman, Inc., 1979), p. 7.

3 Typographic Images

The Fundamental Graphics Tools

How we present information is as important as what we say. Skilled communicators must learn which typographic emissaries carry their messages best to bring the printed word to life. This chapter should help make the formidable task of type selection a bit easier. No lesson is more important to the designer.

Introduction

Typographic symbols are the visual building blocks of language and the fundamental tools of graphic communication. These symbols also are some of the easiest graphics tools to use.

Despite the vast array of sophisticated techniques and media available today for illustration, print designers still often turn to simple type to present their message.

The "Sunrise/Sunset" photography contest announcement shown in Figure 3–2 could easily feature dazzling full-color photographs. Instead, type alone lets readers visualize sunrises or sunsets from their own experiences. "Ahhh . . ." is how we *feel* when viewing such beauty. "Click" is how we respond if we are lucky enough to have our cameras handy.

Why Is it Important to Know Type?

Not only is knowledge of something as historically significant as type personally satisfying—it's also practical. The more we know about any skill or technique, the better we become at applying it. Use of type is no exception. We study typography, the skillful design and use of type, for two important reasons:

First, *to expand our creative options.* Just as a carpenter would be limited to simple construction if he had only a hammer, nails, and saw to use, so would our design creativity be limited if we had only a

rudimentary knowledge about type. Professionals in the use of type can recognize hundreds of typefaces, either by name or by general design characteristics. They also have file drawers crammed with type samples and probably countless more mental images of typographic treatments.

A second reason is *to make professional communications more effective.* Those who set type for a living are accustomed to discussing type with untrained clients. However, they greatly appreciate those customers who show interest in typographic excellence. Quite naturally they tend to provide those customers with higher quality and better service.

If those of us who buy type show clearly that we expect professional care in such things as sizing, spacing, and hyphenation, we are more likely to receive such care.

James Michener, in his popular novel *Chesapeake,* tells of a craftsman who aspired to mastery in shipbuilding, only to be frustrated by his ignorance of the proper technical names for shipbuilding tools and parts. Printers and typesetters, too, use a professional language (and jargon), much of it related to type measurement. It must be deciphered intelligently by anyone who wants to work with type effectively.

Selecting the Right Typeface

We know intuitively that *how* we say something in print is as important as *what* we say. Not only is correct word usage vital to our message, but the

17

B.C.   by johnny hart

FIG. 3–1
By permission of Johnny Hart and News Group
Chicago, Inc.

appearance of our written words also affects the meanings our readers attach to them.

Today we can select from among thousands of typeface designs in choosing those best suited for our printed messages. Some types are designed to enhance a message by evoking its emotional tone. Others are less obtrusive, allowing readers to read through the type and form a mental picture of word meanings. Most professionals who design with type are called upon to use a full range of typefaces—from nondescript sans serifs to the most evocative stylized scripts. Figure 3–3 shows some of the ways types are used to evoke reader involvement.

We can better understand type appropriateness by looking at printed advertisements. It is quite apparent that serious messages—those discussing such matters as interest rates, corporate responsibility, or starving children—are usually presented in serious-looking typestyles. Automobile or movie ads, however, are often splashed with colorful, ornate scripts and so-called designer typefaces. We can be sure that types chosen for an advertisement costing thousands of dollars to produce and print are selected carefully from the bewildering array of type designs available.

Whatever we may call the connotative characteristics of type—*mood, personality, feeling tone, atmosphere,* or *congeniality*—what we usually mean is that the type is right, both emotionally and intellectually, for the subject at hand.

The ability to choose the *one* typeface (or a select group of faces) that best conveys the *essence* of a message is what sets the skilled user of type apart from the novice.

Evocative or psychological characteristics are only part of what we look for in selecting a typeface. Since

FIG. 3–2
Sunrise/Sunset, photo contest poster.

Reprinted with permission from Virginia Commonwealth University
School of the Arts; Richard Rumble, designer; Charles B. Scalin, art director

Transfer Material Collage by Linda Sarro

FIG. 3–3
Setting a mood with type.
Reprinted with permission of *In-Plant Reproductions.*

printed words do not exist in a vacuum, other practical design questions arise. For instance:

• Is the type too *similar* to that used in competitive messages?

• Is the type design *compatible* with corporate symbols, logos, trademarks, or section headings with which it must appear?

• Are the design characteristics of the particular version of the typeface *appropriate* to the overall design? Is the type too large or small, too light or bold, or too tall or slanted to match the size, weight, or angle of other graphic elements?

The Fundamental Elements of Typefaces

Let's remove some of the mystique that surrounds type and makes learning about it seem difficult. Unless you wish to become an expert in the field, the basics discussed here should suffice.

Letterforms and other language symbols have been reproduced in a variety of ways throughout history. No matter how the typographic image is formed—by raised-image impression, photographic exposure, digital scanning, or other means—the end result, the printed *face,* is of most concern to us. Before we look at type measurement and major design differences, let's see what parts make up the printed face. This will help us understand design differences in type.

In Figure 3–4 the words *GATOR* and *gator* are formed by some of the most distinctive letters of the alphabet. The letters are labelled to show essential elements of typefaces. While these elements are derived mostly from a very important category of types called *romans,* other designs have many of these same characteristics.

Letters are formed with the following *main strokes:*

Stem—a thick, straight stroke that forms a major portion of a letter.

Hairline—a thin stroke that balances or connects with the thicker stem stroke. The free end of both a stem and a hairline is called a *terminal,* and it can be straight, tapered, or otherwise modified.

Curve—a thick, rounded stroke that forms a major portion of a letter. Sometimes a *finial* (final or ending) stroke, forming a hook or a ball, is attached to the curve (as in *r*).

Serif—a finishing stroke placed at right angles to the end of a stem or hairline. Often serifs are joined to a stem or hairline by *brackets.*

19

FIG. 3–4
The main parts of a typeface.

Bar—a thin or thick stroke, usually horizontal, that connects stems, hairlines, or curved strokes (as in *H* or *e*).

Cross—a short stroke placed across a stem (as in *t*). The crossbar stroke at the top of a capital *T* forms *arms*.

Important *extension strokes* include:

Ascender—a stroke that extends above the average height (mean line) of the lowercase letters.

Descender—a stroke that extends below the base line on which the letters rest.

Swash—an ornamental, curved stroke usually intended to appear on letters that begin or end a word or heading.

Parts found only on the *G* or *g*, perhaps the two most distinctive characters of all, are:

The beard—a short, thick, stemlike stroke on the capital *G*.

The ear—a small stroke, often finished with a final ball, appearing at the top of the *large bowl* of a lowercase *g*.

The loop—a rounded, usually closed descender on the lowercase *g*.

The link—a short, curved stroke connecting the loop and the *small bowl* of the lowercase *g*.

Figure 3–5 shows how different two of the characters in *regular* (straight-standing) and *italic* (right-leaning) versions of a typeface can be. The italic shown here is a so-called *true italic*, one that is very different in design from its parent. Roman italics are of this type. Some right-leaning types are simply leaning versions of a regular typeface and look very similar to the parent face. We call these *oblique*.

Regular version

Italic version

ITC Century™

FIG. 3–5
Two typeface characters with a strong difference between their regular and italic versions.

Measuring Type Accurately

Relative size, width, depth, and spacing are all important variables in describing type. We use the *point system* to express these values. Its basic elements are simple:

> 1 point = ⅟₇₂ of an inch (or .0138 in.)
> 72 points = 1 inch
> 12 points = 1 pica
> 6 picas = 1 inch

Points are used to express *small* measurements such as type size, leading (interlinear spacing), and thickness of borders and other rules (straight lines).

Picas are used to express *large* dimensions such as widths and depths of type columns, page areas, and illustrations.

We are accustomed to the decimal system, which is based on tens. For examples, 10 cents make 1 dime. The relationship of points to picas is not 10 to 1, but 12 to 1, like inches and feet. We use the 12-to-1 system easily when calculating with a tape measure, but when we encounter the miniature points and picas on a printer's *line gauge* (pica rule), we may become a bit confused.

Take a few minutes now to become famliar with the point system. It is the standard way we measure type and many other graphic elements. Figure 3–6 shows an ordinary ruler and a printer's pica rule so that we can compare the units of measure. Note that 3 picas

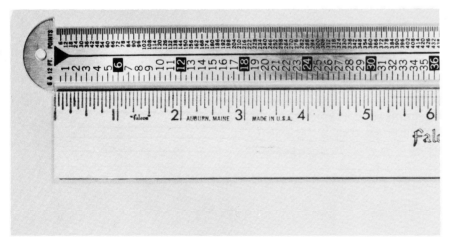

FIG. 3–6
Comparison of a printer's line gauge and an ordinary ruler.

equals ½ inch, but smaller increments do not equate precisely.

How to Determine Point Size of Type

Type size is expressed as height in points. Using a metal or plastic ruler, we measure from the top of the tallest ascender to the lowest descender of the printed typeface.

Figure 3–7 explains this measurement. Note that each letter rests on a *base line*, and that the major portion of most letters rises to the *mean* (average) line. The portions of letters that rise above the mean line are called *ascenders*. Those that drop below the base line are called *descenders*. If we measure the distance from the top of a tall ascender (such as on the lowercase *h*) to the bottom of a long descender (such as on the lowercase *p*), we get the approximate point size of the type.

Why do we say *approximate* point size? Note the drawing of metal foundry type in Figure 3–8. Early typesetting was done entirely with such raised single letters, made of either metal or wood. Later, completely joined lines of type were produced in metal on linecasting machines. Such raised types usually contained a small "shoulder" below the face of each letter. This provided slight separation when lines of type were stacked on top of one another.

Measurement of wood or metal type was a simple matter of measuring the *body* on which the letter rested (the distance from *back* to *belly*). This measurement included the shoulder under each letter.

Since we rarely use raised types today, we cannot measure the body of a letter. Because we can only measure the face of a letter, we cannot accurately account for the small shoulder under each letter. Therefore, our point size measurement is only an approximation.

With some experience you can learn to add a few points to the ascender-to-descender measurement and estimate point size accurately enough for most purposes.

Several manufacturers make clear plastic measuring gauges on which capital letters are printed in different point sizes (Figure 3–9). When placed over the capitals on a printed page, these guides give an easy and fairly accurate estimate of point sizes.

Fortunately, estimates are usually sufficient. If more precision is needed, as when we want to match a previously printed sample exactly, we can consult a type specimen book. There we will find the typeface we need in a full range of point sizes.

Leading

An important aspect of type spacing is *leading* (pronounced ledding). Leading refers to the space between lines of type. In earlier days, this space was formed as printers placed thin metal strips, or slugs, between lines of type. Because these metal strips contained a high proportion of lead, they were called "leads." Today leading refers only to the space beneath type lines that is added *in addition to* space required by the full point size. This space is expressed in points or fractions of a point. The amount of leading is controlled by adjusting the distance the strip of typesetting paper moves after each line is photographed onto it. The more the paper moves after each line is set, the more leading is produced.

When we set type without leading, it is "set solid." Usually, however, type is leaded a little bit for legibility. One or two points of leading is common for reading matter of 12-point size and smaller sizes. If we set 12-point type with 2 points of leading, we mark it 12/14, referred to as "twelve on fourteen," or "twelve-point, two points leaded."

Figure 3–10 shows leading added between the pieces of type forming the word *dog* and the lowercase k beginning the next line. Figure 3–11 shows the visual effect of such spacing, viewed from the top.

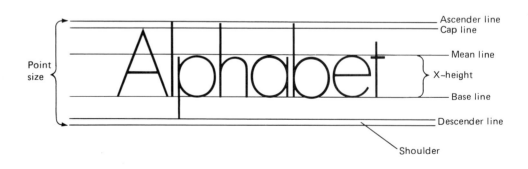

FIG. 3–7
Type measurement.

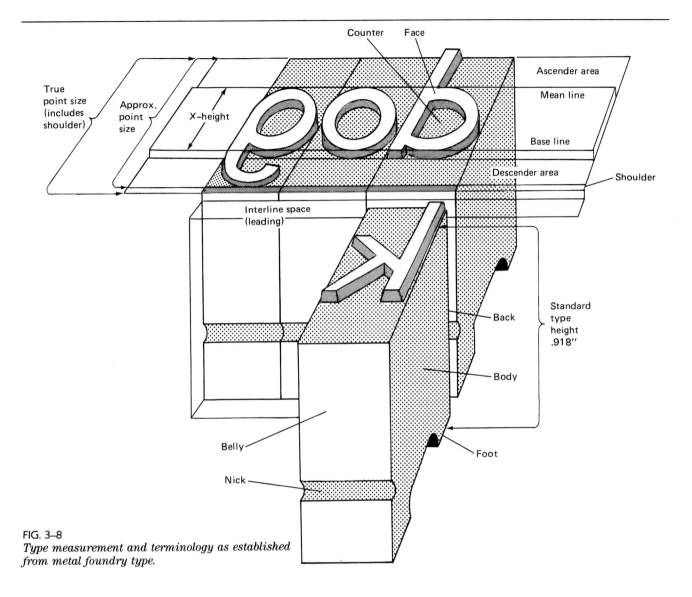

FIG. 3–8
Type measurement and terminology as established from metal foundry type.

Today we often use *negative* or *minus leading*. In photosetting we are no longer restricted to solid type bodies, so we can *decrease* horizontal space between lines as easily as we increase it. We can set type 12/10, or "twelve-point, two points negative leaded." Usually negative leading is reserved for type larger than 12-point, because small lines become too cramped for easy reading.

Today, it is possible to set type in any size from about 4 point to 120-point and beyond, in increments as small as one-tenth of a point. Even the smallest typesetting firms usually offer a range from at least 6- to 72-point sizes. A range of standard sizes from 6 to 36 points is shown in Figure 3–12. Not shown are five other common headline sizes: 42-, 48-, 60-, 72-, and 96-point.

The standard sizes shown here are still the most commonly used today, although photographic typesetting equipment can give us virtually any size we wish. Eventually, as we allow these machines to assume more typefitting tasks automatically, the distinction of standard sizes may blur in a maze of odd sizes such as 16³⁄₁₀- and 70½-point.

Sizes up to 14-point are considered *body type (text* or *reading matter)*. Sizes 14-point and above are called *display (headine* or *title) type.*

If we measure a printed example and obtain a nonstandard point size, either (a) the type was *set* to a nonstandard size to make it fit a layout; or (b) the type was enlarged or reduced photographically *after* it was set.

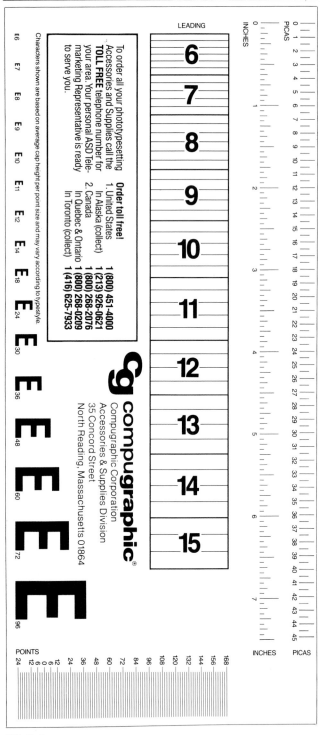

FIG. 3–9
A plastic type-measuring gauge.
Compugraaphic Corporation

Type Design Goes Modern

Today's typography depends heavily upon technology and science. Not only is typesetting almost completely computerized, but typefaces themselves now are being designed with electronic assistance or with electronic reproduction in mind. Figure 3–13 shows the new typeface Shannon, designed principally by Janice M. Prescott of Compugraphic Corporation, with the help of the computer program Ikarus. Ikarus, developed by Dr. Peter Karow of Hamburg, Germany, allows the user to create various weights and styles through analysis of existing forms. Prescott's design was inspired by a classic Irish calligraphic form. Ikarus helped her to isolate recurring shapes in the early typestyle to form the basis of her design.

The typeface Demos (Figure 3–14) was designed by Gerard Unger especially for electronic, vertical-line scanning by the Hell CRT Digiset typesetter. The rounded corners and precise curves of Demos were drawn for minimal loss of sharpness in the scanning process.

Another computer program, Metafont, designed by Donald Knuth, professor at Stanford University, allows a type designer to create completely original typefaces, in many versions, at a computer terminal.

Why Classify Type?

No widely accepted system of classifying typefaces has yet evolved to match the increasing sophistication and precision with which we design and set type. Since typography is still more art than science, this is

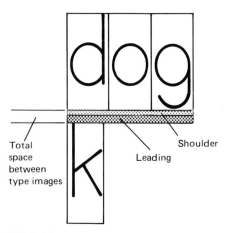

FIG. 3–10
How the outline of the type body affects leading space between printed images.

FIG. 3–11
Effects on line spacing of setting type solid, with leading, and with negative leading.

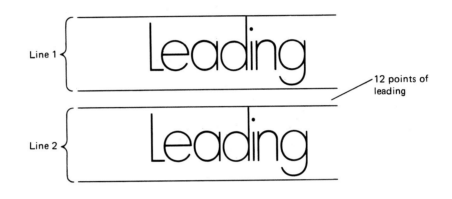

6-pt Serif Gothic
7-pt Serif Gothic
8-pt Serif Gothic
9-pt Serif Gothic
10-pt Serif Gothic
11-pt Serif Gothic
12-pt Serif Gothic
14-pt Serif Gothic
16-pt Serif Gothic
18-pt Serif Gothic
24-pt Serif Gothic
30-pt Serif Gothic
36-pt Serif Goth

FIG. 3–12
Range of standard sizes shown in type books.

119, 686

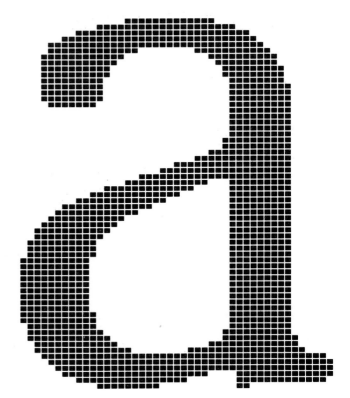

FIG. 3–13
Shannon typeface designed with Ikarus computer program.
Compugraphic Corporation

FIG. 3–14
Demos, a typeface designed for cathode ray tube (CRT) by Gerard Unger.
Reproduced with permission of *Visible Language*, Cleveland, Ohio.

not too surprising. However, we now have an estimated ten to fifteen thousand typefaces available, with hundreds more in various stages of design. We must learn to communicate *about* types as effectively as we communicate *with* them.

Attempts have been made to classify types with great precision. In 1967 J. Ben Lieberman published a Typorama containing 9 categories, 30 classes, and 100 styles of type. His system, based primarily upon metal types, remains one of the most extensive and historically sensitive systems ever devised. The G-wiz system, based entirely on the capital *G* and lowercase *g* of alphabets, was proposed two years later by Henry Borrmann as a way to identify types quickly. A West German standards committee made up of preeminent typographers even proposed a method, much like the

Dewey Decimal System libraries use, to categorize types more precisely.

On a day-to-day basis, few of us need the degree of precision these systems offer. Yet loose systems which divide types into only a few ill-defined categories are not much help in teaching us to distinguish type differences.

A Four-Group Type Classification System

Typefaces, like plants and animals, can be organized into broad-to-narrow classification systems. Typographers, however—unlike botanists or biologists—remain uncommitted to any single system.

Scientists would correctly describe the kind of system presented here as *artificial* (based upon user convenience) rather than *natural* (based upon evolutionary relationships). The sole purpose of the system proposed here is to help students describe and locate typefaces found in many printed sources on the basis of broad design similarities. Many other systems exist. Study as many as you can.

Type classification is roughly based on the following levels: *groups, subgroups, families,* and *fonts.*

Groups (sometimes called races) are major divisions of type. They are formed according to very great design differences, such as the presence or absence of serifs or the consistency of stroke thicknesses.

Subgroups (sometimes called ethnic groups) are major subdivisions of type. Members of these large classes are similar in general design to all other members of their group. However, they also share, with the members of their own subgroup, a substantial common difference from the group.

Families are clusters of related type designs all bearing the names of their parent typefaces. The family members are all very similar in design. They vary only in *weight* (light or bold, for example) and/or *style* (meaning width, stance, or special treatment).

Fonts are the individual branches of a family. They include all possible *alphanumerics* (letters and numbers), *ligatures* (letter combinations), alternate characters, punctuation marks, and symbols that correspond to a single weight, style, and—in most cases—size.

When we use the term *font* to specify type to be *set,* we mean a single size as well as a single weight and style. When we use the term *font* to refer to a *type master* on the typesetting machine, we mean weight and style only, unless the machine is capable of producing only one size per type master.

Note: In the days when type was set by hand, letter by letter, or as lines from a linecasting machine, size was essential in describing fonts. All the pieces of type in a single type case, and all the matrices (letter molds) in a single linecaster storage magazine were the same size. Each case or magazine contained a single font—a branch of a type family in one size only. For example, a printer might have three fonts of Bodoni Bold—24-,

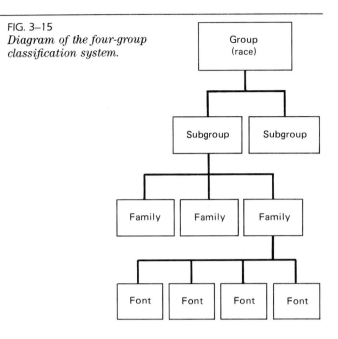

FIG. 3–15
Diagram of the four-group classification system.

36-, and 48-point—each in its own separate type case. Today, we use master fonts on film or magnetic media to produce a complete range (series) of sizes in any given branch of a family.

Through digital or optical modification, we also can produce not only different sizes but different weights and stances from a master font.

If this all seems a little confusing, remember the following example:

a. When we select *Garamond* type for a design, we are referring to a *typeface.*

b. When we further specify our choice as Garamond *Bold Italic,* we are refering to a particular *weight* and *style* of that typeface; and

c. When we call for *12-point* and *24-point* Garamond Bold Italic, we are describing two specific *fonts* of that typeface in a particular weight and style.

Figure 3–15 diagrams the relationships among the four levels of type. We use group or subgroup names to locate type in catalogs. We don't use them when we are providing typesetting instructions. We use the *family name,* preceded by a *point size,* and followed by whatever *font descriptors* identify the branch of the family we want. Below is an example:

36-point Garamond Bold Italic

Size   Family Name   Weight   Style
Font descriptors

Light

Book

Bold

Ultra

Light Italic

Book Italic

Bold Italic

Ultra Italic

Light Condensed

Book Condensed

Bold Condensed

Ultra Condensed

Light Condensed Italic

Book Condensed Italic

Bold Condensed Italic

Ultra Condensed Italic

Outline

Outline Shadow

Contour

FIG. 3–16
The full family of ITC Cheltenham.

FIG. 3–17
The Four-Group Classification System

SERIF	SANS SERIF	HAND-FORMED	SPECIALIZED
Early Roman	Gothic	Text	Novelty
Modern Roman	Monoline	Calligraphic	Decorative
Square Serif		Script	
Round Serif		Cursive	
Sharp Serif			

Figure 3–16 displays the full family of ITC Cheltenham, designed by Tony Stan. Nineteen variations, each with its own weight and style, are shown in 18-point size.

As shown by the table in Figure 3–17, the Four-group Classification System consists of four main groups of typefaces: *serif, sans serif, hand-formed,* and *specialized,* and thirteen subgroups: *early roman, modern roman, square serif, round serif, sharp serif, gothic, monoline, text, calligraphic, script, cursive, novelty,* and *decorative.* We will describe each of these groups and look at representative examples.

The Serif Group

Serif types are defined by the presence of a finishing stroke (serif) at the end of major stem and hairline strokes. There are five subgroups—two for romans and three for other classes of letters.

Roman types vary in three important ways: *serif treatment, contrast* in the thickness of strokes, and *stress* (angle or thrust of curved strokes). Figure 3–18 describes these three characteristics of roman types. These changeable elements help us identify the first two serif subgroups—early romans and modern romans.

Early romans are most numerous and so varied that we would need many subgroups to account for all minor differences. To keep it simple, just remember that *they all have bracketed serifs, contrast of stroke thicknesses, and one or more stressed (left-leaning) bowls on lowercase letters.*

At least four categories are generally recognized: *Venetian, French old style, English old style,* and *transitional.* Examples of each are shown on these pages, as well as updated and stylized roman versions.

Modern romans have straight, unbracketed serifs and strong contrast in stroke thicknesses. The curved strokes are not stressed to the left, but stand upright. Examples shown are ITC Zapf Book and ITC Fenice.

ITC Zapf Book
Light

ITC Fenice
Regular

Romans, as a whole, are the most popular typefaces for reading matter in publications. Their serifs form horizontal lines across each row of type, increasing readability. Almost all newspapers and at least half of all magazines use roman typestyles for body type.

The wide family variations available also make romans suitable for advertising, book publishing, signage, and almost any other printed application.

Three other subgroups complete the serif group. Their members all have serifs and many look quite similar to romans. However, their distinctive serif treatments place them in one of the following categories: *square*, *round*, or *sharp* serifs.

Square serifs traditionally have been described by such terms as slab serif, blunt serif, American square serif, or Egyptian. All have one thing in common—their serifs are squared or blunted rather than completely tapered at the end like those of romans. Look for these features found in three common varieties:

a. *Tall serifs*, thicker than the stem stroke of the letter. Sometimes called American varieties, these

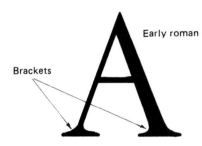
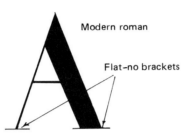

FIG. 3–18
Three distinguishing characteristics of roman typefaces.

1. Serif treatment. Early romans have rounded brackets holding serifs on. Modern romans rarely have brackets.

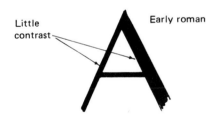
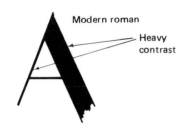

2. Stroke contrast. Early romans have less contrast between stroke thickness than modern romans.

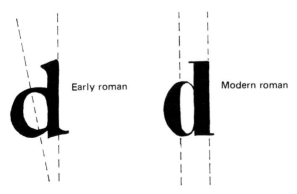

3. Angle of stress. The curved strokes of many early romans lean backwards. Modern roman curves are parallel to the stem.

types conjure up images of circus posters and Civil War headlines. Example shown: Playbill.

Playbill

b. *Moderate serifs*, no thicker than the stem stroke of the letter, and which have *contrast* in stroke thickness like romans. These are often called Egyptian varieties. Example shown: Clarendon.

Clarendon MEDIUM

c. *Moderate serifs with little or no contrast* in stroke thicknesses. These mechanical, monoline types are often referred to as slab serifs. Example shown: ITC Lubalin Graph.

Lubalin Graph Medium

Round serifs may or may not have contrast in stroke thickness, but all have rounded finishing strokes. ITC American Typewriter and Souvenir, shown here, are examples.

American Typewriter Bold

Souvenir Medium

Sharp serifs can also look like romans, with thick/ thin contrast, or they can be uniform in thickness. They all have very sharp, sometimes tiny, finishing strokes. ITC Quorum and ITC Serif Gothic are illustrations.

ITC Quorum Book

Moderate-square, round, and sharp serif types are all popular in advertising and general publications work. They reproduce well, and many new versions are being designed with high degrees of legibility.

The Sans Serif Group

The term *sans* means "without." None of the members of this large group of types have serifs of any kind. Their variety comes from stroke thickness variation and weight differences. There are only two subdivisions: *gothics* and *monolines.*

Gothics are serifless types with some variation in the thickness of strokes. Stroke contrast is not as apparent as in romans. Thickness variations are usually found where curved and stem strokes connect, and may be relatively slight. Gothics are often rather bold and sometimes a bit boxy. Gothics enjoy wide popularity for publication headings, posters, and other places where type must catch the reader's attention but still be serious in tone. ITC Franklin Gothic is one example of a gothic.

Franklin Gothic Medium

Monolines (sometimes called true sans serifs) are serifless types with *little or no variation* in the thickness of strokes. These uniform letters are often round and lightweight compared to gothics. They are highly popular for publication and advertising designs, especially where an elegant, uncluttered look is desired. Futura is a good example of a monoline typeface.

Futura MEDIUM

The Hand-Formed Group

Four subgroups make up this third groups of types: *script, cursive, text,* and *calligraphic.* All look as if they were written or lettered by hand. Some are formal, others quite informal or highly stylized (such as the flowing signature trademark in a perfume advertisement). A variety of lettering or writing instruments would be required to produce such alphabets by hand. These instruments would include broad-tipped lettering pens, fountain or quill pens, brushes, or broad- and fine-tipped markers.

Scripts are types that look like most handwriting. They are slanted to the right. Most lowercase letters are designed so that they can *connect* naturally to adjacent letters. Most capital letters are graceful and flowing, intended to stand alone without connecting to lowercase letters or competing with other capitals.

Some scripts are uniform in thickness (monoline). Others simulate beautiful handwriting done with a quill pen or fine-tipped marker.

Scripts are popular for formal printing, such as wedding announcements and social invitations. They are useful whenever the personal touch of handwriting is desired. Palace Script and Freestyle Script are shown as examples.

Palace Script

Freestyle Script

Cursives are types that slant like handwriting but allow neither lowercase nor capital letters to connect. The ends *(terminals)* of main strokes may be rounded as if produced with a soft, flexible tip, or they may be blunt as if a hard tip was used. Stroke contrast may vary or be uniform throughout.

Note: Don't confuse cursives with roman italics. Italic types have serifs; cursives do not.

Like scripts, cursives may simulate handwritten styles, but many are less formal than scripts. They enjoy popularity in retail advertising, especially where brief slogans or key phrases need emphasis. Examples shown are Dom Casual and Murray Hill Bold.

Dom Casual

Murray Hill Bold

Text types are often called black letter forms. In some type classification systems they are placed in a separate race or group by themselves. We have included them in the hand-formed group because they were originally devised to simulate the practical, though mechanistic, handwritten styles of their day.

It may help you visualize the basic form of text types if you think of Gutenberg's designs for the first movable type (Figure 3–19). These letters are formed almost completely with vertical stem strokes from a hard, broad-tipped pen. The strokes are usually tightly spaced, giving them a condensed look. Stroke contrast is great, coming only from thin strokes formed as the pen tip slides sideways abruptly to connect the thick stems.

Seriflike finishing strokes are sometimes present. These are diamond shaped and unbracketed, however, completely unlike those of the serif type group.

Today we find text types used for newspaper nameplates, certificates, announcements, and other applications where tradition or formality outweighs legibility or readability considerations. Two examples of texts are Old English and Fraktur.

Old English

Fraktur BOLD

Calligraphics are those beautiful alphabets formed from the same free-flowing strokes used when the letters are created by hand with hard-tipped pens. The graceful, sweeping strokes are totally unlike the strict vertical strokes that form most text types. Stroke contrast is strong, flowing gradually from thick to thin like early hand lettering.

These highly individualized forms are used for announcements, posters, publications, initial letters, and any other printing where a classical hand form is wanted. Two examples are ITC Zapf Chancery and Romic.

ITC Zapf Chancery BOLD

Romic BOLD

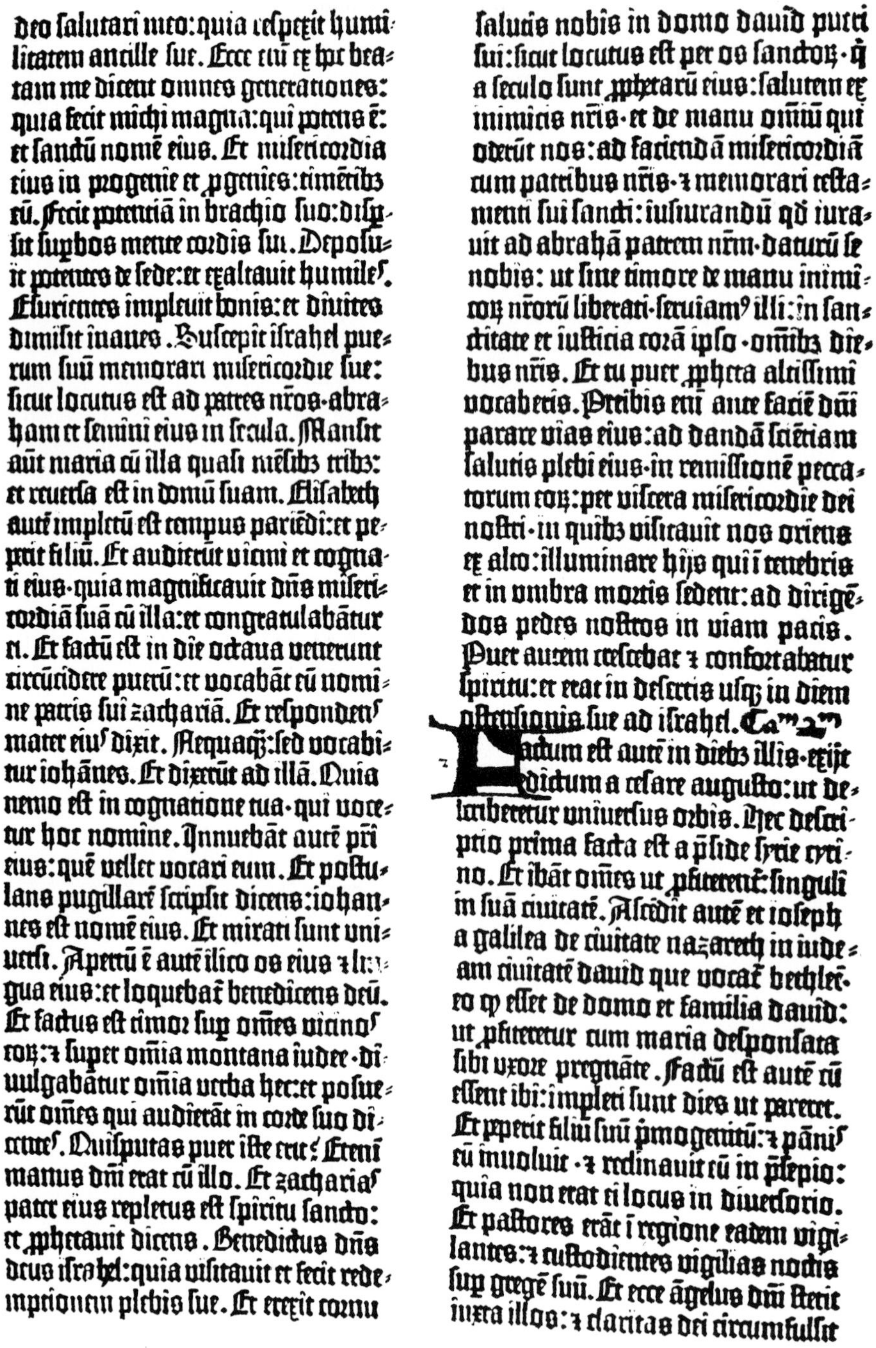

deo salutari meo: quia respexit humi-
litatem ancille sue. Ecce eni ex hoc bea-
tam me dicent omnes generationes:
quia fecit michi magna: qui potens ē:
et sanctū nomē eius. Et misericordia
eius in progenie et progenies: timētibꝫ
eū. Fecit potentiā in brachio suo: disp-
sit superbos mente cordis sui. Deposu-
it potentes de sede: et exaltauit humiles.
Esurientes impleuit bonis: et diuites
dimisit inanes. Suscepit israhel pue-
rum suū memorari misericordie sue:
sicut locutus est ad patres nr̄os abraham et semini eius in secula. Mansit
aūt maria cū illa quasi mēsibꝫ tribꝫ:
et reuersa est in domū suam. Elizabeth
autē impletū est tempus pariēdi: et peperit filiū. Et audierūt vicini et cogna-
ti eius quia magnificauit dn̄s misericordiā suā cū illa: et congratulabātur
ei. Et factū est in die octaua venerunt
circūcidere puerū: et vocabāt eū nomine patris sui zachariā. Et respondens
mater eius dixit. Nequaq̄: sed vocabi-
tur iohānes. Et dixerūt ad illā. Quia
nemo est in cognatione tua qui voce-
tur hoc nomine. Innuebāt autē pri
eius: quē vellet vocari eum. Et postu-
lans pugillarē scripsit dicens: iohan-
nes est nomē eius. Et mirati sunt uni-
uersi. Apertū ē autē ilico os eius et lin-
gua eius: et loquebatur benedicens deū.
Et factus est timor sup omnes vicinos
eorū: et super omnia montana iudee di-
uulgabātur omnia verba hec: et posue-
rūt omnes qui audierāt in corde suo di-
centes. Quis putas puer iste erit? Etenī
manus dn̄i erat cū illo. Et zacharias
pater eius repletus est spiritu sancto:
et prophetauit dicens. Benedictus dn̄s
deus israhel: quia visitauit et fecit rede-
mptionem plebis sue. Et erexit cornu
salutis nobis in domo dauid pueri
sui: sicut locutus est per os sanctorum q̄
a seculo sunt prophetarū eius: salutem ex
inimicis nr̄is et de manu omniū qui
oderūt nos: ad faciendā misericordiā
cum patribus nr̄is et memorari testa-
menti sui sancti: iusiurandū qđ iura-
uit ad abrahā patrem nr̄m daturū se
nobis: ut sine timore de manu inimi-
corum nostrorū liberati seruiamus illi: in san-
ctitate et iusticia corā ipso omnibꝫ die-
bus nr̄is. Et tu puer propheta altissimi
vocaberis. Preibis eni ante faciē dn̄i
parare vias eius: ad dandā scientiam
salutis plebi eius in remissionē pecca-
torum eorum: per viscera misericordie dei
nostri in quibꝫ visitauit nos oriens
ex alto: illuminare hijs qui in tenebris
et in umbra mortis sedent: ad dirigē-
dos pedes nostros in viam pacis.
Puer autem crescebat et confortabatur
spiritu: et erat in desertis usq; in diem
ostensionis sue ad israhel. Ca. ii.
Factum est autē in diebꝫ illis exijt
edictum a cesare augusto: ut de-
scriberetur uniuersus orbis. Hec descri-
ptio prima facta est a preside syrie cyri-
no. Et ibāt omnes ut profiterentur singuli
in suā ciuitatē. Ascēdit autē et ioseph
a galilea de ciuitate nazareth in iude-
am ciuitatē dauid que vocatur bethle-
em eo q̄ esset de domo et familia dauid:
ut profiteretur cum maria desponsata
sibi uxore pregnāte. Factū est autē cū
essent ibi: impleti sunt dies ut pareret.
Et peperit filiū suū primogenitū: et pānis
eū inuoluit et reclinauit eū in presepio:
quia non erat ei locus in diuersorio.
Et pastores erāt in regione eadem vigi-
lantes: et custodientes vigilias noctis
sup gregē suū. Et ecce āgelus dn̄i stetit
iuxta illos: et claritas dei circumfulsit

FIG. 3–19
A page from Johann Gutenberg's famous 42-line Bible, printed between 1452 and 1456.
Courtesy of Richard G. Gray

The Specialized Group

Describing this final group of types is like leading someone to the aviary at a zoo and proclaiming, "Here are the birds." So many splendid varieties exist that it seems almost impossible to begin sorting them out. Indeed, the specialized group is a miscellaneous category; if a type doesn't fit logically anywhere else, it probably belongs here.

Two broad subgroups can be identified: *novelty* and *decorative*.

Novelty types   are those of unique or novel design. A unifying element is present upon which the design of the entire face is based. Some types may suggest specialized functions or hardware (such as computers). Others establish a particularly direct message tone or mood.

Novelty types are either solid in color or designed as illustrations of nontypographic objects, such as a rope or leaves. Data 70 and Quicksilver are examples.

Decorative faces   form the second specialized subgroup. Three-dimensional types (called *shadowed* or *shaded*) are included in this category. Also included are highly ornate or "flowery" designs. In addition, all types with lines added to the outside (called *contours*) or inside (called *inlines*) of the face are included. Finally, those types which are not solid in color or tone, or those which contain light-to-dark or other tonal variations are part of this subgroup.

Uses for either novelty or decorative types are limited only by our imaginations.

Note: Some families normally found in other type groups have special versions or styles that seem to belong in either the decorative or novelty categories. For example, Souvenir, a popular round serif type, also sports an attractive shadowed style. It is better to look first at the *parent* design or family when attempting to locate a type, because specialized versions of type families are more often grouped with their parent family rather than with miscellaneous or specialized types. Most truly specialized faces, however, are designed as individuals, not as families.

The following examples suggest the great variety to be found in the decorative subgroup: Harlow, Sapphire, and Gallia.

Using the Classification System

A good way to become familiar with typefaces is to identify examples in magazines and newspapers and try to classify them according to the table in Figure 3–17. Don't be discouraged if some of the types you find seem to defy pigeonholing, or else seem to fit everywhere. Not everyone agrees exactly where each typeface belongs. It's not the exact classifying that's important here. What counts is that you are noticing and learning to recognize the differences in type.

Selecting Typefaces

The following points are helpful to remember when selecting typefaces.

1.   Avoid getting in a rut with the same old "tried and true" type families (Helvetica, Souvenir, Times Roman, or Goudy, to name a few popular ones). From a choice of thousands, you can probably find exactly the typeface you need.

2.   Remember that typeface names can be misleading. For example, the typeface Smoke may look more like ink blots to most people. Be sure you haven't been more influenced by the name of a face than by its actual appearance.

3.   Select type based on actual need. Remember that your audience will see only the letters, numbers, etc. contained in the actual message. Thus the beautiful capitals in the typeface Vivaldi will be lost if your message is in lowercase.

4.   Check for alternate characters (different versions of certain letters, such as the lowercase *e* and *a*), ligatures (letter combinations such as *fi, fl, ffl*), and other special font characteristics. Be sure you use alternate characters consistently throughout your printed design.

Additional Resources

The nondesigner can help make up for a lack of
typographic experience by accumulating and studying
type reference materials. Ask local typesetters and
printers for type specimen books. When you visit art
and graphics supply retailers, request catalogs
containing transfer type samples. Buy type reference
books or borrow them from school or public libraries.

Begin collecting advertisements, brochures,
newspapers, and magazines with examples of effective
or innovative typography. Soon you will gain some of
the confidence you need to design with type
creatively.

4 TYPOGRAPHY

From Simple Symbols to Professional Messages

Picking the right typefaces is just the beginning in making type work for you. Sensitive blending of typographic symbols with other graphics and choosing appropriate combinations of size, configuration, space, and special effects without sacrificing clarity and purpose—these are the ultimate challenges.

Introduction

Typography today is both art and science. Alphabet designers must understand how their images will reproduce electronically. Those who design with type must appreciate its unique effect on our ability to read and understand the printed word.

It is more important now than ever that we learn to use type wisely. With the advent of electrographic printing and other high-speed imaging processes offices, schools, libraries, even homes are attaining systems with sophisticated type-imaging capacities. No longer are consumers limited to the typestyles, sizes, and formats of professional typesetting houses. As surely as we have come to expect our education to include some experience with computer terminals, so do we also expect more freedom to control our own communications. With this newfound freedom comes the challenge to create effective graphic and typographic communications.

After almost half a century of substantive typographic research, we still rely as much upon professional trends, traditions, even fads, as we do upon scientific data. Much research on type was done before many of today's best typefaces even existed. And, because of the number and diversity of today's type designs, it is increasingly hard for us to say how much the results of one type study accurately predict effects when other typefaces and formats are considered. We simply haven't adequately tested most of the countless combinations of new typefaces and typographic formats.

This chapter discusses commonly accepted traditions, research generalizations, and innovations in modern typography. As we learn more about type in use, no doubt some of the rules we list here will be modified.

Why Typeset at All?

What are the advantages of typesetting as opposed to typing?

1. Typesetting saves page space. Depending upon size, style, and format, typeset material may reduce the space needed by half. This means savings in paper, printing, mailing, and storage.

2. Typeset material can be read faster. Estimates vary, but reading speed and comprehension may improve by up to one-fifth.

3. Typesetting can be cheaper. Standard typewritten material must be retyped when revisions are necessary. Text processing systems, designed for typesetting applications, store character keystrokes electronically. These characters can be recalled, revised, and printed out easily in many forms. Word processing systems (see Chapter 7) offer similar advantages if linked (interfaced) to typesetters.

4. Typeset material looks more professional. It is more attractive, more readable, and typefaces can be coordinated to reinforce an organization's identity in print. With professionalism also comes credibility. We tend to have more faith in permanent-looking print.

5. Typewritten material sometimes is used to emphasize the timeliness and personal nature of

newsletters and special reports. This practice may diminish as more and more people begin to take both high-speed and high-quality type imaging for granted.

6. Crude, distorted typestyles common with some computer printers are rapidly giving way to the typeset look.

The Dimensions of Typography

Gutenberg's wondrous invention, movable type, came into the world with a bonus—he also showed us how to use it sensitively. His 42-line Bibles remain priceless models of beauty and utility in print.

We know that many typographic messages, even without other visual complements, have the power to attract attention, to persuade, to crystallize thought, and to initiate action.

What sets effective printed messages apart from the ineffective? Let's examine typographic variables that help determine how people perceive printed words, along with some practical rules on how to use them to the best advantage. As with most rules, these have exceptions. Be on the alert for creative ways to bend or break them once you understand their basic importance.

Contrast Variables

Size

None of the contrast variables discussed here is as important or as easy to use as size. Varying point sizes, as in the headlines of a newspaper, help establish a reading priority by "grading" the importance of a message in relation to the words around it. In an advertisement, size contrasts help maintain reader interest in each separate portion of the message. Common size guidelines include:

1. Body type sizes for newspapers, magazines, and newsletters normally should be from 8- to 11-point. Ten-point type has scored consistently well in the research on reading ease. Some brochure designs require larger body type, but usually only for emphasis. Very young readers and elderly readers appreciate sizes larger than 8-point.

2. Consider setting captions one point size smaller than the body type size used, even if another typeface is used.

3. When contrasting two display type sizes, make one about half the size of the other—for example, a 48-point magazine title would contrast well with a 24- or 30-point subtitle. Don't be afraid to use a much greater size difference for special impact, however.

4. Always consider x-height when sizing body type. Types with large x-heights appear larger than those with smaller ones, even though they are the same point size (Figure 4–1).

Type Groups

Selecting and mixing type from major groups or races of type requires care. Major differences in design between groups provide easy contrasts. However, contrast may be less important than compatibility. Here are some guidelines for choosing and mixing type groups:

1. Avoid mixing type groups at all, if possible. The safest approach is to use weight and style variations of the same type family—for example, Souvenir in extrabold, bold, and bold italic.

Large Medium Small
} Same point size

FIG. 4–1
Differences in x-height for three typefaces of the same point size.

2. If type group mixing is desired, try to use only two compatible subgroups, one from each main group. For example, an unusual novelty type, used with a plain sans serif, will be enhanced by the contrast. Publication designers often use types from two groups to break up monotony or to "index" content.

3. Generally, avoid mixing roman typestyles, especially early romans. Romans are too similar in stroke and serif design to compete with one another. Mixing romans may be fine, however, if highly contrasting weights and styles are chosen.

4. Don't worry too much about mixing sans serif display with roman body type (and vice versa). This used to be taboo, but magazines and newspapers ignore this prohibition today. Text and headlines often are linked typographically with initial letters that match the headline typeface.

5. For body type, consider serifed type first. Studies have confirmed that readers generally prefer romans and other serif types, especially for lengthy stories. Serifs at the top and bottom of letters form horizontal lines that keep the eye moving easily across the page. The vast majority of newspapers use roman body type. Magazines are more likely to seek sophistication in sans serifs or nonroman, serifed designs.

6. Don't use script or cursive typestyles in all capitals. The capital letters in most of these faces were designed to stand alone with lowercase letters, not fight with one another for attention.

7. Use italics and boldface designs sparingly in body type. Italics are good for emphasis, quotations, captions, and wherever a little extra attention is needed.

Spacing

We consider five kinds of space when working with type:

- space taken up by the letters themselves (*unit widths* or values)
- space between words (interword or simply *word spacing*)
- space between letters (intercharacter or *letter spacing*)
- space between lines of type (interline space or *leading*)
- surrounding space (framing or *negative space*)

Three increments of space are horizontal: space between words, space between characters, and space required for individual characters.

Today's horizontal spacing originated from the system used for metal type. This system is based on the em (or quad), a two-dimensional square of the type size in question. An em is about as wide as a capital M.

An em space in 24-point type is 24 points tall and 24 points wide. A 10-point em is 10 points tall and 10 points wide.

An old-fashioned metal em space is three-dimensional, square on top and not quite as tall as a regular piece of metal type. It is used to fill large areas of space in hand-set lines as they are being composed in a hand-held tray called a type stick.

Figure 4–2 shows the use of the type stick and the California Job Case in which an entire font of type and its appropriate spacing material is contained. Figure 4–3 shows spacing material other than the em, all based on fractions of the em. Each kind of spacing is contained in a separate compartment of the Job Case.

Spacing of Words. Traditionally, a 3-em-space (one-third of the width of an em) was used for hand-set word spacing. Today's sophisticated typography often demands tighter spacing—about one-fourth of the em width (a 4-em-space).

Let's see what the actual difference in word spacing would be: We use a 3-em-space between words for traditional typesetting. In 36-point type, this is one-third of the width of a 36-point em, or 12 points between words. If we want a more modern look, we use a slightly thinner 4-em space. This is one-fourth of the 36-point em, or 9 points of space between words.

Of course, we no longer place metal spaces between words. Spacing is now controlled by computer in increments called units—discussed later in this chapter. When ruboff or transfer type is used, spacing is estimated visually.

FIG. 4–2
*Line of type set from California Job
Case.*
Photo by Pat O'Donnell

For quick reference when you are using transfer type, the width of a lower case *t* is about right for traditional spacing. A width slightly greater than the capital *I* is about right for modern spacing.

Consistency in word spacing is important for reading ease. It is affected most by two things: line format or mode, the manner in which lines of type are set; and how well compensations are made for differences in the shapes of letters appearing on either side of word spaces.

The three columns of type in Figure 4–4 show the effects of line width and column format on word spacing. The first is set with justified (even right and left) margins in a narrow width. The second is also justified, but set much wider. The third is set unjustified, with flush left and ragged right margins.

In justified typesetting, space is used to spread words evenly across each line. The result is straight margins on both sides. The lines in example 1 are fairly short,

so the number of words and spaces per line varies widely. Note how the spaces between words form irregular patterns down the column. These "rivers" of space detract from horizontal eye flow and inhibit reading. They also produce unattractive breaks in the even "texture" of body type.

Example 2, set wider, allows more words and spaces per line; therefore, word spacing is more even. Note that a few lines still have noticeably wider spacing.

Sophisticated hyphenation programs for typesetting minimize word space problems in moderately-short line lengths. However, some word space variation is to be expected in justified typesetting.

When type is set unjustified, as in example 3, all word spacing is exactly the same. An ideal word spacing value is programmed into the typesetter. As the machine sets each line, spacing each word identically, any remaining space is placed at the end of the line.

Example 3 is set with even left and ragged right

Early metal quads (EMS) were used to space whole lines in a type stick. Today, "quadding" refers to the positioning of type—left, right, or center.

EM (quad)

ENS are simply half-EMS. Previously used for spacing lines, they are now used—along with EMS—to indicate paragraph indentions.

$\frac{1}{2}$ $\frac{1}{2}$

EN ($\frac{1}{2}$ of EM)

This space is *traditional* for word spacing. The lowercase t provides a good visual estimate of this amount of space.

$\frac{1}{3}$ $\frac{1}{3}$ $\frac{1}{3}$

3-EM space ($\frac{1}{3}$ of EM)

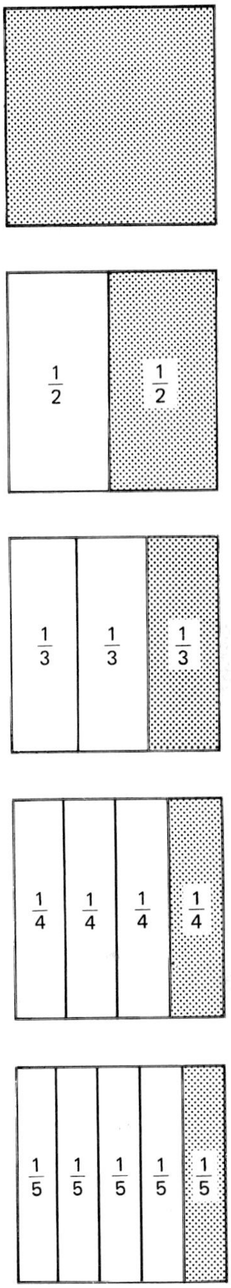

FIG. 4–3
The em and its simple divisions. We still find these useful for discussing type spacing even though we no longer use metal type.

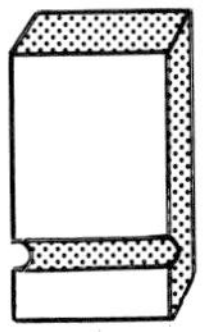

This space is better for the more *contemporary* word spacing popular today. The width of a lower case i (with serifs) gives a workable space estimate.

$\frac{1}{4}$ $\frac{1}{4}$ $\frac{1}{4}$ $\frac{1}{4}$

4-EM space ($\frac{1}{4}$ of EM)

Side view of a metal quad—the square of any type size. Years ago, printers gambled with quads, rolling five of them like dice and counting the number of nicks that landed face up.

When type is set very tight, word spacing may be reduced to this width or less. It is best to avoid such spacing, since it can hamper reading.

$\frac{1}{5}$ $\frac{1}{5}$ $\frac{1}{5}$ $\frac{1}{5}$ $\frac{1}{5}$

5-EM space ($\frac{1}{5}$ of EM)

1. Type justified in a narrow column width produces great variations in word spacing. Result: rivers of space down the column.

2. Type justified in a wider column is more pleasing since word spacing is divided over more words per line.

3. Unjustified type provides even word spacing, since all extra space is placed at the end of each line.

FIG. 4–4
Effects of column width and justification on word spacing.

margins. Notice how the consistent word spacing eliminates rivers and provides an even gray texture to the ragged right column.

Modern typesetters can be programmed to set with or without hyphens, and to allow no more than a

predetermined amount of space at the end of any line. This is called "maximum rag." If more than the desired amount of space appears at the end of the line after hyphenation and justification (H & J), the lines are identified by the computer so the operator can reword copy or hyphenate manually.

Letter Widths and Letterspacing

Individual letter widths and space between letters are two other important horizontal spacing variables. They are determined by a system of units. Units are simply very fine vertical fractions of the em or quad—much finer than the thirds or fourths of a quad we used on page 37 to describe word spacing. The Linotype, the best-known hot metal linecasting machine, used a popular system based upon 18 units per em. Today's sophisticated computers are more likely to use 36, 54, or even more units per em.

Each letter in a type font is designed to be a certain width in units. This width includes a little extra space on each side of the letter to prevent its touching adjacent letters in typesetting.

Figure 4–5 shows an em divided into 54 units. The point size of the em doesn't matter here. There would be 54 units regardless of the em's size. The capital W, a 54-unit character, is shown within the em. Two letters, the capital T and the lowercase j, also are shown to indicate their widths relative to the 54-unit em. Note that capital T is 39 units wide, while lowercase j is only 15.

The unit system is just a refinement of the simpler word spacing system discussed earlier. For example, in a 54-unit system we can produce traditional ⅓-em spacing by putting 18 units, or one-third of the width of a 54-unit em between words.

Letterspacing

Control of letterspacing also is based on unit system increments. Tighter fitting of letters has become a strong trend in professional typesetting, especially for large display type. Tighter-than-normal letterspacing is created by programming the typesetter to subtract a certain number of units automatically from between each letter combination. This is known as minus-unit letterspacing or tracking. Wider-than-normal (plus-unit) letterspacing also can be used, although this effect is rarely pleasing in body type.

Figure 4–6 indicates examples of normal, minus-1,

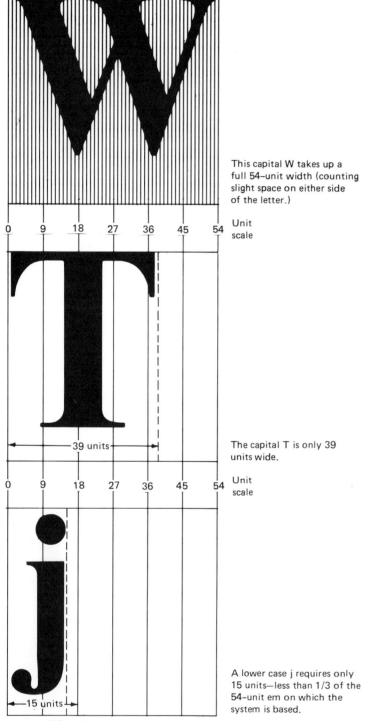

This capital W takes up a full 54–unit width (counting slight space on either side of the letter.)

Unit scale

The capital T is only 39 units wide.

Unit scale

A lower case j requires only 15 units—less than 1/3 of the 54–unit em on which the system is based.

FIG. 4–5
Letter width values based on 54-unit em.

Four score and seven years ago our fathers brought

Four score and seven years ago our fathers brought forth

Four score and seven years ago our fathers brought forth

Four score and seven years ago our fathers brought forth on

FIG. 4–6
Examples of 10-point type set with units subtracted from between each letter (based on an 18-unit system).

minus-2, and minus-3-unit spacing. The type shown is based on an 18-unit system. Notice that, as more units are subtracted from between each letter combination, the closer they come to overlapping. The last line, with minus-3-unit tracking, results in most letters touching, an effect that could impair reading.

Normally, minus-1-unit letterspacing (on an 18-unit system) is about right for most body type.

Typesetting houses often describe letterspacing in general terms such as loose, tight, or touching, but they usually can provide any degree of precision you specify. Find out what unit systems their typesetters use and what their actual letterspacing limits are.

Kerning

Space adjustment between individual letters is known as letterfitting or kerning. This is space adjustment beyond that done equally between all letters. Certain pairs of letters leave too much space between themselves when set normally. Extra units must be subtracted from between these character pairs. Examples are VA, Yo, and LY.

Figure 4–7 shows several examples of letter pairs both kerned and unkerned.

Most computer typesetting devices today can be programmed for automatic kerning. Sophistication of kerning programs varies widely. Many systems now offer user-programmable features, allowing kerning software to be designed for each separate typeface in all sizes.

Figure 4–8 is a typical table showing kerning pairs and the unit values that must be subtracted for a single typeface.

A recent kerning innovation is the universal kerning program. It allows a computer to memorize the topographical profile of each letter to determine how close to its edges adjacent letters may be set.

For example, the computer would be programmed to account for the wide-to-narrow shape of a capital *A*. It also would account for the opposite shape of a capital *T*. A width value for each letter, at its base, top, and several points in between, is stored in memory. By rapidly comparing the width values of both letters, the computer would subtract the number of units required to produce a pleasing fit. The same logic would be applied to letter pairs such as *L* and *Y*.

Universal kerning eliminates the need for separate kerning tables for each typeface.

The diagram in Figure 4–9 indicates how the computer assigns a numerical value to each portion of a letter's height, according to its topographical profile.

Vertical Spacing

In an earlier chapter, we discussed a very important vertical type measurement—type size. We said that the measurement from the highest ascender to the lowest descender of a typeface approximated its point size. We also discussed the measure of interline spacing.

Interline spacing, or leading, is any space placed between lines of type beyond that required by the point size. When we set type without leading, we are "setting solid" with no extra space between lines. Example 12/12, or 12-point type on a 12-point body. If we add two points of leading, we are now setting 12/14, or 12-point type on a 14-point body.

Interline spacing actually refers to space placed beneath each line of type. In the days of hot metal typesetting, leading was done by inserting a thin strip of metal (called a lead) between lines. Type on linecaster slugs also could be cast with a thin shoulder to provide built-in leading under each line. Today, we add leading in the computer by advancing the photographic paper slightly more than the height of the point size as each line is set (See Figure 4–10). This is somewhat like double- or triple-spacing typewritten lines to make them more readable.

One of the most prominent typographic trends of the past decade has been the great increase in the use of negative leading. Negative or minus leading is created by reducing rather than increasing the space allowed for each line of type.

Negative leading is indicated by a fraction in which the point size is larger than the body. For example, the fraction 36/32 calls for 36-point type set on a 32-point body. This indicates four points of negative leading.

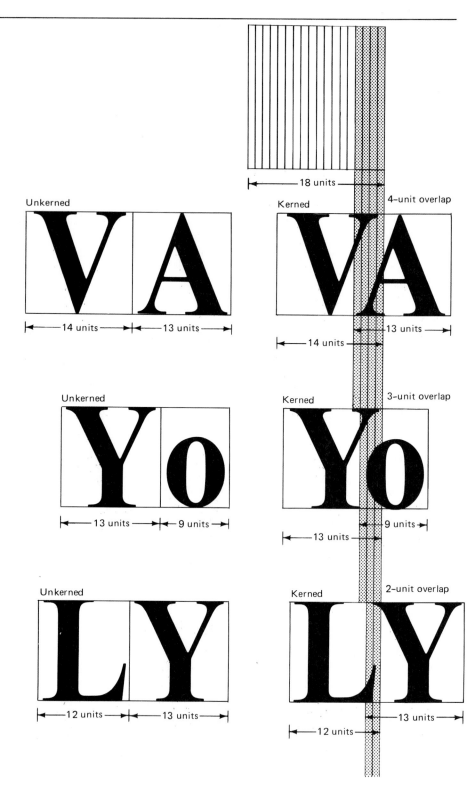

FIG. 4–7
Three letter pairs with and without kerning (18-unit system).

```
KERNING  RECORD  =         1
  6   62    5  †)      6   64    5  †`     18   27   -5  r.     18   28   -5  r,     22   27   -5  v.     22   28   -7  v,     23   27   -5  w.
 23   28   -5  w,     25   27   -5  y.     25   28   -7  y,     27   64   -7  .`     28   64  -10  ,`     33   22  -15  Av     33   35   -5  AC
 33   39   -5  AG     33   47   -5  AO     33   49   -8  AO     33   52  -10  AT     33   53  -10  AU     33   54  -12  AU     33   55   -8  AW
 33   57  -10  AY     33   64   -5  A`     34   33   -5  BA     34   57   -8  BY     35   33   -2  CA     36   33   -5  DA     38    1   -5  Fa
 38    5   -5  Fe     38   15   -5  Fo     38   27  -10  F.     38   28   -8  F,     38   33   -8  FA     44    5   -2  Le     44   25   -5  Ly
 44   28   -2  L,     44   47   -3  LO     44   52   -5  LT     44   53   -5  LU     44   54  -10  LU     44   55   -8  LW     44   57   -5  LY
 44   64   -5  L`     47   27  -10  O.     47   33   -5  OA     47   52   -3  OT     47   54   -5  OU     48    1   -5  Pa     48    5   -5  Pe
 48   27  -10  P.     48   28  -10  P,     48   33  -10  PA     52    1   -8  Ta     52    5   -5  Te     52   15   -8  To     52   21   -8  Tu
 52   27  -10  T.     52   28   -5  T,     52   33   -8  TA     52   47   -3  TO     53   33  -10  UA     54    1  -10  Va     54    5  -10  Ve
 54   15  -10  Vo     54   21   -8  Vu     54   27  -10  V.     54   28  -10  V,     54   33  -12  VA     55    1   -8  Wa     55    5   -8  We
 55   15   -8  Wo     55   21   -8  Wu     55   27  -10  W.     55   28  -10  W,     55   33  -10  WA     57    1   -8  Ya     57    5   -8  Ye
 57   15   -8  Yo     57   21   -8  Yu     57   22   -8  Yv     57   28  -10  Y,     57   33  -10  YA     64   27   -7  `      64   28   -7  `,
```

The printout shown is a Kern record for Baskerville, based on standards set up for the Compugraphic 8600 phototypesetter for kerning specific character pairs. Each column block is divided into three columns: the first two are the font locations of the two characters to be kerned, and the third column shows the character pair and the amount of space to be added or taken away from between the characters. For example, the third combination of characters in the first column block is AG. The A is in position 33 on the type master. The G is in position 39. Each time these two letters occur together in typesetting, 5 units of space are subtracted automatically. In this example, an em is 54 units wide. Once a kern record is loaded into the system, whenever the family is called for—Baskerville in this case—the system will automatically set the type based on that record. 500 kerned pairs are available per font, and these pairs are user definable. No two fonts need to be the same.

FIG. 4–8
Kerning table for one typeface.
Penta Systems International Inc.

FIG. 4–9
Topographical model for capitals A, T, and O in a universal kerning program.
Bert De Pamphilis/World Typeface Center

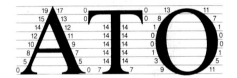

1. Letters are exposed to typesetting paper.

2. Paper advances far enough to allow next line to be exposed.

3. Next line is exposed. The amount of leading is determined by how far the paper moves.

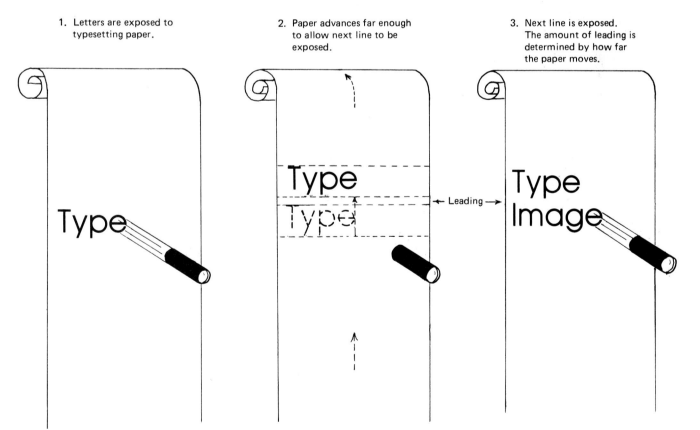

← Leading →

FIG. 4–10
Advancing the paper after each line is set to provide leading.

Many of the most attractive advertisements we see in today's publications are produced with some negative leading, especially in their all-capital headlines.

Figure 4–11 gives two examples of negative leading. Notice what can happen when ascenders and descenders overlap.

Framing or Negative Space

Space that surrounds type creates a field within which the image is displayed. This space can be used creatively to enhance a typographic message.

The use of framing space is important in page layout and design. This nonimage or "white" space, discussed more thoroughly in Chapters 8 and 9, influences how our eyes move within a layout.

The shapes of letters themselves determine space irregularities or profiles. Individual letters, when combined to form words, produce a distinctive word profile which aids reading. Rather than pausing to recognize each letter, a reader learns to recognize word shapes.

Legibility is a term applied to the ease in which we can recognize type in blocks. It usually refers to display type. The term readability—when applied to typography, not writing—refers to how easily we read masses of body type. Figure 4–12 shows why lowercase letters usually are preferred over capitals for legibility. Note the distinctive profile of the

lowercase letters. Words like "the" form unique shapes—in this case, an "old shoe." All-capital words form a nondescriptive box.

The color of the background—black, white, tones of gray, or colored hues—affect both the legibility and readability of type. Studies have shown that when body type is reversed (printed as white on a dark background), readability drops about 15 percent.

A background for reverse type can be tinted from black to various shades of gray. The gray tint must be dark enough to provide contrast for the white type. Black type should be printed over light gray background tints. Tints are produced by screening solid black into hundreds of tiny dots. Type, either printed over those dots or reversed within them, should be neither too small nor contain fine hairlines and serifs.

Reversed and overprinted (surprinted) type is used extensively in brochures, magazines, and printed advertisements of all kinds. Where these techniques add emphasis and impact to the design without being overused, they are very effective.

Background colors, rather than black and gray, enhance typographic effectiveness if chosen carefully for contrast. Care also must be taken to avoid the same destructive effects from color dots that black dots can have on delicate hairlines and serifs.

Where type is reversed from a background made up of several colors of screened dots, precise printing (registration) of dots is essential. Otherwise, a faint

FIG. 4–11
ITC Bookman™ Demi with swash, set 48/42 (6 points negative leading).

A rectangle

A distinctive shape
(like an old shoe,
in this case)

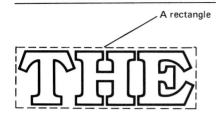

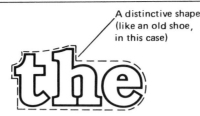

All capitals can form
nondescript rectangles
that make rapid reading
difficult.

Lowercase letters
have height differences
which give them a
distinctive profile. This
generally makes them
easier to read.

Capitals are popular for display typesetting.
They are neither boring nor difficult to read
if set in special faces such as the ITC^TM
Bookman Swash letters above.

FIG. 4–12
Caps versus lowercase letters.

tint of color will be visible at the edges of the white letters. This may create a "muddy" look, or hopelessly blur the letters.

Space between blocks of type affects meaning and impact. An important rule with type, as with most graphic elements, is that related units should be closer to one another than unrelated elements.

While research has provided no perfect formula, several guidelines have evolved for estimating an appropriate column width for body type.

Column width (line length) determines how many words will fit per line. Typographic research has shown that we read type in short takes or "fixations"—several words in a phrase or group—not one word at a time. The fewer pauses we must make to absorb meaning, the faster and more thoroughly we read and comprehend.

Short column widths require many breaks in phrases, word groups, even words themselves, since only a few characters will fit on a single line. Very long lines require fewer breaks; however, they also force us to search harder for the beginning of the next line each time we return to the left edge of the column. Moderate line lengths are a good compromise.

Typographic researchers M. A. Tinker and D. G. Patterson found years ago that, of the body type formats they studied, the most readable overall was

10-point type on a 19-pica line length. The typical column format of today's magazines and newspapers—about 8- or 9-point type set on a 14- or 15-pica width—comes very close to this size-to-width ratio.

A commonly accepted rule of thumb for finding the best line length is this: multiply the lowercase alphabet length by 1.5. Give the answer in picas.

Example: abcdefghijklmnopqrstuvwxyz

The line of 8-point Eras Bold, above, measures 9 picas from end to end. Multiplying 9×1.5, we get $13\frac{1}{2}$. Therefore $13\frac{1}{2}$ picas would be a good line length to consider.

The above formula, as with all guidelines given here, provides only a starting point. Lines somewhat shorter or longer than $13\frac{1}{2}$ picas usually would be fine.

Here is an easier formula for finding approximate line length: multiply the point size of type to be used times two, and give the answer in picas.

Example: 9-point type $\times$ 2 equals 18.
            Answer: Use an 18-pica line length.

Another popular rule of thumb that is independent of point size suggests that the number of characters per line should range from about 40 to 60, and should never exceed 75.

Typesetting Modes

Several common typesetting modes are used today. Lines of type can be justified, unjustified (ragged left or right margins), centered, staggered, run-around or contoured to fit illustrations, or molded into a special shape.

Most newspapers and books use justified columns of type. Blocks of type with even margins are easy to read and give structure to a page.

Studies performed in the late-1960s showed that most readers neither knew nor cared when newspapers changed from justified to ragged columns. However, most papers have stayed with justified modes.

Magazines, annual reports, and brochures are more likely to use ragged right columns for a contemporary look.

The advantages of even word spacing and the setting of large type in narrow columns makes unjustified typesetting highly popular today. Ragged (rag) right is used most often because it allows an even left-hand margin for easy reading. Ragged left usually is reserved for special applications, such as when a caption is placed to the left of a photograph.

Thin lines (still called column rules) are sometimes placed between columns of ragged right type to add structure to the page. Also, paragraph indentions may be omitted in ragged setting.

The centering mode is used mostly for display type or body type in advertisements. While centered type appears balanced, it presents the problem of two irregular margins which can hamper reading.

Staggered type is rarely used except for poetry and other special typographic effects.

Indention of type can be done in several ways. A hanging indention mode is used where several items are placed under one heading, but each one must stand out. The first line is full length, then all lines beneath it are indented one or two ems.

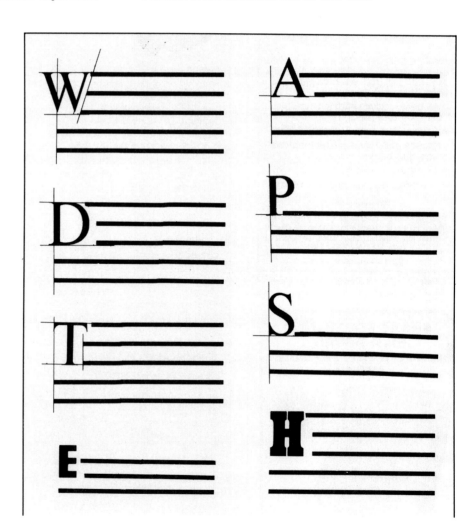

FIG. 4–13
Eight popular ways to use initial letters.
© 1982 Graphic Arts Monthly, Technical Publishing

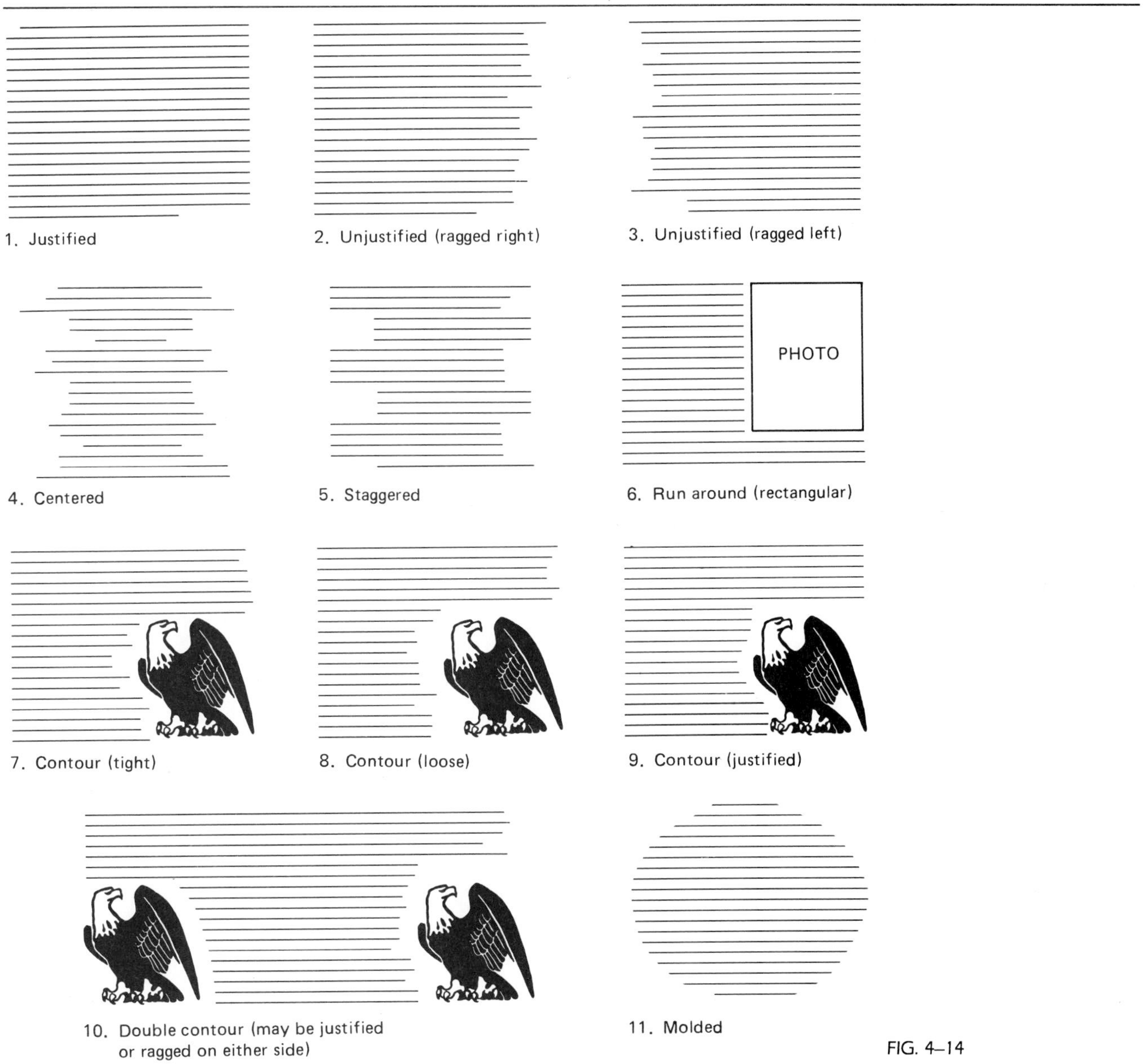

1. Justified

2. Unjustified (ragged right)

3. Unjustified (ragged left)

4. Centered

5. Staggered

6. Run around (rectangular)

7. Contour (tight)

8. Contour (loose)

9. Contour (justified)

10. Double contour (may be justified or ragged on either side)

11. Molded

FIG. 4–14

Paragraph indentions are normally one-em wide for average column widths, but may be two or more ems for wide columns. When paragraph indentions are not used at all, a blank space equal to the thickness of one line is left between paragraphs.

Initial letter indention is done to accommodate an oversized character that begins the first word of a story. The first few lines of body type must be indented to form a shape compatible with the large letter. The indention may be either rectangular or contoured, depending on the shape of the letter. Unless this letter forms a complete word such as A, the first line of body type must be placed close enough to it to complete the first word. Figure 4–13

shows several ways initial letters can be integrated into body type, as suggested by *Graphic Arts Monthly* phototypography columnist Carl P. Palmer.

A run-around or right-angle "wrap" is a temporary change in line length in order to allow body type to be set around an image such as a photograph. This abrupt break in body type works best with rectangular images. When used with illustrations having irregular edges, a run-around can look too mechanical.

Along with the increase in complex, irregularly shaped illustrations in publications today, there has been a strong move toward contoured typesetting. It has been bolstered by the ability of newer typesetting systems

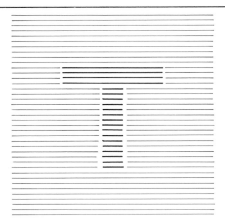

FIG. 4–15
Typographic texture is shown by the contrast of the capital T—set in bolder type—against a background of medium-weight type.

Typographic Texture

Texture (or "color") of body type refers to its apparent density when viewed as a unit. Weight, size, and spacing all affect typographic texture. Figure 4–15 shows how texture can provide contrast on a page. The capital *T* is formed by a block of type with heavier texture than that surrounding it. Pages on which light-textured body type is printed tend to require more elements in order to provide relief from the monotony of gray columns.

Typography Today

The remainder of this chapter is devoted to examples of the more common effects that designers and typographers seek to achieve through the beauty and versatility of type.

to manipulate characters on terminal screens. A person operating a sophisticated terminal can wrap a type block around any illustration or photograph in a fraction of the time it would take by earlier trial-and-error methods.

Type contours can be tight—within about a pica of the image—or loose (several picas away). They can be curvilinear, with the edge of the type in "flowing justification" with the edge of the illustration, or they can be ragged. Type can also be double-contoured, with two curvilinear edges. Double-contours are required where type must run between two irregular shapes simultaneously. Most double-contours are set with a ragged right edge.

Molding is simply the creation of an illustration graphic by forming body type into a distinctive shape. This mode of typesetting is very time-consuming unless executed on a terminal which allows previewing the image in actual size and justified form.

The illustrations in Figure 4–14 show the major body typesetting modes.

NEGATIVE LEADING
By tightening interline spacing, the designer pulls the top four lines together as a unit and emphasizes the last line.
Knoth & Meads Company

Consumer Orientation
No. 20 in a Series
of Technical Papers
Subject: Introduction of the 928S.
New Power. New Performance. New
Parameters of Comfort and Luxury.

Porsche 928S

At Porsche, our philosophy is to
design, test, produce, and constantly improve. The new 928S em-
bodies this tradition and is the proud
successor to the 928. Consider its liquid-
cooled, fuel-injected, aluminum-alloy V-8
engine. Displacement has been increased to
4.7 liters. And output has been raised to 234 hp.
On the track, with manual transmission, the 928S
accelerates from 0 to 50 mph in 5.2 seconds. It reaches
the ¼-mile mark from a standing start in 15.2 seconds at a
speed of 90 mph. And it has a maximum speed of 146 mph. The
928S transaxle design places the engine in front and transmission in
back. It produces a nearly-perfect 50-50 front-to-rear weight distribution for
improved cornering and balanced braking. And it creates a high polar moment of
inertia that reduces pitching, resists cross-winds, and increases directional control.
The 928S' unique Weissach rear axle optimizes rear-wheel alignment during deceleration
or braking and while cornering. A kinematic effect changes toe-out to toe-in in no more than 0.2
seconds to control oversteer. The 928S' aerodynamic design includes integral front and rear
spoilers to reduce lift and improve road holding. To optimize driver performance, standard equipment
includes: An adjustable-tilt steering column and instrument cluster. Power-assisted, variable-boost, rack-and-
pinion steering. Four-wheel, internally-vented, power disc brakes. A power seat on the driver's side. Automatic cruise
control. Automatic climate control. Electrically-heated and adjustable outside rearview mirrors. Retractable halogen
headlights with a power-spray washing system. And a choice of 5-speed manual or new 4-speed automatic transmission. Priced
at $43,000: the new 928S is Porsche's finest. For your Porsche + Audi dealer, call toll-free: (800) 447-4700. In Illinois, (800) 322-4400.
Manufacturer's suggested retail price. Title, taxes, transportation, registration and dealer delivery charges additional. © 1982 Porsche Audi.

PORSCHE · AUDI
NOTHING EVEN COMES CLOSE

CREATIVE MODES

A dramatic block of body type is produced when each line is centered. Here, the three-dimensional perspective of a road is created above the water. Excellent overall design and consumer interest help overcome potential reading problems presented by the unusual shape, irregular left margin, and extralong lines.
Porsche Audi Division

The Delta Wing

It's best that you use light-weight cloth for this kite.
Tightly-woven cotton (not silk) — the more
air-tight the better.
You can see the dimensions
on the pattern.

And you'll
need three
pockets for battens
(sticks that go in
between the folds of cloth A).
If the cloth is light enough you
can stitch the vane (B) into place
at the same time you're stitching
down the short (31½") central
batten that forms the "spine," or "keel,"
of the kite (as shown in C).

Use thin, lightweight strips of wood
(bamboo, pine, cedar, etc.) or plastic for
the battens; and then a piece of bamboo,
or some other springy piece of wood (or
plastic) for the spar (D) that spreads the
wings apart. The spar is inserted on the
outside — and top-side — of the kite at
points (F) and its ends are constructed
as shown in (E), using short lengths of
wire similar in gauge to a coat hanger.

The triangular vane (B) becomes the "bridle" of this
particular kite — the line or tether (the string that
connects you to your kite way up there) is fastened
directly at the apex of the triangle to eyelet (F).
(Note: All three positions marked "F" should be
secured solidly with metal eyelets.)

Metallic blue.

Normally, we avoid breaking up the flow of body type to prevent reader confusion. Here, the kite wing splits the column in two but does not break the natural reading pattern of the column. It also presents a pleasing line against which the type may be contoured.
S. D. Warren Paper Company

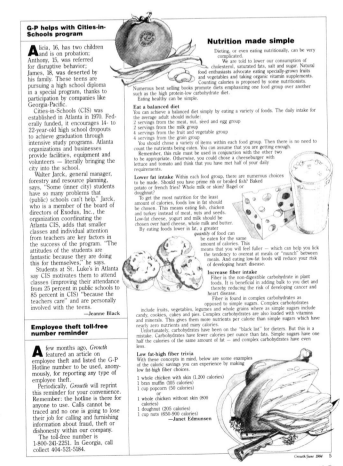

G-P helps with Cities-in-Schools program

Alicia, 16, has two children and is on probation; Anthony, 15, was referred for disruptive behavior; James, 18, was deserted by his family. These teens are pursuing a high school diploma in a special program, thanks to participation by companies like Georgia-Pacific.

Cities-in-Schools (CIS) was established in Atlanta in 1970. Federally funded, it encourages 14- to 22-year-old high school dropouts to achieve graduation through intensive study programs. Atlanta organizations and businesses provide facilities, equipment and volunteers — literally bringing the city into the school.

Walter Jarck, general manager, forestry and resource planning, says, "Some (inner city) students have so many problems that (public) schools can't help." Jarck, who is a member of the board of directors of Exodus, Inc., the organization coordinating the Atlanta CIS, adds that smaller classes and individual attention from teachers are key factors in the success of the program. "The attitudes of the students are fantastic because they are doing this for themselves," he says.

Students at St. Luke's in Atlanta say CIS motivates them to attend classes (improving their attendance from 25 percent in public schools to 85 percent in CIS) "because the teachers care" and are personally involved with the teens.
—Jeanne Black

Employee theft toll-free number reminder

A few months ago, *Growth* featured an article on employee theft and listed the G-P Hotline number to be used, anonymously, for reporting any type of employee theft.

Periodically, *Growth* will reprint this reminder for your convenience. Remember: the hotline is there for anyone to use. Calls cannot be traced and no one is going to lose their job for calling and furnishing information about fraud, theft or dishonesty within our company.

The toll-free number is 1-800-241-2251. In Georgia, call collect 404-521-5184.

Nutrition made simple

Dieting, or even eating nutritionally, can be very complicated.

We are told to lower our consumption of cholesterol, saturated fats, salt and sugar. Natural food enthusiasts advocate eating specially-grown fruits and vegetables and taking organic vitamin supplements. Counting calories is proposed by some nutritionists. Numerous best selling books promote diets emphasizing one food group over another such as the high protein-low carbohydrate diet.

Eating healthy can be simple.

Eat a balanced diet
You can achieve a balanced diet simply by eating a variety of foods. The daily intake for the average adult should include:
2 servings from the meat, nut, seed and egg group
2 servings from the milk group
4 servings from the fruit and vegetable group
4 servings from the grain group

You should chose a variety of items within each food group. Then there is no need to count the nutrients being eaten. You can assume that you are getting enough.

Remember, this rule must be used in conjunction with the other two to be appropriate. Otherwise, you could chose a cheeseburger with lettuce and tomato and think that you have met half of your daily requirements.

Lower fat intake Within each food group, there are numerous choices to be made. Should you have prime rib or broiled fish? Baked potato or french fries? Whole milk or skim? Bagel or doughnut?

To get the most nutrition for the least amount of calories, foods low in fat should be chosen. This means eating fish, chicken and turkey instead of meat, nuts and seeds. Low-fat cheese, yogurt and milk should be chosen over hard cheese, whole milk and butter.

By eating foods lower in fat, a greater *quantity* of food can be eaten for the same amount of calories. This means that you will feel fuller — which can help you lick the tendency to overeat at meals or "munch" between meals. And eating low-fat foods will reduce your risk of developing heart disease.

Increase fiber intake
Fiber is the non-digestible carbohydrate in plant foods. It is beneficial in adding bulk to your diet and thereby reducing the risk of developing cancer and heart disease.

Fiber is found in complex carbohydrates as opposed to simple sugars. Complex carbohydrates include fruits, vegetables, legumes and whole grains where as simple sugars include candy, cookies, cakes and pies. Complex carbohydrates are also loaded with vitamins and minerals. This gives them more nutrients per calorie than simple sugars which have nearly zero nutrients and many calories.

Unfortunately, carbohydrates have been on the "black list" for dieters. But this is a mistake. Carbohydrates have fewer calories per ounce than fats. Simple sugars have one half the calories of the same amount of fat — and complex carbohydrates have even less.

Low fat-high fiber trivia
With these concepts in mind, below are some examples of the caloric savings you can experience by making low fat-high fiber choices.

1 whole chicken with skin (1,200 calories)
1 bran muffin (105 calories)
1 cup popcorn (50 calories)
 or
1 whole chicken without skin (800 calories)
1 doughnut (205 calories)
1 cup nuts (650-900 calories)
—Janet Edmunsen

Growth/June 1984 5

This page utilized several eye-catching type contours that flow around the food items in a natural way. Note how the type in the middle of the page forms a justified contour around the muffins while loosely contouring the cheese.
Georgia-Pacific Corporation

These Time-Zero cameras not only focus themselves, they give you the time of day.

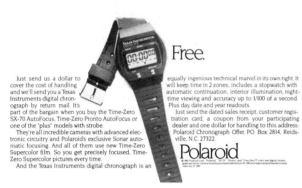

Free.

Just send us a dollar to cover the cost of handling and we'll send you a Texas Instruments digital chronograph by return mail. It's part of the bargain when you buy the Time-Zero SX-70 AutoFocus, Time-Zero Pronto AutoFocus or one of the "plus" models with strobe.

They're all incredible cameras with advanced electronic circuitry and Polaroid's exclusive Sonar automatic focusing. And all of them use new Time-Zero Supercolor film. So you get precisely focused, Time-Zero Supercolor pictures every time.

And the Texas Instruments digital chronograph is an equally ingenious technical marvel in its own right. It will keep time in 2 zones, includes a stopwatch with automatic continuation, interior illumination, night-time viewing and accuracy up to 1/100 of a second. Plus day, date and year readouts.

Just send the dated sales receipt, customer registration card, a coupon from your participating dealer and one dollar for handling to this address: Polaroid Chronograph Offer, P.O. Box 2814, Reidsville, N.C. 27322.

Polaroid

The body type around the watch has been set as a tight, fully justified contour on both sides of the image.
Polaroid

A lot of people thought we were ready to cry "Uncle."

News about International Harvester® hasn't been too good over the past few years. Like a lot of companies, we've had to struggle harder than ever to make it in this tough economy. And lately, quite a few people were beginning to think we'd have to throw up our hands and give in.

Instead, we're going to do everything we can to go on.

We know it won't be easy. Because that means trimming our entire operation from top to bottom. Closing plants that aren't profitable. And consolidating others to improve production capacity.

It also means putting all our resources behind the operations with the most potential to make it in this tough economy.

Trucks and farm equipment.

Because these two operations have proved

they can make it. During the first six months of this year we outsold everyone in medium and heavy trucks. And in the past seven months alone over 33,000 farmers bought International farm equipment.

So, we're planning to make International Harvester a trimmer ...and tougher... company than before. A company with a better chance to succeed than before. We know it will be some time before we completely turn the corner and head into a profitable future.

But we're doing everything we can to assure that we do indeed have a future.

We're not giving in. We're going on.

IH

International Harvester

Contours are often set ragged on both sides of an image. Here the soft, vignetted edge of the illustration is matched by a loose, ragged type contour to maintain the soft edge. Note how the illustration projects above the body type to prevent it from becoming lost in a mass of type.
International Harvester

(Left) A similar accenting effect is achieved when type is contoured around the top of a subject. More space between illustration and type should be allowed to prevent a cramped look.
Kaiser Permanente

KAISER PERMANENTE
Medical Care Program
Southern California Region

Planning for
Health

Spring 1982

Your
LIFESTYLE
can make a difference

No one can control all the factors that contribute to good health. Some risks are beyond the control of the individual—environmental pollution and unsafe highways, for example. However, you do have control over your behavior and habits. Improving those aspects of your lifestyle can make a difference in both the quality and quantity of your life.

Health experts now consider lifestyle to be one of the most important factors affecting health. It is estimated that as many as seven of the 10 leading causes of death in the United States could be reduced through common sense changes in lifestyle.

What type of changes should you make? The most important are: stopping smoking, moderation in drinking, developing good eating habits, proper handling of stress, regular exercising, using safety measures in your car and around work and home and having a positive outlook.

Unfortunately, too few of us make the effort. Maybe you will after considering the following:

Smoking
Cigarette smoking is the single most important preventable cause

of illness and early death. It is especially risky for pregnant women and their unborn babies. Persons who stop smoking reduce their risk of getting heart disease and cancer.

So if you're a smoker, think twice about lighting up that next cigarette. But if you choose to continue smoking, at least try decreasing the number of cigarettes you smoke and switch to a low tar and nicotine brand.

Drinking
Alcohol produces changes in mood and behavior. Most people who drink are able to control their intake of alcohol and to avoid undesired and often harmful effects. Heavy, regular use of alcohol can lead to cirrhosis of the liver—a leading cause of death.

If you do drink, do it wisely and in moderation. Statistics clearly show that mixing drinking and driving is often the cause of fatal or crippling accidents.

Use care in taking drugs too. Today's greater use of drugs—both legal and illegal—is one of our serious health risks. Even some of the drugs prescribed by your doctor can be dangerous if taken when drinking alcohol or before driving.

Using or experimenting with illicit drugs can lead to a number of damaging effects and even death. Physical and mental problems can develop from the excessive continued use of tranquilizers and "pep pills".

Continued on the next page.

(Right) This creative magazine spread presents a variety of body type modes built into a basic justified column format: (a) justified contours; (b) justified run-arounds; and (c) a molded circle of type completely enclosing an illustration.
CH2M Hill Reports, 1980, Vol. 18, No. 3

LET'S STOP INVENTING THE WHEEL OVER AND OVER AND OVER AND OVER AND OVER AND OVER AND OVER AND

FADE-AWAYS

Screen tinting of type from dark to light is easily achieved by the printer and often appealing. Here, the descending dark-to-light image reinforces the message "over and over."
Donnelley-Directory Record

Type fitted to rectangular shapes is called a run-around. Unjustified type creates an irregular run-around to the left of the image and an even margin to the right. Here, this effect helps to soften an otherwise formal layout. If the body type had been justified (even left and right margins), the geometric shape of a capital H would have been accented. Such alphabetic shapes can prove distracting to readers.
U.S. Navy

focuses on projected national energy needs, evaluates traditional and alternative energy sources, and details our abilities to help municipalities, agencies, utilities and industries deal with the development, supply and cost aspects of the energy problem.

In this study, we paid particular attention to the two factors that form the major part of any energy solution: energy management and the development of new, non-conventional energy sources.

Energy Conservation and Management

One of the basic energy problems facing industries and utilities alike is illustrated by the experience of a potato processing plant in the Northwest. During 1973, the monthly natural gas bill for this plant averaged $17,000; by the end of 1979, however, operating with only a slight increase in production, this same plant was paying over $100,000 a month for fuel.

These drastically increasing energy costs and their resultant effects on profitability have prompted many firms to institute a program of energy conservation in order to reduce consumption of fuel.

Energy conservation is often thought of in terms of lowering the thermostat and turning off unused lights. On a larger scale, though, it involves the design and implementation of energy-efficient systems.

With modern engineering abilities, existing processes that normally use tremendous amounts of oil, coal or natural gas can be redesigned to burn fuel more efficiently, to recover and reuse process byproducts, and to operate at reduced levels of energy consumption.

Such energy management approaches are also important to the development of new facilities. In addition, comparisons can be made between the premium cost of high-efficiency electric motors and the cost of electricity. This type of study can help determine whether, in the long run, special equipment will save sufficient energy dollars to justify the initial expense.

Another significant aspect of energy management is the energy audit. Both public agencies and private industries can benefit from this in-depth technical analysis of various energy conservation schemes.

Non-Conventional Energy Sources

Fossil fuels and natural gas cannot be relied upon to meet our energy needs indefinitely. So it is imperative that we

reduce our dependence on conventional fuels, production of which in the United States is expected to peak during the 1980s, and pursue the development of non-conventional energy sources.

Of the non-conventional energy sources being investigated throughout the world, some are readily adaptable to large-scale use, others require years of intensive study and research. Among the most prominent of the sources that are likely to offer widespread applications are: solar energy in all its direct and indirect forms—thermal, photovoltaic, ocean thermal, wind, biomass and hydropower; geothermal energy; and synthetic fuels, including coal gasification and liquefaction, oil shale, tar sands and heavy oil.

CH2M HILL's Role in the Energy Solution

Beginning with our hydropower work in the 1940s, CH2M HILL has been bringing clients the required planning and design skills needed to complete a wide range of energy-related projects. Energy-efficiency considerations have long been part of our standard design procedure and are an integral part of the advanced energy programs in which we are participating today.

The Energy Systems Report is one more step in our continuing commitment to provide state-of-the-art services and to remain current with the needs, technology and direction of the intensifying energy situation that affects all of us. This study has enabled us to identify our strengths and to proceed with the expansion of our already extensive involvement in the energy field.

The following capsule reports highlight some of the energy projects we have been associated with recently.

GEOTHERMAL
Boise's Alternative Energy Program

In one of the most ambitious applications of an alternative energy source, Boise, Idaho, has entered a five-year, $10 million geothermal development project.

Located above a major geothermal reservoir, Boise intends to provide geothermal heating for the downtown business district and private homes.

Already benefiting from the program is an Idaho Department of Agriculture and Health office and laboratory building. Since converting from natural gas to geothermal, the building has been able to cut its heating costs in half.

The entire Boise system, which will save an estimated 50,000 barrels of oil per year, will deliver energy equivalent to the production of a 28-megawatt electric power plant.

Cogeneration at California WWTP

The Point Loma Wastewater Treatment Plant at San Diego is on its way to becoming completely energy-self-sufficient. In the process, it will not only eliminate its consumption of purchased energy, but will produce an excess of electricity for sale to a local utility.

Using its more than 1 million cubic feet per day of digester gas, much of which is currently being flared, the plant will fuel two 1,350-kilowatt engine generators which have the capacity to supply nearly three megawatts of power.

In addition, waste heat from these generators will be recovered to heat the plant's digesters.

This project is scheduled for completion by the end of 1981.

Bioconversion Yields Renewable Energy Resource

The dream of efficiently generating a long-term supply of energy from a renewable resource is about to be realized in Lamar, Colorado.

The $14 million project undertaken by the Lamar Utilities Board involves con-

verting manure from 50,000 cattle in Lamar area feed lots to 1 million cubic feet per day of methane gas. The gas will be burned at the city's electric plant, supplying one-third of Lamar's energy needs.

A byproduct, 130 tons per day of protein supplement, will be sold as livestock feed. In addition, the plant will use hot wastewater from the existing Lamar power plant as a source of heat.

This project, expected to be operational by late 1982, won the 1978 American Consulting Engineers Council Grand Conceptor Award for Engineering Excellence.

Solar Study for Northwest

What are the barriers and opportunities involved in developing large-scale solar electric power systems? This is

the question being investigated in the Northwest Regional Assessment Study (NWRAS), one of six nationwide studies sponsored by the Solar Energy Research Institute.

These studies examine both the technical considerations of solar development and the legal, social, institutional, environmental and economic issues.

The NWRAS covers Alaska, Washington, Oregon, Hawaii, Idaho, Montana, Nebraska, North and South Dakota, and Wyoming.

CH2M HILL is participating in the NWRAS, assisting JBF Scientific Corporation. Our work entails development and implementation of public involvement programs and liaison between JBF and portions of the electric utility industry.

AIR FORCE "ATTACKS" ENERGY PROBLEM

McClellan Air Force Base at Sacramento, California, is mounting an all-out assault on the nation's energy problem.

Sponsored jointly by the Department of Defense and the Department of Energy, McClellan's Advanced Energy Technology Demonstration Program will include fuel cells, Stirling and advanced diesel engines, a biomass Brayton cycle, cogeneration, fluidized bed combustion of coal, energy conservation, solar energy, windpower and centralization of airbase utilities.

Within seven years of the recently completed feasibility study, the Air Force expects to have invested

$200 million in a power plant, utility improvements and working examples of advanced energy systems. Information derived from the solid fuel, solar and wind-powered projects, as well as from the advanced building total energy systems, will be made available to the public.

Additional studies are planned to evaluate the feasibility of a 49,000-kilowatt power plant at McClellan Air Force Base.

Through the Advanced Energy Technology Demonstration Program, similar energy initiatives will be stimulated in the public and private sectors throughout the country.

KINGSLEY DAM TO GO HYDROPOWER

Since its construction in the late 1930s, Nebraska's Kingsley Dam has been used primarily for irrigation, flood control and downstream power generation. However, its original design did consider the later addition of a hydroelectric generating plant at the dam itself. The Central Nebraska Public Power and Irrigation District, owners of the dam, initiated formal planning for this step in 1973.

Currently in the licensing phase, the project will result in a 50-megawatt hydroelectric plant adjacent to the dam's spillway. At completion, the approximately 89 million kilowatt-hours of electricity generated annually will be sold by the District to the Nebraska Public Power District, thus supplying energy which otherwise would consume 150,000 barrels of oil.

NOW WE KNOW WHAT IT'S LIKE TO BE ALONE.

It wasn't always this way.

It used to be hard to decide among all the 35mm SLR's. Then we created the new Minolta XG-M. A camera so extraordinary it stands alone in its class. With an unrivaled combination of creative features.

First, it's automatic. So it's easy to get sharp, clear pictures. You just point, focus, and shoot. It even has electronic features that keep you from making mistakes.

As your skills advance, you'll appreciate advanced features like metered manual and exposure override for full creative control.

To further separate ourselves from the competition, we built in the option of professional motor-drive. Something normally found only on more expensive cameras.

It lets you shoot a blazing 3.5 frames per second. So you can catch a baseball as it comes off the bat. Or halt a horse leaping a hurdle.

But to fully grasp the XG-M's advanced design, you have to hold it.

The body feels rugged yet light. With a built-in textured grip that's sculpted to fit snugly in your hand. And oversized controls that make it easy to adjust to changing conditions.

As your creative potential develops, you'll have access to over 45 interchangeable, computer-designed Minolta lenses. As well as the Minolta system of SLR accessories.

All in all, the XG-M is a remarkable achievement. But then, we have over 50 years of remarkable achievements to draw on.

The new Minolta XG-M. Now we know how it feels to be alone. But we're not complaining.

WAIT 'TIL YOU SEE HOW GOOD YOU CAN BE.

MINOLTA

(Left) Body type contoured to the shape of irregular display headlines may present problems. Any time the left edge of the body type is contoured, paragraph indentions may disrupt the even flow of the type unit. Here, three consecutive indentions keep the type from conforming to the shape of the W at the top of the type area. Extrawide indentions minimize the problem in the rest of this otherwise dramatic ad.
Minolta Corporation

T/A HIGH TECH™ RADIALS

Objective: Utilize the most advanced technology to create a line of radials meeting almost every driving and performance requirement.

Solution: BF Goodrich introduces T/A® High Tech™ radials. Truly, the State of the Art.

ANGLING FOR ACTION

(Above right) Italic and oblique typestyles—those that lean to the right—lend impact when bold, active messages are needed. Diagonal alignments of type columns add even more punch. Notice how the right-leaning body type blocks shown here suggest movement.
B.F. Goodrich Company

(Right) This entire ad has been given a diagonal stress to match the angle of the oblique typestyles. This has the effect of standing the normally-leaning type upright for easier reading. Column margins also stand upright.
Nissan Motor Corporation, USA

52

NEWS You Can Use®

IN YOUR PERSONAL PLANNING

2300 N Street, N.W.
Washington, D.C. 20037

Rumors about jobs in other parts of the country--often the sun belt--have
sent many unemployed workers off in their cars without much planning. This
frequently leads to problems as bad as unemployment.

TRAVELING FOR WORK. When reports of available jobs prove to be untrue,
job seekers and their families may find themselves stranded in unfamiliar
places, with little money and few friends or contacts. To avoid this, the
Travelers Aid Association of America suggests a few precautions:
One day's drive. If you can't find work where you live, look within one
day's drive from home. That way, you can investigate a large area without
giving up the benefits of home. If you find work in that area, you may not
have to move at all--or at least not as far.
Long distance. If you look outside your region, conduct a long-range
search before you travel. Check advertisements in out-of-town newspapers at a
public library, and write or call potential employers. It's cheaper to call
long distance than to travel to a distant city.
Employment services. Investigate the out-of-town listings at your nearest
state-employment-service office or contact the employment service in the cities
where you want to relocate. Your local library may also have out-of-town
telephone directories to help you find correct phone numbers and addresses.
Verify job offers. If you're offered a job, make sure it's valid before
moving. Ask for a letter assuring your employment and giving the starting
date, salary and pay period. Seek references about the employer from the local
Chamber of Commerce, Better Business Bureau or local trade union.
Advance planning. Determine beforehand all costs of the trip, including
motel rooms en route and at your destination, as well as food and fuel to last
until your first paycheck. Remember that many nonreimbursed moving costs will
be tax-deductible. Take money for emergencies--or to return home if things
don't work out. Use traveler's checks, since carrying a lot of cash can be
risky, and it may be difficult to cash a personal check in a strange city.

AIR FARES. Travelers bemoaning the end of those $99 one-way air fares
between the East Coast and California--take heart. Those fares were supposed
to end in mid-February, but some airlines are extending them through April 1 on
many busy routes. Most bargain air fares are available only in limited
numbers, for short periods or with restrictions. Check with your travel agent
or watch for newspaper advertisements for the most up-to-date information.

TIPS AND TAXES. Bars and restaurants must refund any money that has been
incorrectly withheld from employes' tips this year, says the Internal Revenue
                                                                        (over)

TYPEFACE UNITY
*This page shows how type can be
selected to reinforce a subject. The
typewritten look of a newsletter is
stressed in this newsmagazine
feature, both for display and text.*
From U.S. News and World Report, Feb. 28, 1983.
Copyright 1983, U.S. News & World Report, Inc.

INTEGRATED TYPE UNITS

When type forms are thoughtfully selected, spaced, and connected in a particularly pleasing way, they often form single reading units. The following examples exhibit a special sensitivity to making type a design element in its own right.

GENERAL MOTORS CORPORATION

From Autorama USA, a special advertising section. TIME Magazine, April, 1983.

Lucy Lucid was a bright and beautiful bride. The trouble was she'd never learned to cook. Knowing her limitations, she bought a packaged pie-mix for the cherry pie dessert she was planning to serve at her first dinner party.

The instructions were short and explicit. A minute with the mixer, a couple of quick passes with the rolling pin, add the cherries and bake for <u>exactly</u> nine minutes. Simple as pie!

Just as she was ready to slide her masterpiece into the oven, the hectic party preparations drove her kitchen clock cuckoo—it quit cold. Knowing it was the only clock in the house, her husband thought "Bye, bye, pie."

But not Lucy. Without hesitation she set up the two "hour-glasses" she had received as wedding presents. One measured four minutes; the other seven. "Just what I need," she said. And nine minutes later she took her pie from the oven. It was baked to perfection.

"How did you do it?" marveled her husband — but Lucy was smart. She kept him guessing.

Can you figure out how Lucy, using these two old-fashioned timers, measured exactly nine minutes?

Copyright © Westvaco Corporation 1975

Westvaco Corporation

and the Early Days

St. Joseph Lead Company was formed in 1864 for "manufacturing, mining, mechanical and chemical purposes." A combination of expertise and tenacity carried the domestic mining company through some very tough times.

Throughout history, writers have referred to the smelting of a malleable, bluish-gray metallic element—lead. This serviceable metal was used for everything from water pipes in ancient Pompeii to sling-ejected projectiles on the battlefields of Greece.

Today, besides its continued use in the manufacture of pipe, solder, cable sheaths and radioactive shields, lead is the major component of electric storage batteries. In the United States, batteries consume some 60 percent of domestic lead production.

Lead mining and smelting began in the U.S. in the 1700s when numerous French expeditions were dispatched to find new sources of the versatile metal. What they discovered was a very rich corridor of lead-bearing ore located in Southeastern Missouri. This geological formation later became known as the *Lead Belt.*

Those early mines were merely open pits with few, if any, smelters located nearby. The miners would simply take their accumulated cache to a convenient point where it was sold to a smelting operation.

In those days, smelting techniques were very crude. Lead was processed in log furnaces, which were holes in the ground covered by a smooth coating of clay.

Over the top of the hole, men placed two or three large logs and set smaller logs and sticks on top. They then scattered the crude mineral over this pyre.

When the fire had roasted and oxidized the lead, the resultant molten metal trickled into the cavity below. They repeated the process until the hole was filled with impure lead; then they pulled the slug and cut it into convenient-sized pieces to sell.

Time passed and technology improved. It wasn't long before the Lead Belt was drawing much attention. Lead demand grew.

But there was a problem: how to transport product from the mines to the major population centers?

In 1809, after a concentrated search, a group of lead producers picked a site alongside the Mississippi River that afforded easy river shipment. Roads were built and a town sprang up–a town they named Herculaneum.

The district grew and, by 1819, Herculaneum became the county seat. By 1834, however, larger towns were built along the river and the legal administrators moved west.

Herculaneum began to lose impor-

7

Fluor Corporation

TYPE AS ILLUSTRATION
Type forms often become a vital part of the message itself.
In this last group, the illustrations and type become one.

Crime Prevention Coalition and the Ad Council

Central Magazine, South Central Bell

THERE'S NO TURNING BACK

Early in April, AAUW sent letters to a list of people who had spent long years at the forefront of the struggle for passage of ERA, asking them to set down their thoughts about the movement and what it meant to them. Some were unable to respond, but among those who did, one theme soon became evident. As succinctly stated by Liz Carpenter, "the inalienable rights of women are irreversible."

People who have been active in the cause of equality can see, as perhaps their opponents are incapable of doing, that the most important product of the years of struggle is the vitality of the movement itself. The process has made women strong and determined, skeptical of power but trusting of one another, less naive about political realities and more sympathetic of each other's difficulties. Women have learned to build on their defeats as well as their triumphs, and they will not stop striving for full participating citizenship under the US Constitution.

There is no turning back.

Mary H. Purcell, President, American Association of University Women

Liz Carpenter, Former Assistant Secretary for Public Affairs, US Department of Education

The ERA Woman? First she was the zealot feminist, noisy in the caucuses, taking to the streets with placards.

Then, the organization woman—the clubwoman—marched hand-in-hand with the feminist movement, effective and well-organized in coalitions, lobbying the state legislators. From the blue jeans of the '60s, the ERA woman of the '70s moved into ultrasuede, more determined and gaining power, rallying her forces behind ERA. She was in Houston, at the first national convention ever held in this country open to the concerns of women. With 34 states ratified, there were only three more to go—so close and still so far.

In 1980, she battled the Republican platform committee, joined by pro-ERA Republican men—governors and vice-presidential candidates—marching through Detroit, angry, stunned, defeated but undaunted.

Today, ERA still wins every national poll, and by bigger margins. The ERA woman has a new recruit in the '80s—yesterday's "babydoll," the prototype of the beauty pageant queen, the last bastion of the bigoted "Old Boys" system has joined her sisters. She knows when she is being ripped off.

The anti-ERA constituency is dead. Politically, so are those who would deny equal rights to women of America, because the inalienable rights of women are irreversible.

It is extremely irritating, just at the present moment, to be a feminist in the United States. It is mortifying to contemplate the headlines all over the world saying "American Women Unequal." It is depressing to think what such headlines will do to the hopes of other women living in less smugly

10   GRADUATE · WOMAN

The American Association of University Women

WINE
NEW FLAVOR FOR MISSISSIPPI'S ECONOMY
by Sandra Bearden

If Jane Wyman and the rest of the "Falcon Crest" cast ever get tired of the vineyards of central California, they might move their location shooting to Mississippi.

Mississippi?

Yes, Mississippi. The state, legally as dry as the Sahara until 1966, is now developing a thriving grape-growing and wine-making industry that promises to enrich and diversify the state's economy. Hub of this new industry is the A.B. McKay Food and Enology (wines and wine-making) Laboratory at Mississippi State University, the only academic research facility of its kind in the U.S., outside of California.

Hostesses from Corinth to Gulfport are now able to team native red and white table wines with the tasty Edam and cheddar cheeses they've been buying for years from the dairy science department at Mississippi State. They've found that home-produced wines and cheeses are delectable party fare.

Wine-making research at Mississippi State is the brainchild of Dr. Louis Wise, vice president, Division of Agriculture, Forestry and Veterinary Medicine, and Dr. Boris Stojanovic, now head of the enology laboratory.

"Actually, it all started in Boris's basement," confesses Wise, a genial gentleman with an easy manner and smile-crinkled face. "He's a native of Yugoslavia, and members of his family were winemakers. After four years as a German prisoner of war, he earned a doctorate in chemistry and emigrated to the States. He learned enough English to attend Cornell University and earn another Ph.D., this time in microbiology. Then he came to Mississippi State."

Stojanovic, who had brought wine cultures with him to Mississippi, started making wine in his basement. Another neighbor and Wise joined him, and they labeled the homemade concoctions SEFAM, a cryptic acronym for "Mississippi Agricultural and Forestry Experiment Station" spelled backwards.

"Then we decided to make it legal," smiles Stojanovic, an affable Henry Kissinger look-and-sound-alike. He and Wise took samples of wine, mostly made from native muscadines, to members of the legislature, suggesting that both grape-growing and wine-making could help boost the state's economy.

They and their samples were convincing. In 1976, the legislature passed the Native Wine Law, which permits the establishment of vineyards and wineries in Mississippi. Legislators also appropriated $150,000 to the experiment station for wine-making research and education the first year. The funding has continued under the leadership of Dr. Rodney Foil, director of the experiment station.

"We appreciate the legislature's having the foresight to establish and develop this program," says Wise. "I believe there are several reasons we continue to receive their support. First, our research program is stringently controlled and academically

6

Central Magazine, South Central Bell

"The most audacious, strange, deeply imagined and, at last, most accomplished book that Mailer has ever written."
—*Richard Poirier*

ANCIENT EVENINGS

NORMAN MAILER

"A great work.... Mailer, a bold gambler, wins all the chips with this one."—*Publishers Weekly*

At bookstores now
A Literary Guild Dual Main Selection   LITTLE, BROWN and COMPANY

Little, Brown and Company

5 Text Preparation

Editing, Marking and Copyfitting

Anyone who has ever worked with type would agree that the more carefully we prepare our copy for typesetting, the better our finished result will be. And, the work will be done faster, cheaper, and with fewer errors than if we are careless. Enough said.

Introduction

Every verbal communication we read originates as an idea. When ideas are recorded on paper in a form suitable for typesetting, they become *copy*. Nontypographic originals—photographs, slides, and art—are not referred to as copy. Instead, terms such as *glossy, transparency,* or *separation* are used that refer to processes or materials required to capture and reproduce such images. Typewritten copy is quite literally material to be redone or copied—usually in a typesetting process.

Today's technology has changed the nature of copy. Instead of typing, then rekeyboarding copy at a typesetting device, we input original matter directly using a video display terminal (VDT). The VDT can help us store, process, and output (typeset) information electronically. Output, in typeset *galley* or page form, may be the only printed record of the original information.

Handwritten notes or typed copy are still often produced prior to electronic keyboarding. Copy can also be produced in printout (hard copy) form after terminal keyboarding. When used only for error detection, such printouts are more accurately called *proofs,* not copy. No overall rekeyboarding takes place before the file is sent directly to typesetting. When substantial rekeyboarding from printouts is required, those printouts become simply electronically produced copy.

Word processors make electronic copy preparation very efficient. Even though much word processing today is interfaced (connected) directly with typesetting systems, a large volume of word-processed hard copy is still rekeyboarded in the typesetting process.

Optical character recognition (OCR) devices, which allow typesetters to scan typed pages electronically and convert copy to typesetting codes, are also still in use today. They offer the advantage of inexpensive input from familiar typewriters instead of terminals.

Not everyone has convenient access to electronic terminals. Many who have such access still prefer to create copy at a typewriter, or use electronic means only to produce a paper original of their work. Since paper copy will still be with us for some time, we need to understand how to prepare it, how to estimate the space it will require when typeset, and how to give typesetting instructions.

Copy Preparation

Procedures for copy preparation vary widely, but some rules always apply:

1. *Be neat.* Type on one side only, double-space, and use clean, white, 8½ × 11 bond paper. If color-coded copy sheets are used, pick light pastels which contrast well with typed characters and are easy on the eyes. Never use erasable bond—it smears.

2. *Allow marking space.* Leave extra room at the top of the first page for typesetting specifications and heading information. Preprinted forms make marking such information simple. Figure 5–1 shows a preprinted copy sheet containing typical typesetting specifications and common copy-editing symbols. Generous side margins allow room for any special instructions.

3. *Keep editing clear and concise.* Copy that leaves your hands should leave no questions unanswered. Copy-editing symbols should be used to mark corrections and modifications on the copy. These symbols may be found in any basic editing text or style manual. Don't assume that everyone uses exactly the same symbols you do. If you have the slightest doubt that a rarely used symbol may be misinterpreted, *spell out the problem in plain*

language. Special instructions should be written in blue or another color which contrasts with copy typewritten in black.

4. *Establish a style, then follow it.* For a fee, some typesetting houses will modify your copy to conform either to generally accepted style or to your own special requirements. However, you will probably find it cheaper and safer not to delegate the task of checking copy carefully for style errors *before* it goes to typesetting.

Whatever the case, insist upon absolute consistency in abbreviations, spelling, titling, and language usage. Many systems today allow the programming of special dictionaries that automatically correct not only common spelling errors, but also specialized terminology, names, and titles.

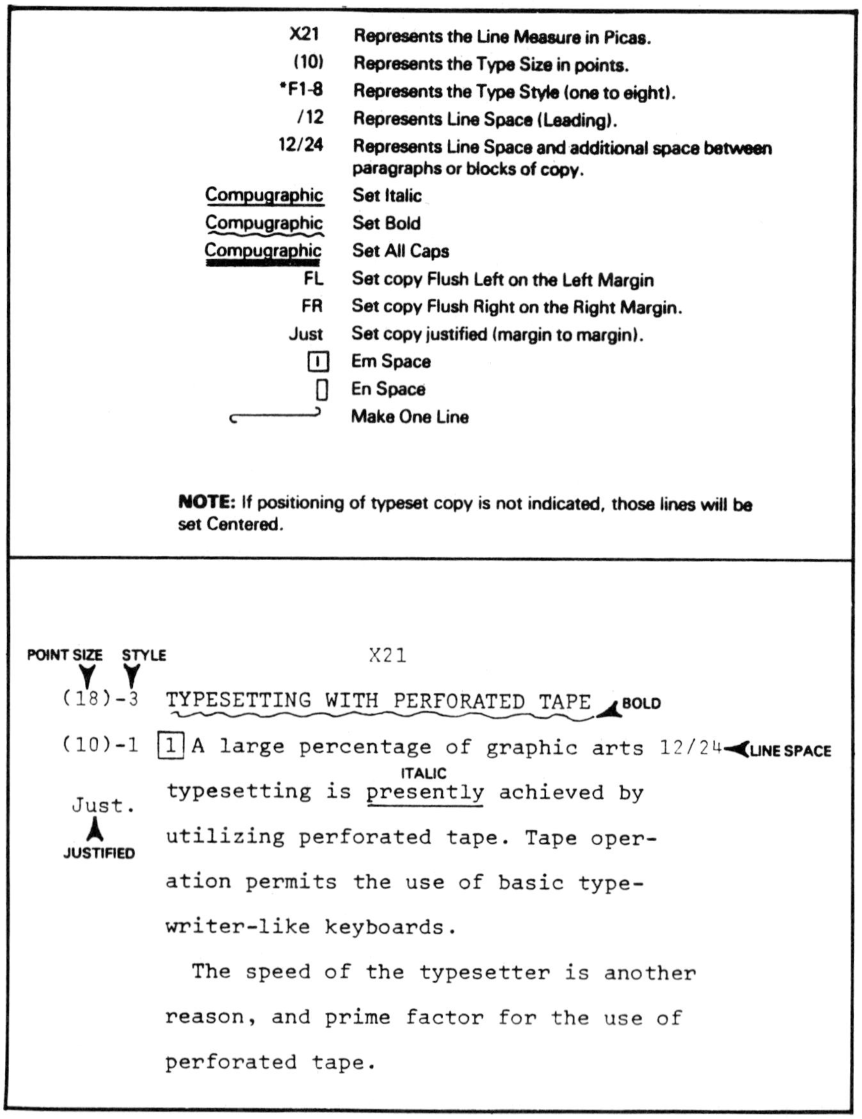

FIG. 5–1
Preprinted copy sheet containing editing symbols and typesetting instructions. On opposite page, typemarked copy for an advertisement is shown, and how it appears when typeset.

X21	Represents the Line Measure in Picas.
(10)	Represents the Type Size in points.
*F1-8	Represents the Type Style (one to eight).
/12	Represents Line Space (Leading).
12/24	Represents Line Space and additional space between paragraphs or blocks of copy.
Compugraphic	Set Italic
Compugraphic	Set Bold
Compugraphic	Set All Caps
FL	Set copy Flush Left on the Left Margin
FR	Set copy Flush Right on the Right Margin.
Just	Set copy justified (margin to margin).
[1]	Em Space
[]	En Space
⌐___⌐	Make One Line

NOTE: If positioning of typeset copy is not indicated, those lines will be set Centered.

POINT SIZE   STYLE   X21

(18)-3   TYPESETTING WITH PERFORATED TAPE   BOLD

(10)-1   [1] A large percentage of graphic arts   12/24 LINE SPACE

ITALIC

Just.   typesetting is presently achieved by

JUSTIFIED   utilizing perforated tape. Tape oper-

ation permits the use of basic type-

writer-like keyboards.

The speed of the typesetter is another

reason, and prime factor for the use of

perforated tape.

X30

(30)F3 The Gilstraps aren't moving. /28) FL
        They're being robbed.

X18

(12)F1 The Gilstraps aren't home today. They're in Toledo on vacation. /13/26
        And these moving men aren't movers at all. They're crooks. They
JUST.   think they're pretty smart. They think no one will notice.
        They're wrong.

        Across the street, the neighbors are calling the cops. Because the
        neighbors know, if they don't call the cops now, the Gilstraps
        will have to call them later.

        This neighborhood has learned how to prevent crime. Find out what
        you and your neighbors can do. Write to: Crime Prevention
        Coalition, Box 6600, Rockville, Maryland 20850. That'll help. /38

(24)F3 TAKE A BITE OUT OF   /50
(60)   CRIME

The Gilstraps aren't moving. They're being robbed.

The Gilstraps aren't home today. They're in Toledo on vacation. And these moving men aren't movers at all. They're crooks. They think they're pretty smart. They think no one will notice. They're wrong.

Across the street, the neighbors are calling the cops. Because the neighbors know, if they don't call the cops now, the Gilstraps will have to call them later.

This neighborhood has learned how to prevent crime. Find out what you and your neighbors can do. Write to: Crime Prevention Coalition, Box 6600, Rockville, Maryland 20850. That'll help.

PHOTO

TAKE A BITE OUT OF
CRIME

Consult standard style manuals (AP, UPI, or University of Chicago style books, for example). These can be used in tandem with any supplement you prepare which shows special usages appropriate to your organization.

5. *Proofread thoroughly.* After type is set, it must be checked carefully for errors. A set of common proofreading marks is shown in Figure 5–2 with examples of how these marks are used on a galley proof. Far fewer people know proper proofreading symbols than know copy-editing symbols, so take no chances. If a mark seems at all unclear, *spell out the problem in simple language.*

Don't proofread alone. Have another person read the copy out loud—including all punctuation, capitals, quotes, and special spellings—while you check the proof. All errors in typesetting mechanics or lapses in general typographic aesthetics (such as careless letterspacing), should be corrected free of charge by the typesetter. You pay for changes in content or style

	Marginal Marks	Example
Delete; take out	*dl*	Draw a diagonal line thus; show the symbol in margin.
Left out; insert	*ar*	Use a caret and write in margin.
Insert a question mark	?	Use a caret. What
Insert a colon	:⊙	Use a caret and circle. As follows
Insert an exclamation mark	!	Use a caret in text. Write in margin. No
Insert an apostrophe	⋁	Proofreaders marks. Insert it in margin.
Insert a semicolon	;/	Use a caret in text, write in margin.
Insert a hyphen	=/	Checkout. Show two parallel lines in margin.
Delete and close up		Counter clockwise.
Insert en dash	⊥/n	Counterclockwise
Insert em dash	⊥/m	Counterclockwise
Insert a comma	⋀	Use a caret in text, write in margin.
Insert a period	⊙	Use a caret in text. Draw circle around it in margin.
Insert quotation marks	⋎⋎ ⋎⋎	He said, I will not.
Insert brackets	[/]	The result: H_2O
Insert parentheses	(/)	The result: H_2O
Stet; let it stand	*stet*	Do not ~~make~~ correction. Place dots under crossed-out word.
Insert space	#	For better reading.
Equalize spacing between words	eq #	Too many spaces are not good.
Spell out	(sp)	The U.S. government.
Transpose		Written for a purpose. Or, in order reverse
Align text or columnar matter	‖	43879 76120 34827 63001

FIG. 5–2
Proofreading marks and their proper use.

	Marginal Marks	Example
Change capital letters to lowercase	*lc*	Do not use ~~CAPS~~.
Change lowercase to capitals	*caps*	Williston, North Dakota
Change to small capital letters	*sc*	Small capital letters.
Use boldface	*bf*	Draw a wavy line under the word. Write in margin.
Use roman letters	*rom*	Use (roman) letters.
Use italic	*ital*	Use italic.
Raise copy; move up	⌐¬	Use caution in drawing angles.
Lower copy; move down	⌐¬	Use caution in drawing angles.
Move copy to left	⊏	The word "the" will be moved left to the vertical line.
Move copy to right	⊐	The word "the" will be moved right to the vertical line.
Make a new paragraph	¶	Use the symbol with a caret. ∧This sentence will then begin a new paragraph.
No paragraph	*run in*	Use a line. Connect the sentences.
Insert a superscript	∨°	32∧F.
Insert a subscript	∧₂	H∧O
Bad letter; change	X	Circle letter.
Letter is inverted	⊙	Underline and use symbol in margin.
Straighten jumbled type	=	The letters are uneven.
Push space down	⊥	Draw a diagonal through it; mark in margin.
Indentions	☐☐☐	☐☐☐ This indicates an indention of three ems.
Wrong font; change	*wf*	The letter is not of the same face or size.

(author's alterations, or AAs). These can be extremely expensive and time-consuming.

Insist upon revised proofs of the corrected page or galley. If revised proofs are not obtained, a final proof of the entire job can be checked.

Manual Copyfitting

Copyfitting is the accurate estimation of the space that copy will occupy when it is set into type. It is one of the least understood and most necessary tasks connected with the use of type.

Today, many electronic text processing systems provide precise copyfitting at the VDT screen with relative ease. These systems are discussed in Chapter 7. We should, however, understand the general procedures for calculating copy-to-type relationships when electronic means are unavailable or inconvenient to use.

Two general methods are commonly used for fitting copy. The *units-per-column* method is applied where display (headline) type of 14-point and above is required. The *characters-per-pica* method is suited for text (body) type in all sizes under 14 point.

Units-Per-Column Method for Headlines

This relatively unsophisticated method for estimating display type fitting is still used today on some newspapers, magazines, and other publications—especially small ones. With practice, we can apply it easily and with sufficient accuracy.

This method sacrifices some precision to gain simplicity. Each letter of copy to be set into type is assigned a numerical value that describes its width in relation to that of average lowercase letters. Letters of average width, such as *e*, *a*, or *o*, are assigned a base value of 1; narrow letters, such as *l*, *i*, *f*, and *t* and thin punctuation marks are ½; most capitals, like *A* or *Z*, are 1½; and the widest capitals, *M* and *W*, are given a value of 2. A typical unit width chart is shown in Figure 5–3. To use this chart, editors first determine how many units will fit into each column width of their publication. Each type size and column width combination requires a different number of units. The smaller the type or the wider the column, the more units the editor can fit.

In writing headlines or titles, editors use the unit counts as guides to determine the maximum number of units they have for their wording. They know that a headline that exceeds the maximum count for a particular size and column width will not fit properly, and may cause typesetting or page layout problems. Headlines with very short unit counts will leave excessive space in the page, although with today's formats employing more ragged-right setting and more white space, this is less of a problem.

Figure 5–4 compares four lines of typed copy with the same lines in typeset form. Since the number of keystrokes (letters or spaces) is the same for all four typewritten lines—26, the lines in the left column are equal in length. However the typeset lines on the right vary in length because of their varying unit counts. Line 2 sets the longest because it is made up of capitals; line 3 is the shortest because it contains many thin letters. Lines 1 and 4, although identical in unit count, are not exactly the same length when typeset because this system of unit counting is not exact.

The difference in actual typeset length of lines 1 and 4 helps remind us that the units per column system is based on a simplification of letter widths. It only provides an *estimate* of widths—selected from only four possible choices, ½, 1, 1½, and 2. The unit system, discussed later on, is based upon *precise* letter widths, expressed as fractions of the width of an *em* (square of the type size). Don't confuse these em-fraction units with the crude but adequate units we are using in the present discussion to show headline estimations.

Publishers usually establish master *head schedules* that show examples of headlines set in all type sizes,

FIG. 5–3
Unit System for Counting Headlines and Titles

Count each character or keystroke according to these values:

Capital M, Q, W	2 units
Capital I, J	½
All other capitals	1½
Lowercase l, i, f, t, j	½
All other lowercase	1
Figure 1	1
All other figures	1½
Punctuation	½
$, %, &, ?	1½
Dash (—)	2
Space between words	½

Example:

This is a headline

T h i s   i s   a   h e a d l i n e

1½ 1 ½ 1   ½ ½ 1   ½ 1   ½ 1 1 1 1 ½ ½ 1 1   = 15 units

		No. of chars.		Unit count
1.	a b c d e f g h i j k l m n o p q r s t u v w x y z	26	abcdefghijklmnopqrstuvwxyz	24
2.	A B C D E F G H I J K L M N O P Q R S T U V W X Y Z	26	ABCDEFGHIJKLMNOPQRSTUVWXYZ	39½
3.	She lifted a little kitten	26	She lifted a little kitten	18½
4.	The mother cat moved again	26	The mother cat moved again	24

FIG. 5–4
Character count versus unit count.

styles, and column widths they use. The unit count for each example also is shown to aid headline writers.

Sometimes allowances must be made for differences in type design when using the unit count method. For instance, types without serifs often produce shorter typeset lines than serif types, even though the two may have the same unit counts. This is because many of the lowercase letters in sans serif styles are quite thin. Problem letters usually include the *r*, which may require a ½-unit count, and the *f*, *j*, and *t*, all of which may be somewhat less than ½-unit.

Characters-Per-Pica Method for Fitting Body Type

As we have seen, each letter can make a big difference in the width of a line of large type. Small body type is a different matter. Since so many characters fit on each line, the width of a single character is not too important. Instead of counting the number of units in a given column width, we use an average number of characters per pica to find out how many characters will fit on a whole line. For example, if a 10-point Futura Light type allows us to fit 3.0 characters per pica (⅙ of an inch), a line 15 picas wide will allow $3.0 \times 15 = 45$ characters per line in typesetting. We now can either *fit* 45 characters we have already typed onto each printed line, or we can *type* 45 characters for each line of type we need in our finished layout.

Typewritten characters, except for typewriters with proportional spacing, are all the same width. A *pica* typewriter produces a consistent 10 characters per inch; an *elite* typewriter yields 12. Typeset characters vary greatly in width. They give a more professional appearance to finished pages. *Copyfitting is simply*

the task of determining how much space typed copy will require when it is typeset.

Two simple tools are needed in copyfitting: a small hand calculator and a type gauge marked in various point size increments. Figure 5–5 depicts the Haberule, a popular, slotted-plastic scale used for copyfitting.

Copyfitting by Character Count

Whether we are copyfitting space to copy or copy to space, we always need to count our typed characters. The more accurately this is done, the better our copyfitting will be.

Abraham Lincoln supposedly composed his Gettysburg Address on the back of an envelope. We've typed it here (Figure 5–6) to make it easier for us to count the characters in his famous message. Every keystroke is counted, including punctuation and word spaces. Spaces between sentences count as only one character, even if two spaces were typed. Because the last line of each paragraph varies in length, a total for *each paragraph* is computed.

Any one of the following three methods may be used to count characters, but the vertical line method and character scale counting are more accurate.

Average line method

Multiply the number of characters in the average typed line by the total number of lines. Example: margins set for average 60-character line times 40 lines equals 2,400 characters. This method is easy but crude, since it doesn't account for short lines at the end of paragraphs. It also assumes that the typist actually averages 60 characters per line, although

FIG. 5–5
The Haberule.

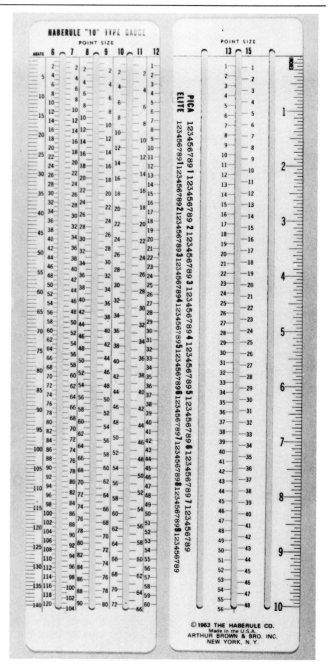

many people type short of or beyond their set margins. It also may fail to account for copy editing in the manuscript.

Vertical line method

Draw a vertical line down the right side of the typed page, at the *actual* average length of typed lines. For example, the vertical line shown in Figure 5–6 is drawn after the 60th character. Every typed line that stops at this vertical line is exactly 60 characters long.

In paragraph one, the first line is 60 characters long, the second line is 58 (2 short of the vertical line), and the third line is 60 again—for a total of 178 characters.

Note that the paragraph is indented 4 spaces—the amount needed for a fairly wide (2-em) paragraph indention in typesetting. Two spaces of typed indention would allow for a smaller (1-em) paragraph indention, about the same as a paragraph of type in a newspaper. Spaces used for indenting typewritten matter are included in the paragraph total.

With the vertical line method, count all typed lines but the short ones at the end of the paragraph as full lines. Then add or subtract characters on either side of the line. Finally, add the characters in the last, short line of the paragraph. Character counts, done easily with a calculator, are shown to the right of each paragraph in Figure 5–6:

Paragraph one: 3 long lines × 60 characters each = 180 characters. Subtract 2 characters (the amount that the second line is short of the vertical line), for a total of 178.

Paragraph two: 6 long lines × 60 characters each = 360 characters. The third, fifth, and sixth lines are each one character short, so subtract 3 characters, for a total of 357. Add the 18 characters in the last, short line of the paragraph, for a paragraph total of 375.

Paragraph three: 15 long lines × 60 characters each = 900 characters. Add the 5 total characters that cross *over* the vertical line, then subtract the 20 total characters that fall *short* of the line. To this total of 885, add the 6 characters in the final line—for a total of 891 characters in the paragraph.

The calculations here are accurate, but tedious. Fortunately, there is another counting method that is both accurate and rapid.

|←——————————— (60 keystrokes) ———————————→|

Paragraph 1

Four score and seven years ago our fathers brought forth

on this continent, a new nation, conceived in Liberty, and   −2

dedicated to the proposition that all men are created equal.

60
×3
180
−2
Total = 178

Paragraph 2

Now we are engaged in a great civil war, testing whether

that nation or any nation so conceived and so dedicated, can

long endure. We are met on a great battlefield of that war.   −1

We have come to dedicate a portion of that field, as a final

resting place for those who here gave their lives that that   −1

nation might live. It is altogether fitting and proper that   −1

we should do this.   +18

60
×6
360
+18
− 3
Total = 375

Paragraph 3

But, in a larger sense, we can not dedicate--we can not   −1

consecrate--we can not hallow--this ground. The brave men,   −2

living and dead, who struggled here, have consecrated it, far +1

above our poor power to add or detract. The world will little +1

note, nor long remember what we say here, but it can never   −2

forget what they did here. It is for us the living, rather,   −1

to be dedicated here to the unfinished work which they who   −2

fought here have thus far so nobly advanced. It is rather for +1

us to be here dedicated to the great task remaining before   −2

us--that from these honored dead we take increased devotion   −1

to that cause for which they gave the last full measure of   −2

devotion--that we here highly resolve that these dead shall   −1

not have died in vain--that this nation, under God, shall have +2

a new birth of freedom--and that government of the people,   −2

by the people, for the people, shall not perish from the   −4

earth.   +6

60
×15
900
+11
−20
Total = 891

4-char.
indent
(equals a
2-em
indent
when set
in type)

FIG. 5–6
How to count characters.

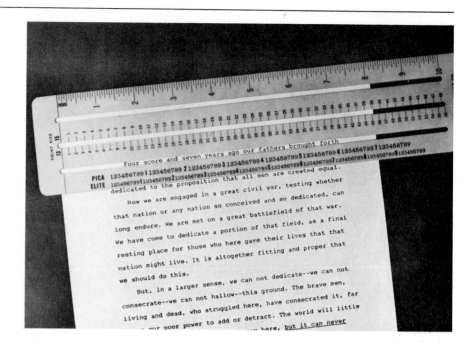

FIG. 5–7
Measuring a typed line with the Haberule.

Character scale counting

Use a ruler or gauge with a typing scale at the edge. It will show 10 characters per inch for a pica typewriter, 12 characters per inch for elite. (Don't confuse pica typewriting with the pica we use in type measurement.) Figure 5–7 shows how to measure a typed line with the Haberule. The count for each line of a paragraph can be added up quickly on a calculator.

Look for shortcuts. A quick glance tells us that the second line in the first paragraph of Figure 5–6 is two characters shorter than the first line of 60 characters, so why measure it?

Remember to allow for editing. *If a word is added or deleted, account for those characters.*
Publications often use preprinted copy forms that show the number of characters to type so that each typed line will yield one typeset line. Figure 5–8 shows such a form.

Copyfitting by Characters Per Pica

For this method, we use the average number of characters per pica, not unit counts, to gauge how many characters will fit in a typeset line. Both letter design widths and point sizes can vary in typesetting; therefore the number of characters that can fit in a one-pica width also varies. Each type style, in each possible point size, has its own per pica character count.

The table in Figure 5–9 gives the average character counts per pica for five popular typefaces in five common body type sizes. Each character count represents the *average* number of typeset characters per pica for that size/style combination. Note the large differences in counts, even for similar point sizes. For example, Futura Light's 10-point count is 3 (or 3.00) characters per pica, while 10-point Helvetica—because of its wider design (set width)—allows only about 2½ (exactly 2.43) characters per pica. This means that Futura Light would allow about 19 percent more characters on a line.

Copyfitting situations require that we answer one or more of the following questions:

1. How many typeset *lines* will a given number of typewritten characters produce?
2. How many typewritten *characters* will it take to fill a *rectangular* area with typeset material?
3. How many typewritten *characters* will it take to fill an *irregularly-shaped* area with typeset material?
4. Given a certain number of typed characters and a type area to fill, which *type style* will produce the best fit?

To keep things simple, we'll use the following abbreviations in our examples:

TC = The *total* number of typewritten *characters* in the copy.
C/P = the number of *characters* that will fit in a *1-pica* width (each style and size of type is different).

```
CAPTION COPYFITTING SHEET

Photo/art title_____ Position_____

Typesetting code_____ Written by_____

1234567890123456789012345678901234567890123456789012345678901234567890
For 1/3-page col. type to here--x
For 1/2-page col. type to here---------------------x
For 2/3-page col. type to here------------------------------------x
_____

Write caption below (each line typed equals one typeset caption line):

. . . . . . . . . . . . . . . . . . . . . . . . . . . . . . . . . . .
. . . . . . . . . . . . . . . . . . . . . . . . . . . . . . . . . . .
. . . . . . . . . . . . . . . . . . . . . . . . . . . . . . . . . . .
. . . . . . . . . . . . . . . . . . . . . . . . . . . . . . . . . . .
. . . . . . . . . . . . . . . . . . . . . . . . . . . . . . . . . . .
_____

Special instructions:
```

FIG. 5–8
Typical preprinted copyfitting form.

FIG. 5–9

TYPE STYLES	CHARACTERS PER PICA				
	8-pt.	9-pt.	10-pt.	11-pt.	12-pt.
Bookman	3.10	2.90	2.60	2.40	2.20
Futura Light	3.75	3.34	3.00	2.73	2.50
Helvetica	3.03	2.70	2.43	2.21	2.02
Optima	3.24	2.92	2.59	2.38	2.15
ITC Souvenir Light	3.50	3.11	2.80	2.54	2.33

C/L = the number of *characters* that will fit in one *line*.

WL = The *width* of one typeset *line* (in picas).

DL = the *depth* of one typeset *line* (in points).

TL = the *total* number of typeset *lines*.

DA = the *depth* of a type *area* (in points).

If you don't like mathematical symbols and formulas, just read the verbal explanation for each example.

You'll find the figuring to be quite easy, especially with a small hand calculator.

SOLVING FOR THE NUMBER OF LINES — This is the most common kind of copyfitting. It is applied in all phases of publications and general graphics work.

Use this kind of problem when you have determined . . .
- *the total number of typewritten characters
- *the type size, leading, and style
- *the desired width of typeset lines

. . . and you want to know . . . how many lines of type will be produced?

Steps to follow:
1. Find the *characters* per line

Multiply the per pica character count of the desired type size and style *times* the width of the desired line (in picas). This will give the number of characters that will fit on each typeset line.

chars/pica × width of line = *chars/line*

2. Find the total number of *lines*

Divide the total number of typewritten characters in the copy by the number of characters that will fit on one typeset line. This will give the total number of typeset lines that will be produced.

chars/line)total chars = *total lines*

This formula in words . . .

$$\text{total lines} = \frac{\text{total characters}}{\text{char. count per pica} \times \text{width of a line}}$$

or symbols . . .

$$TL = \frac{TC}{C/P \times WL}$$

may help you recall the steps.

SOLVING FOR THE NUMBER OF LINES	A WORKED EXAMPLE Suppose we wish to copyfit the 375 characters in the *second* paragraph of the Gettysburg Address into the body type area of a magazine advertisement. We have selected 9-pt. Bookman with 1-pt. leading, and we wish to use a 15-pica line width.
We have already specified . . .	*the total number of typed characters: 375 *the type size, leading, and style: 9/10 Bookman *the desired width of typeset lines: 15 picas
We also have found that . . .	*9-pt. Bookman has a per pica character count of 2.90 (see Figure 5–9).
We want to know:	How many lines of type will be produced?
Steps: 1. Find the characters per line	Multiply the characters per pica for 9-pt. Bookman (2.90) *times* the desired line width of 15 picas to get 43 characters per line: $2.90 \times 15 = 43.5$ $= 43$ (round off *characters per line* to the lower whole number)
2. Find the total number of *lines*	Divide 43 into the total number of typewritten characters (375) to find how many lines will be produced. The answer is 9 *lines*. $43\overline{)375} = 8.72$ $= 9$ (round off total number of lines to higher whole number)
The formula helps us see how the steps are related	$TL = \dfrac{TC}{C/P \times WL}$ $TL = \dfrac{375}{2.90 \times 15}$ $TL = \dfrac{375}{43.5 = 43}$ $TL = 8.72 = 9$

We have found that the 375 characters in Lincoln's second paragraph will produce 8.72 (rounded to 9) lines of 9/10 Bookman type, when set on a 15-pica line width. Figure 5–10 shows the typed copy and how it would look when typeset.

Marking the depth of a type column is important in layout. Notice that the plastic type-gauge in Figure 5–10 indicates that there are, indeed, 9 lines of 9-pt. Bookman, each line measuring 10 points in depth. We can use the 10-point scale on the type gauge to mark an accurate type block depth on a layout. In effect, we can measure in 10-point-deep lines. Nine lines, each 10 points deep, take up a depth of 90 points. If we need

to express our answer in picas or inches, we can measure the type (if it has been set), or measure the 90-point type area on the layout (if the type has not yet been set).

To convert points to picas or inches without using a plastic or metal scale, use simple division: Divide the 90-point type depth by the number of points in one pica (12) to convert to picas:

$$12\overline{)90} = 7\tfrac{1}{2} \; picas \text{ of type depth}$$

Divide the 90-point type depth by the number of points in one each (72) to convert to inches:

$$72\overline{)90} = 1\tfrac{1}{4} \; inches \text{ of type depth}$$

SOLVING FOR THE NUMBER OF CHARACTERS	This second kind of copyfitting problem is called *writing to fit* a given space. It may be encountered where an advertisement or brochure copy block has been designed before copy has been written.
Use this kind of solution when you have determined . . .	*the type size, leading, and style *the desired width of typeset lines *the desired depth of the type area
. . . and you want to know . . .	How many characters will it take to fill the given type area with typeset matter?
Steps to follow: 1. Find the characters per line	Multiply the per pica character count of the desired type size and style *times* the desired line width (in picas). This gives the number of characters that will fit on each typeset line. chars/pica × width of line = *chars/line*
2. Find the total number of lines	Divide the depth of the type area (in points) *by* the depth of each line (in points) to get the total number of lines that can be fit into the type area depth. Depth of lines$\overline{)}$depth of area = *total lines*
3. Find the total number of characters	Multiply the number of characters per line *times* the total number of characters needed. chars/line × total lines = *total characters*
This formula in words . . .	$\dfrac{\text{total}}{\text{chars}} = \text{chars/pica} \times \text{width of line} \times \dfrac{\text{depth of type area}}{\text{depth of each line}}$
or symbols . . .	$TC = C/P \times WL \times \dfrac{DA}{DL}$
may help you recall the steps.	

Now we are engaged in a great civil war, testing whether
that nation or any nation so conceived and so dedicated, can
long endure. We are met on a great battlefield of that war.
We have come to dedicate a portion of that field, as a final
resting place for those who here gave their lives that that
nation might live. It is altogether fitting and proper that
we should do this.

Now we are engaged in a great civil war, testing whether that nation or any nation so conceived and so dedicated, can long endure. We are met on a great battlefield of that war. We have come to dedicate a portion of that field, as a final resting place for those who here gave their lives that that nation might live. It is altogether fitting and proper that we should do this.

10-pt scale
Type
gauge
- 0
- 1
- 2
- 3
- 4
- 5
- 6
- 7
- 8
- 9

FIG. 5–10
Second paragraph of Gettysburg Address set in 9/10 Bookman 15 picas wide.

We have found that 1,260 total characters must be typed to fill an area 18 picas wide and 15 picas (180 points) deep with 8/9 Souvenir Light. Figure 5–11 shows the ad layout and why it would be important to fit copy into it accurately.

	A WORKED EXAMPLE
SOLVING FOR THE NUMBER OF CHARACTERS	Suppose we are designing an advertisement that calls for 8/9 Souvenir Light type, and we wish to fit an area 18 picas wide by 15 picas (180 points) deep.
We have already specified . . .	*the type size, leading, and style: 8/9 Souvenir Light *the desired width of typeset lines: 18 picas *the desired depth of the type area: 15 picas (or 180 points)
We also have found that . . .	*8-pt. Souvenir Light has a per pica character count of 3.50 (see Figure 5–9)
We want to know:	How many characters will it take to fill the given type area with typeset matter?

Steps:

1. Find the characters per line	Multiply the per pica character count for 8-pt. Souvenir Light (3.50) *times* the desired line width of 18 picas to get 63 characters needed per line. 3.50 × 18 = 63 chars/line
2. Find the total number of lines	Divide the depth of the type area (180 points) *by* the depth of each line (9 points) to get the total number of lines that can be fit into the type area depth. The answer: *20 lines.* 9)180 = 20 total lines
3. Find the total number of characters	Multiply the number of characters per line (63) *times* the total number of lines needed (20) to get the total number of characters needed. The answer: 1,260 characters. 63 × 20 = 1,260 total characters

The formula helps us see how the steps are related	$TC = C/P \times WL \times \dfrac{DA}{DL}$ $TC = 3.50 \times 18 \times \dfrac{180}{9}$ $TC = 63 \times 20$ $TC = 1,260$ *total characters*

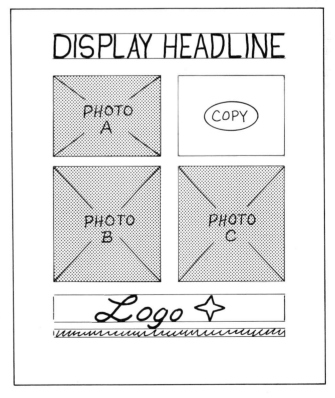

FIG. 5–11
A simple ad layout that requires copy to be written to fit a specified area.

In the problem just completed, we determined that 1,260 total characters were needed. The copywriter might simply set up the margins of a typewriter to produce 63 characters per typewritten line. Every typed line would then equal one typeset line. This is known as typing *line for line*. By typing 20 lines, each with 63 characters, the writer would produce the 1,260 characters needed.

Solving for Irregularly Shaped Type Areas

The third kind of problem we face in copyfitting is that of figuring out how type can be set to flow around an irregular shape. This kind of problem can be either a variation on the *solving for lines* or the *solving for characters* problems we have just examined.

Two basic kinds of type irregularity are common:

The *run-around*, where type is set to fit around a rectangular (square finish) area such as a photograph.

The *contoured wrap*, where type is set to fit around a curved or shaped area, such as line art.

Run-Arounds

Figure 5–12 shows a type area where typesetting is to be run around a small photograph. Notice that we really have two copyfitting tasks here: calculating the number of characters for line width in area A, then calculating again for line width in area B.

Let's copyfit the last paragraph of the Gettysburg Address—891 characters—around the Lincoln photograph shown in Figure 5–12. We'll use 9/10 Bookman. The lines of type to be placed to the *left* of the photograph are to be 12 picas wide, and the lines that fall *below* the photograph are to be 22 picas wide. The picture is 1½ inches wide by 2¼ inches deep.

Calculations. This problem is a combination of both types of problems we have just worked. First, we must solve for *total characters* that will fit to the *left* of Lincoln's picture. The space is 12 picas wide. Although the photo is 2¼ inches deep, our type block must be a bit deeper to bypass the photo before we begin setting the wider lines that pass beneath it. Therefore, *type area A* should be 2½ inches, not 2¼ inches deep.

Type area B, the lines that pass *below* the photograph, is 22 picas wide. Now we must solve for the *number of lines* that will be produced from those characters which would not fit to the left of the photo.

SOLVING FOR RUN-AROUND TYPESETTING	A WORKED EXAMPLE Copyfit the last paragraph of the Gettysburg Address around a 1½ × 2¼ in. photograph of Lincoln.
Specifications for type area A—the type to the left of the photo	*total characters: 891 *type: 9/10 Bookman (characters per pica: 2.90) *line width: 12 picas *type area depth: 180 points (2½ inches)
Problem	How many of the 891 characters will fit in this type area?
Steps: *area A* 1. Find the characters per line	2.90 × 12 = 34.8 = 34 characters per line (C/L)
2. Find the total number of lines	10)180 = 18 total lines (TL)
3. Find the total number of characters	34 × 18 = 612 total characters (TC)
Specifications for type area B—the type that passes *below* the photo	*total characters: 279 (891 − 612 in area A) *type: 9/10 Bookman (characters per pica: 2.90) *line width: 22 picas
Problem	How many lines of type will be produced from the 279 characters remaining after area A is filled?
Steps: *area B* 1. Find the characters per line	2.90 × 22 = 63.8 = 63 characters per line (C/L)
2. Find the total number of lines	63)279 = 4.4 = 5 total lines (TL)

To summarize the problem, we found that 612 of the 891 characters in the last paragraph of Lincoln's address would fit to the *left* of his picture—in a space 12 picas wide by 180 points (2½ inches) deep. Then we found that the remaining 279 characters would produce 5 lines of type, each 22 picas wide, *below* his picture.

Figure 5–13 shows the type when set into position around the photograph.

Contoured Wraps

In the previous run-around example, we only had to calculate for two line widths. With a curved or contoured shape, as in Figure 5–14, the problem is very similar to the run-around, except that we are working with several different line widths.

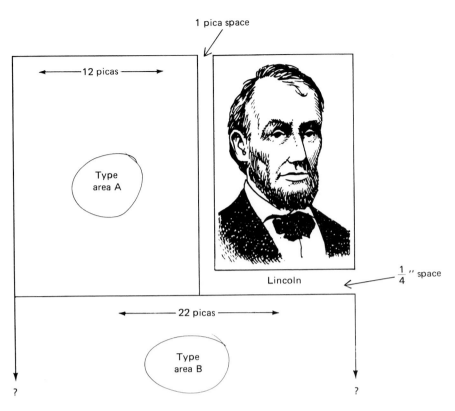

1 pica space

12 picas

Type area A

Lincoln

$\frac{1}{4}$" space

22 picas

Type area B

?   ?

FIG. 5–12
Type areas where run-around must be set.

FIG. 5–13
Type set for run-around.

Type
area A
18 lines,
612 chars.

Type
area B
5 lines,
278 chars.

Four score and seven years ago our fathers brought forth on this continent, a new nation, conceived in Liberty, and dedicated to the proposition that all men are created equal.

Now we are engaged in a great civil war, testing whether that nation or any nation so conceived and so dedicated, can long endure. We are met on a great battlefield of that war. We have come to dedicate a portion of that field, as a final resting place for those who here gave their lives that that nation might live. It is altogether fitting and proper that we should do this.

But, in a larger sense, we can not dedicate—we can not consecrate—we can not hallow—this ground. The brave men, living and dead, who struggled here, have consecrated it, far above our poor power to add or detract. The world will little note, nor long remember what we say here, *but it can never forget what they did here.* It is for us the living, rather, to be dedicated here to the unfinished work which

LINCOLN

Line	Pica Width	Character Estimate	Actual Characters
1	19	55	58
2	19	55	58
3	19	55	56
4	19	55	58
5	19	55	59
6	19	55	57
7	19	55	57
8	13	37	37
9	12.5	36	36
10	12	34	33
11	11.5	33	35
12	11	32	30
13	11	32	34
14	11	32	29
15	11	32	33
16	11.5	33	36
17	12.5	36	40
18	12	34	36
19	11.5	33	35
20	11.5	33	35
21	11.5	33	37

FIG. 5–14
Calculating a contour.

A contour often requires careful calculations. Shown in the example to the left are the number of lines, the pica width of each line, the number of characters estimated per line (2.91 x each pica width), and the actual number of characters that fit on each line. In this example, all letterspacing was decreased slightly. This effect, known as *tracking*, allowed a few more characters than normal to fit on some of the lines.

But, in a larger sense, we can not dedicate--we can not consecrate--we can not hallow--this ground. The brave men, living and dead, who struggled here, have consecrated it far above our poor power to add or detract. The world will little note, nor long remember what we say here, <u>but it can never forget what they did here</u>. It is for us the living, rather, to be dedicated here to the unfinished work which they who fought here have thus far so nobly advanced. It is rather for us to be here dedicated to the great task remaining before us--that from these honored dead we take increased devotion to that cause for which they gave the last full measure of devotion--that we here highly resolve that these dead shall not have died in vain--that this nation, under God, shall have a new birth of freedom --and that government of the people, by the people, for the people shall not perish from the earth.

But, in a larger sense, we can not dedicate—we can not consecrate—we can not hallow—this ground. The brave men, living and dead, who struggled here, have consecrated it far above our poor power to add or detract. The world will little note, nor long remember what we say here, *but it can never forget what they did here.* It is for us the living, rather, to be dedicated here to the unfinished work which they who fought here have thus far so nobly advanced. It is rather for us to be here dedicated to the great task remaining before us—that from these honored dead we take increased devotion to that cause for which they gave the last full measure of devotion—that we here highly resolve that these dead shall not have died in vain—that this nation, under God, shall have a new birth of freedom—and that government of the people, by the people, for the people, shall not perish from the earth.

Copy typed line-for-line so that person doing typesetting will know approximately where to break each line. Note that actual characters in each line do not match exactly the calculation shown above.

Let's take the same copy we used for the run-around and do a contoured wrap around a picture of Lincoln. This time we will use 9/10 Optima. We will begin with a 19-pica line width, then go to shorter lines as we wrap type around the picture.

For this kind of problem it is easier to sketch the area to be fit rather than write it out as we have other problems.

In Figure 5–14, the lines of type have been sketched in around the shape of the photo, and the pica width of each line is given. It is a simple matter to multiply the 9-point Optima per pica character count (2.91) times the pica width of each line to see how many characters fit in each one. Of course, each line is 10 points deep just like in the previous problem.

In line one, we find that 55 characters will fit by multiplying 2.91×19. When we get to line 8, we have a shorter line width to multiply: $2.91 \times 13 = 37$ characters. By line 21 we have used up all 891 characters of Lincoln's last paragraph.

Figure 5–14 also shows how the copy could be typed line-for-line to help the typist and the typesetter see how many lines to set and how.

Ragged Composition

Unjustified or *ragged* composition is used regularly today. While it is not a form of irregular typesetting, it should be copyfit a bit differently than justified matter.

Ragged right or ragged left typesetting is not difficult to copyfit. All we need remember is that the line width marked on the copy for typesetting is the *maximum* width possible, not the average width. The average width falls *between* the *maximum width* of lines and the *maximum rag* parameter (that is, the shortest line permitted by the typesetting program).

For example, suppose we wish to set type ragged right on a 20-pica line width. If the maximum rag parameter is set for 3 picas on our typesetter, the shortest possible line would be 17 picas. The *average* line we set will be between the two extremes of 20 and 17— that is, about 18½ picas. When we calculate copyfitting, we would calculate for lines averaging about 18½ picas in length.

Solving for Type Style

This final copyfitting situation is not as common as the others, but your knowing how to handle it easily without trial and error might amaze even some professionals who have been copyfitting for years.

Sometimes, because of rigid formats or changes in copy, we find that the only thing left to decide is which type style to use. Everything else is fixed.

For example, let's imagine that we wish to typeset the last paragraph of Lincoln's Gettysburg Address. Our character counting showed that it contains 891 characters. Suppose that we must fit these characters into a type area 20 picas wide and 168 points deep. Finally, we must use 10-point type, leaded 2 points. The only thing we don't know in this case is *which* 10-point type would produce the *best* fit.

Since we are now familiar with general copyfitting steps, only the worked example and formula, without elaborate explanation, follows.

	A WORKED EXAMPLE
SOLVING FOR BEST FITTING TYPE STYLE	We wish to fit the final paragraph of the Gettysburg Address into a given area, and the only remaining specification is the exact type style to use for an ideal fit.
We have already specified . . .	*the total number of characters: 891 *the type size and leading: 10/12 *the desired width of typeset lines: 20 picas *the desired depth of the type area: 180 points
We want to know:	what type style, in the given 10-pt size, will provide the best fit?
Steps: 1. Find the total number of lines	$12\overline{)180} = 15$ total lines (TL)
2. Find the number of characters that must fit on each line	$15\overline{)891} = 59.4$ $= 59$ characters per line (C/L)
3. Find the characters per pica	$20\overline{)59} = 2.95$ characters per pica (C/P)
4. Match a type style to the characters per pica	Look at a characters-per-pica chart to find the 10-pt. style that most closely matches the 2.95 character count. From the table in Figure 5–9, we see that Futura Light has the closest character count: 3.00, and therefore would provide the best fit.
This formula summarizes the steps we took	$C/P = \dfrac{\dfrac{TC}{DA}}{DL} \div WL$ $C/P = \dfrac{\dfrac{891}{180}}{12} \div 20$ $C/P = 59 \div 20$ $C/P = 2.95$ characters per pica

As we have seen, we work *backwards* to determine how many characters must be placed on each line, then how many characters must be put in each pica of each line. We found an *ideal* character count of 2.90 to fit our 890 characters.

We should be sure that the type style selected, is appropriate in *design* as well as ideal in fit for characters per pica. A perfectly fitting type style which does not match the nature of the message is no solution.

To supplement Figure 5–9, Appendix B lists 50 popular type styles and their character counts.

Type manufacturers always provide copyfitting tables for their types. Typesetting professionals can provide you with the copyfitting tables they use, or can give you selected counts as you need them. Several good type reference books, listed in the bibliography, also contain copyfitting tables and aids.

How to Get a Quick Character Count

Occasionally, we are unable to obtain a particular character count readily. A good character count estimate can be made by counting the total number of characters in 20 lines of a printed sample. Multiply the line width in picas times 20, then divide this figure into the total number of characters. This figure will be fairly close to the count provided by the type manufacturer.

Some Final Tips

Don't expect hand-calculated copyfitting to be perfect. We are dealing with *average* characters per pica in body type. If our copy contains more than the average number of extrawide or extranarrow letters, our fitting may be off slightly. In most cases, however, a one-line error is not critical.

Counting and fitting by *paragraph* is more accurate than counting all characters as a single total. Calculate the number of lines produced by each paragraph, then sum up the *line* totals to get full copy depth in lines.

Allow for variations in typesetting. When type is *minus-unit set* (see Figure 4–6), slightly more characters can be fit per line. Example: 12-point Optima yields 2.15 characters per pica when letterspaced normally. When tightened to minus 2-unit spacing, however, its character count increases to 2.38. (That is, if 12-point Optima is designed on an 18-unit system, and we subtract 2 units—⅑ of an em—per 12-point em, we gain a *full em* of space for every 9 picas of line width.) The gain in this case amounts to 5 characters, from 38 to 43, in an 18-pica line width.

Just remember that, if you specify the tighter letterspacing popular today, adjust your copyfitting accordingly by adding a few characters per line. Your typesetting house can help you determine proper counts.

6 Graphic Images

Powerful Partners of Words in Print

Today's information consumers demand more than words; they also want pictures and other graphic images that help explain what they read. With illustrative imagery, as with type, our choices seem limitless. Graphic images have a unique power to enhance—or to impede—understanding, depending upon how well we select them. Certainly, we should not forget how much both words and pictures depend upon one another for their meanings.

Introduction

Sometimes type alone is enough to convey a printed message. Most of the time, however, we need the special communicative powers of other graphic images to give life to our designs. Such nontypographic images include line art, continuous tone art or photographs. Photomechanical or electronic images derived from other images or generated in original form are also included in the term graphic image.

Those who manage the creation and use of images in publications go by many titles, including: art or design directors, who work for publishers and agencies; graphics, layout or picture editors, who work for newspapers; and graphic designers, who work independently or for organizations in a variety of settings.

Some of these people serve as their own artists and photographers, originating most or all of the images they use. Most, however, concentrate their energies and talents on selecting and blending images from many sources into complete packages of visual and verbal information.

Skillful planning and supervising of art and photograph production is basic to graphics. High-quality, original illustrations must be appropriate to the message, clear and concise. They must be used effectively to retain their communications value.

After images are produced, we edit them by altering their sizes and width-to-depth relationships. We sometimes cut away (crop out) unneeded portions. We also may retouch or delete distracting information. Finally, we position the images on our pages for greatest impact.

Why Use Graphic Images?

1. Graphic images get a message noticed. Printed communications rely on design to attract a reader's attention, and images are a large part of design.
2. Graphic images give additional information. They

NOTE: The author wishes to thank Douglas C. Covert, Assistant Professor of Journalism, Indiana University, Indianapolis, who contributed much to the writing of this chapter.

supplement the printed word by adding visual detail and explain relationships that are difficult to describe with words.

3. Graphic images provide impact. They show drama, action, emotion—any mood or setting desired.

4. Graphic images save words. They communicate many ideas more efficiently than type alone.

5. Graphic images add beauty. Even when they largely repeat typset information, they give an aesthetic atmosphere or special quality to the message.

Illustrative images range in visual content from highly abstract to extremely realistic. Simple, side-view line drawings, such as cartoons, may be quite abstract, leaving the reader to fill in or ignore missing details. Art with dimensional perspective and shading, and black and white photographs provide more details and convey a higher level of realism. Full color photographs, screened to 200 or more dots per inch and reproduced on gloss-coated paper are the ultimate in printed realism.

The choice of how realistic or abstract to make illustrations depends greatly upon how such images will interact with accompanying material. It also depends upon the cognitive level, topical interest, and sophistication of intended readers. Subject matter often dictates which images must be literal and which may be more symbolic. For example, a brochure published by a major manufacturer to promote one of its 35mm camera models might contain many realistic halftone images in full color to demonstrate picture quality. It might also contain simplified line drawings to explain settings discussed in the text, as well as technical charts and graphs. The cover might be a rather abstract color photo to symbolize fine photo-illustration.

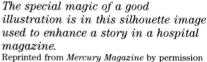

The special magic of a good illustration is in this silhouette image used to enhance a story in a hospital magazine.
Reprinted from *Mercury Magazine* by permission of Memorial Medical Center of Long Beach.

Labor and delivery had gone well. The eight and a half pound baby boy was everything they wanted.

Thirty-six hours later the parents' joy had faded. The baby was blue—a heart catheterization had to be performed to see what was wrong. His tiny heart was failing and he might die if something was not done—soon.

The parents had waited—for what seemed like days. But it was only an hour—an hour in which, the doctor explained, a spaghetti-like tube was threaded through the infant's umbilical cord vein to the heart. A special "dye" had been injected through the catheter which entered the heart, so it could be seen on a special x-ray machine.

The problem was found and temporarily remedied—until the baby could have corrective surgery. Without it, the baby could have been dead.

Rare?

Not today.

"Until the early forties the only thing we could tell parents whose child had a serious congenital heart defect was that we were sorry but there was nothing we could do," Richard Wittner, M.D., a pediatric cardiologist at Miller Children's Hospital Medical Center explained.

When symbolism is used as illustration, we must again consider the audience. Just how literal our visual message must be depends upon how familiar the reader is with the subject and what we intend the reader to do with the information we provide. A "how-to" booklet must be long on detail. An album cover can be a masterpiece of abstraction.

The Statue of Liberty is one of our most familiar symbolic images. Figure 6–1 (pp. 78–79) depicts a variety of its recent illustrative uses in advertisements and covers.

Editing Images

A prime function of image editing is the selection of illustrative matter appropriate for the purpose of a publication and the needs of its audience. Line or halftone, art or photography, multicolor or monochrome, colored paper or ink, are all visual decisions that should be made or anticipated early in reproduction planning. Close cooperation between production workers, artists, and photographers will improve the effectiveness of graphic images.

Images usually are selected to achieve certain objectives. An industrial publication intended to document a manufacturing process for the benefit of engineers probably would not fulfill its purpose by using a dramatically moody, retouched photograph.

Angle of view, lighting effects, scale, and amount of detail must all be considered by the designer. From a variety of miniature thumbnail sketches or photographic proofs, one or more pictures with appropriate content may be chosen. In editing, we must keep in mind that presentation of a picture's content is as important as the content itself.

Words and Pictures

Pictures rarely stand alone in publications. Art may be used to illustrate text or text may be used to enhance art. Whether they contribute to a message equally, or one is more important that the other is often difficult to determine. However, we know that people do respond to the way in which words and pictures interact.

Decorative images may simply attract a viewer's attention, offering little information but creating an atmosphere or mood. Some illustrations pack great

quantities of information into concise form, as with charts and graphs. Photojournalists try to tell a story about people, places, or events in each picture. Other photographs may be highly detailed records of things, events, or processes. Still others are intended to be persuasive. Whatever their purpose, pictures do not often work well without words to establish a context for viewing or to explain their significance.

We can group words that accompany pictures into three categories: *headlines, captions,* and *text.* Each is considered a separate element in layout.

Headlines. A headline may be subordinate to a picture. If so, the typeface and size should not compete with the picture. Instead, it should support its dominance. When the headline is intended to dominate, its size, its style, and especially its position in the layout should establish that dominance. White space often serves well to set off and emphasize a headline. On the other hand, the headline needs to be physically close to other elements in the layout for a unified appearance.

Superimposing a headline on art frequently creates visual conflict, especially for attention. With such designs, the headline must clearly dominate, usually by using a plain but large typeface—serif or sans serif—with strong contrast in tone and color.

Captions. A caption separated from its picture is lost. It is distracting to a reader to have to seek out a caption or to be directed from a caption back to "top left," "opposite," or "right center, below." *National Geographic* magazine is one of the few publications with content so excellent and audience so enthusiastic that captions can be separated from photographs— even placed within the text.

A caption may be placed above a picture ("capping" it). A caption also may be placed alongside or underneath the picture, where it is called a cutline. Visually, the caption should appear to be united with the picture, separated by not more than one or two picas of space.

Stacked captions are acceptable when the layout excludes other options. They should be separated from each other by two or more picas of space. Beginning each caption unit with a few words in boldface may be helpful.

When both headline and caption are used with a single illustration, the headline sometimes is placed above the photo and the caption below. The most acceptable order when a photograph accompanies a story is (from the top) picture, caption, headline, and

Polychrome Corporation

Bausch & Lomb makes sunglasses that make America look good.

American Express

Hammermill Paper

Look to Liberty for life insurance, too.

For generations, people have looked to Liberty Mutual for value.
In workers' compensation insurance. Business property insurance. Automobile insurance. Homeowners insurance.
Now, you can look to Liberty for value in life insurance as well.
In 1963, we organized the Liberty Life Assurance Company of Boston. Its sole purpose, to offer you an excellent value in life insurance.
Life insurance worth looking for.

LIBERTY MUTUAL

Public Relations Society of America

FIG. 6–1
This spread of images shows the limitless variety of visual effects possible from a single general theme.

Enco Printing Products

Reprinted from *U.S. News & World Report*, Feb. 28, 1983. Copyright 1983 U.S. News & World Report. Photo courtesy Karl Kummells/Shostal Associates, N.Y.C.

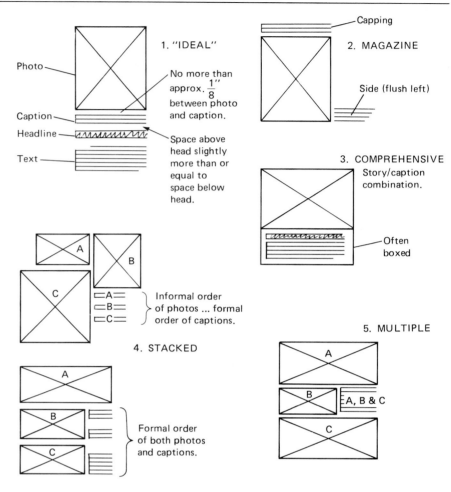

FIG. 6–2
Common caption styles.

story. Occasionally, a longer headline may serve as an abbreviated caption, generally appearing above the picture. Various caption styles are illustrated in Figures 6–2 and 6–3.

Text. Layout of text blocks is discussed in Chapter 9, but some combinations of text and pictures should be noted here. An illustration which clarifies text should be adjacent to text. A picture may protrude (be notched) into a text block or may separate blocks of text. Such insertions are most satisfactory when picture tones are middle- or high-key rather than dark. This helps avoid breaking the visual flow too severely. Low-key or boldly colored illustrations usually work better when placed outside text blocks. The preferred position for these more obtrusive pictures is below the text, unless the picture is intended to dominate all text on the page.

Surprinting text onto a picture may be effective and also may make efficient use of available space. Strong contrast with the background is needed for clarity.

Contouring text around the shape of a major picture element also may be effective if not overdone. This tends to add emphasis to the text, reducing the dominance of the picture around which it flows.

Type reversed within a picture requires special consideration. Type selection, sizing, and problems with screen patterns are discussed in Chapter 4.

Whenever type is to be superimposed on a picture, we should be sure the background will be the right tone or color for contrast and properly located within the frame.

Line Images

The most basic form of graphic image is line art. This includes any high-contrast image rendered in noncontinuous tone. The drawing in Figure 6–4 is an example of a line image.

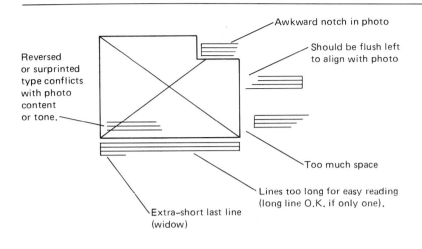

Awkward notch in photo

Should be flush left
to align with photo

Reversed
or surprinted
type conflicts
with photo
content
or tone.

Too much space

Lines too long for easy reading
(long line O.K. if only one).

Extra-short last line
(widow)

FIG. 6–3
Troublesome caption styles

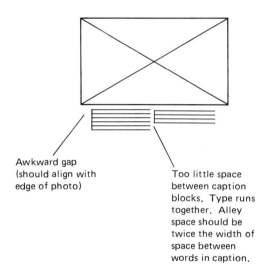

Awkward gap
(should align with
edge of photo)

Too little space
between caption
blocks. Type runs
together. Alley
space should be
twice the width of
space between
words in caption.

Line art should be done in black drawing ink or another medium that produces an image that can be photographed. Drawings, today, usually are made with technical pens, although quill- or broad-tipped inking pens are satisfactory for some work. Since originals are often drawn larger than the desired reproduction size, it is important to remember that fine lines may disappear as the image is reduced photographically.

Border tape, type, shading films, and many other materials are used to produce line images. Shading or fine dots produced mechanically by the artist, not photomechanically by the printer, are line images, not true continuous tone. Figure 6–5 depicts several types of line work.

Photoillustration

Today's virtual explosion of exciting photography—both in moving- and still-image media—has made us

more aware than ever before of how we can apply the camera to graphic illustration. Most of us have at least a passing acquaintance with photographic processes. When photographs are used in publication, special care must be taken to ensure that image quality and overall impact are preserved.

Evaluating Photo Images

If you have not studied photography before, the following brief discussion of films, papers, and reproduction quality of photographs may be a little confusing. If so, skip to the sections on Enlargement and Reduction. If you want to know more about photographic reproduction, take a basic photography class or read one of the many excellent texts available at most book stores.

A photographer judges the basic technical quality of a photograph on three characteristics: grain, gradation, and sharpness. Grain, more accurately called granularity or the appearance of granularity, is an inherent characteristic of the film. The granular appearance may be controlled by choice of film, degree of enlargement, handling of the film during its development, and to some extent by manipulation while making the photoprint.

Films which are most sensitive to light exhibit the greatest granularity. Continuing improvements in films keep reducing the problem of grain so that a photographer's choice of film usually is based on other characteristics. Moderate enlargements—up to about 7 times (7×)—from commonly used high-speed films rarely exhibit noticeable granularity. Grain in enlargements to about 12× is not usually objectionable and depends largely on the subject matter. Indeed, some grain clumping has a desirable effect for it tends to cut off images between grain clumps, enhancing an appearance of sharpness. This same cut-off effect often improves the appearance of sharpness in coarse halftones (those of less than about 120 dots per inch).

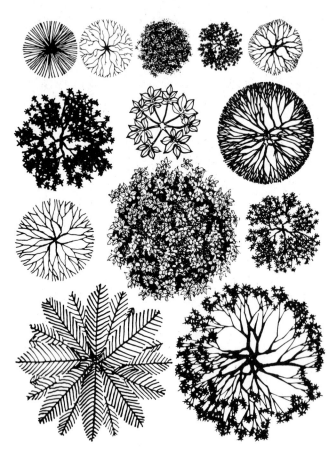

FIG. 6–4
Examples of line drawings.
Graphic Products Corporation (Formatt)

Gradation is the modulation or change of tones within a photograph. It is more than simple contrast. A photoprint exhibiting a full range of tones, from photographic black to highlights with form and substance, may still not appear bright, clean, and vigorous. The internal contrast or tonal modulation between similar grays in a photograph provides the impression of vigor—the "snap"—which enhances reproduction.

Medium-speed films show more internal contrast or modulation than either fast or slow films. This characteristic often improves reproduction of photographs in publications printed on lower grade, uncoated papers. However, it sometimes adds an impression of harshness on high quality, coated papers, depending upon subject matter and the skill of the printer.

Sharpness in a photograph depends on a photographer's skill, film characteristics, and photoprint quality. Fast (more light-sensitive) films permit faster shutter speeds, which promote sharp image recording. Slow films are inherently sharper but may require slower shutter speeds. Selective focus in the photograph may permit high shutter speeds with slow films. Visual contrast between points of detail in the photograph will also give an impression of sharpness. A photographer usually will select the slowest film compatible with the subject matter and the amount of light available.

The balance between grain, gradation, and sharpness in modern films and the characteristics of printing processes generally favors the use of higher speed films. This would include those with ISO ratings of about 400 for black-and-white, 64 to 100 for color. Color prints from ISO 400 negative-color films usually reproduce about as well as those from slower materials.

Black-and-white photoprints made from nonsilver negatives often exhibit excellent overall contrast characteristics. Internal tonal modulation may, however, lack the vigor desired for good reproduction. The image-forming dye clouds provide smooth gradation but this favorable characteristic may be offset by a slight softness of the image.

Black-and-White from Color

Black-and-white reproduction from color-negative original photos may be approached in several ways. A halftone negative may be made directly from a color print. However, this is usually the least satisfactory

FIG. 6–5
*Examples of line images containing
tones and textures obtained through
the use of self-adhesive, hand-cut films.*

Graphic Products Corp.

Letraset USA

Chartpak

Artist John Creed

Artist Terry Russell

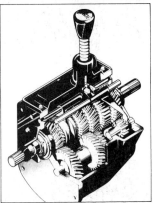

Artist John Thompson

method for faithful reproduction of all tones, even if the printer uses panchromatic film (film that is sensitive to all colors of light). Some loss of sharpness also may result.

Special panchromatic black-and-white photo papers allow not only improved sharpness and good tonal rendition but also manipulation of the tonal values when making the photoprint.

Some photographers have discovered that for most of their work an ordinary graded or variable contrast photo paper is satisfactory, especially since it allows adjustment of print contrast where panchromatic papers do not.

Black-and-white reproduction from a color transparency (slide) is a special problem. Different colors may have the same density in the transparency, producing the same gray values in the ordinary halftone. Panchromatic halftone films eliminate this problem. The most satisfactory method is to make a black-and-white transfer internegative, then a regular photoprint, but most photographers are unfamiliar with this technique.

Enlargement and Reduction

As a general practice, the best reproduction of continuous tone images is made from a same-size original. A halftone enlarged from an original photoprint will lose overall contrast, internal tonal modulation will be suppressed, sharpness will reduce, and detail in both highlights and shadows will tend to disappear. When the original is larger than the halftone, contrast and apparent sharpness will increase, modulation will become harsh, and again detail will be lost in the highlights and shadows. While photographs may be enlarged or reduced considerably more and be generally accceptable, the best results will be obtained when reproduction size is kept between about 80 to 125 percent of original size (see Figure 6–6).

Line art originals usually should be larger than the desired reproduction size, since no tonal contrast is present—only solid black or white. Reduction of line art usually sharpens detail and eliminates minor imperfections.

When retouching is required, or artwork is added to a photoprint, a large original may be necessary. Both the photographer and the artist must consider the contrast gain and detail loss inevitable during the subsequent reduction.

Four-color process work originating from a 35mm transparency also may suffer from extreme enlargement. Strong colors in the original will help, as will slight underexposure. When the four separation negatives are produced (see Chapter 12), contrast may be increased but the printed image may look harsh. Larger original transparencies yield better quality, but 35mm is used widely for color in general publications work.

The Process Camera

Both line and continuous tone images are photographed for reproduction on a device called a process camera. Figure 6–7 depicts a modern vertical camera and its essential parts. Figure 6–8 shows a large horizontal camera with an automatic film cassette which simplifies mass production.

FIG. 6–6
Continuous tone versus line copy. High-contrast images are better able to withstand changes in size.

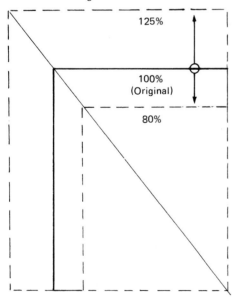

Continuous tone copy.
May reduce or enlarge moderately
without affecting quality.

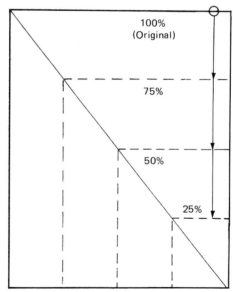

Line copy.
Always reduce, if possible,
for sharper image.

FIG. 6–7
*Modern vertical camera with fully automatic sizing,
focusing, and exposure control.*
Agfa-Gevaert, Graphics Systems Division

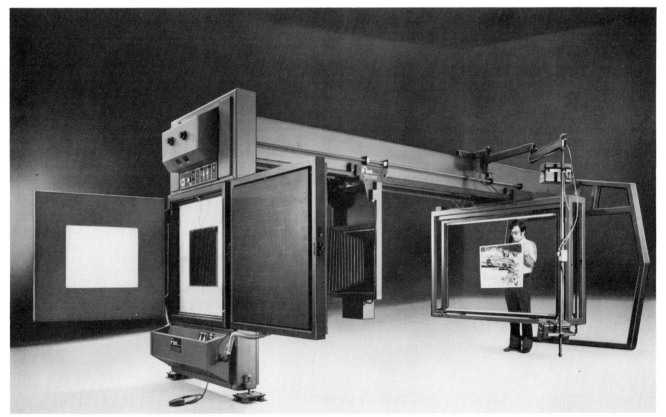

FIG. 6–8
Overhead type horizontal process camera with roll-cartridge feed.
ACTi Products Inc.

Film or paper positives and negatives can be produced on most cameras, and many special effects are possible.

Diffusion-transfer materials are highly popular today for process work. They require only one chemical for developing and can produce line and halftone paper positives in minutes with a simple processor instead of developing trays. Reverses can be made, and some systems can produce paper negatives. These materials have largely replaced the "velox" print—a photographic paper proof made from a negative. Fine halftones still require film, however.

Figure 6–9 shows how diffusion-transfer materials are processed.

The Halftone

A photograph or other continuous tone image may have a great many shades of gray between black and white. However, the printer is limited to a condition of ink or no ink, so gray is simulated by printing large or

small dots. The halftone is the printer's primary tool for reproducing the many tonal variations in a photograph from a single ink color. One printing process, gravure, requires a different procedure for simulating tone (see Chapter 11).

Line images are simple to photograph in the process camera. They are placed on the copyboard, and their images are exposed to the film or diffusion transfer paper in the back of the camera. Figure 6–10 shows the steps required.

Halftones are a bit trickier. After the film or paper is positioned in the camera, a special contact screen is placed over it before the image is photographed. When the image is exposed through the screen, a delicate pattern in the screen's emulsion (Figure 6–11) acts as thousands of tiny lenses, breaking up any continuous tone image into dots of different sizes. Dot sizes vary according to the light and dark areas of the image being photographed.

The number of halftone dots per inch is referred to as lines per inch or screen ruling. This varies with the type of screen used. Lines per inch usually range from 65 to 150 when produced on process cameras.

1

2

3   4

1. Expose KODAK PMT Negative Paper through a contact screen from original continuous-tone print to make a screened paper print.

2. Feed the PMT negative paper into a diffusion-transfer processor emulsion-to-emulsion with KODAK PMT Receiver Paper.

3. Separate the two sheets 30 seconds later—the result: a screened paper print of noteworthy quality.

4. Strip the screened paper prints into position on a mechanical, along with appropriate line copy.

FIG. 6–9
How a diffusion-transfer print is made
Reprinted courtesy of Eastman Kodak Company

5

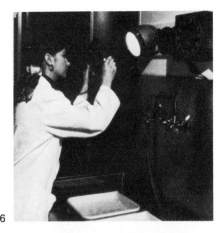

6

5. From the mechanical, make a line negative.

6. Process the negative in a tray . . .

7

8

7. Or process it in a film processor, such as a KODALITH Processor.

8. Result is a finished, full-page line negative, ready for platemaking.

A 100-line contact screen produces 100×100 (10,000) dots per square inch. Much finer rulings can be produced with electronic scanning devices which are discussed in Chapter 12.

Several dot shapes are possible, including square, round, and elliptical, depending upon the type of halftone screen that is used. Elliptical dot screens produce very smooth middle-tone variations. Halftone dots produced on scanners are actually groups of tiny dots which combine to form the size and shape of each larger dot. Figure 6–12 depicts a halftone image with shadow, highlight, and middle-tone dots.

Both the printing process and the paper to be used determine how fine a screen may be used for a halftone. A coarse, absorbent paper such as newsprint encourages dots of ink to run together into a nearly solid mass. A high-quality coated paper allows dots to retain their size and shape. Clear, sharp, fine-screen dots reproduce better detail.

Continuous Tone Originals and Reproduction

Nonphotographic continuous tone originals may be treated in much the same way as photographs. Rendering of shades in media such as charcoal, pencil, ink wash, or watercolor requires the halftone process to reproduce tones faithfully. As with photographs, tonal smoothness of nonphotographic continuous tone improves as the number of halftone dots per inch increases.

A glossy photographic print contains a longer range of tones than any major printing process can produce. Printed black may appear as dark as the black in the photograph. However, reflection characteristics of photographic paper are different from paper used for printing; therefore, the photoprint usually produces brighter whites. In addition, there will be less distinction between dark tones on the press sheet, and

a.

b.

c.

d.

FIG. 6–10
Steps in the operation of a process camera.
Photographed by Pat O'Donnell, with permission from Smith Printers &
Lithographers, Tustin, CA.

e.

f.

g.

h.

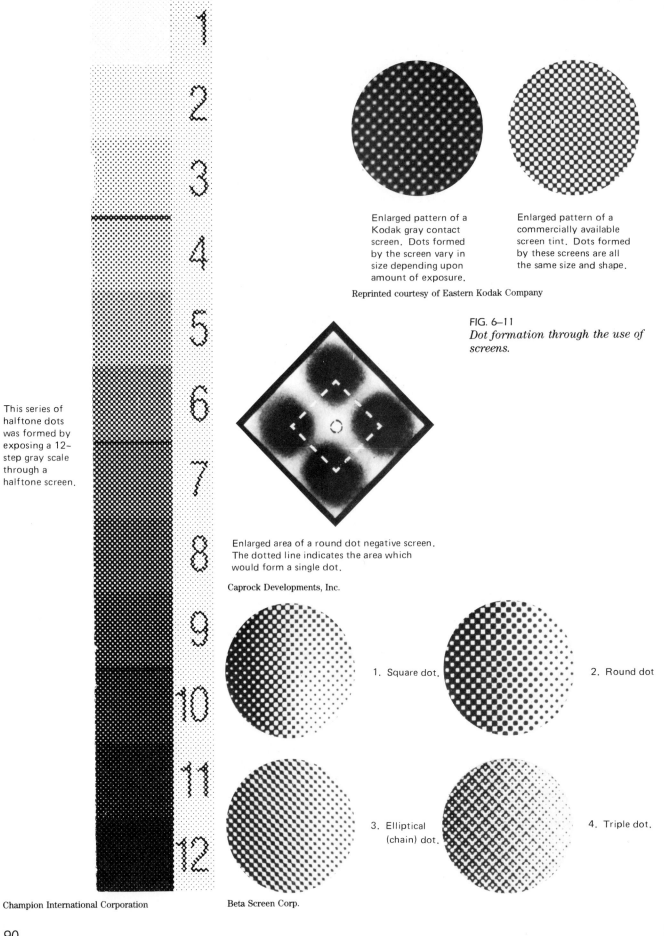

This series of halftone dots was formed by exposing a 12–step gray scale through a halftone screen.

Champion International Corporation

Enlarged pattern of a Kodak gray contact screen. Dots formed by the screen vary in size depending upon amount of exposure.

Enlarged pattern of a commercially available screen tint. Dots formed by these screens are all the same size and shape.

Reprinted courtesy of Eastern Kodak Company

FIG. 6–11
Dot formation through the use of screens.

Enlarged area of a round dot negative screen. The dotted line indicates the area which would form a single dot.

Caprock Developments, Inc.

1. Square dot.

2. Round dot

3. Elliptical (chain) dot.

4. Triple dot.

Beta Screen Corp.

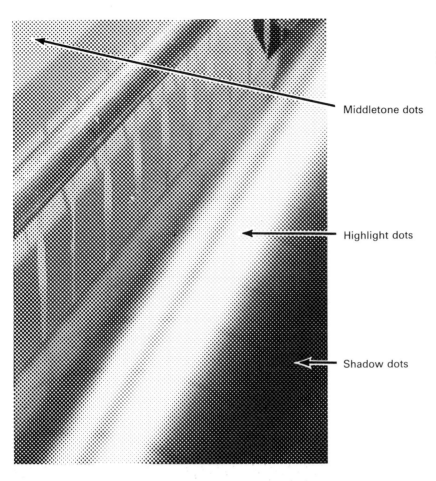

Middletone dots

Highlight dots

Shadow dots

FIG. 6–12
A halftone, enlarged to show dots.
Hammermill Paper Company

subtle differences in the lightest highlight areas will disappear.

Publications photographers usually are well aware that their images, when transferred from photoprint to press sheet, will be reduced in overall tonal range and that details at extreme ends of the tonal scale will be subdued or lost. They make an effort to produce originals with full ranges of tone and well-defined shadows and highlights.

Screened positive prints (either diffusion transfer or velox halftones) tend to gain contrast when rephotographed with the page on which they are pasted. Therefore, screened prints may need to be somewhat flat in contrast. Photographers should not attempt to compensate for contrast increase. They should provide full-tone photographs and let the process camera operator compensate for contrast increase in printing by adjusting halftone exposures.

Preparation of Photographs for Reproduction

When preparing an original photoprint or piece of art for reproduction, we can compare the brightness ranges and reflection densities of the original to those of the printed reproduction.

A well-ferrotyped (glossy) photoprint may have highlights that are 75 times as bright as the darkest shadows. This brightness range is commonly expressed as 75:1. The brightness range of a well-printed halftone on reasonably good paper will be about 25:1. Newsprint commonly produces a brightness range of only 5:1.

Photographs which depend for their effectiveness on details or subtle tonal gradations in either the highlights or shadows will be unsatisfactory for reproduction on newsprint. If printed on a higher grade uncoated paper, the halftone still may not satisfy either the photographer or the client since reduction in brightness range is still evident.

Even when printed on smoother, coated papers, the halftone reproduction may be disappointing because of another quality—the comparative reflection density. A sheet of coated paper printed with black ink in a single impression has a reflection density range of about 1.6. The glossy photoprint may have a reflection density of 2.0. This means that the tonal distinction between black and white will not appear to be as great, and tonal separations within the picture will be reduced.

The photographer will find reproduction more closely matches the original if the photoprint is made on regular glossy paper or on glossy-surfaced stabilization paper without ferrotyping. The reflection density of these papers is about 1.6 or 1.7, about the same as coated printing paper.

Reflection density of the printed reproduction can be improved through the selection of higher quality papers or the addition of varnish to the sheet. These solutions can add substantially to printing costs.

A good photograph for reproduction will have a full range of tones from photographic black (the blackest the photoprint can produce) through a variety of grays to highlights at which, at their lightest, still have a sense of substance and form.

Many photographs do not reproduce well because information in the shadow areas—less commonly in the highlights—is not tonally distinct. The most common cause is simply underexposure of the original negative by the photographer. To adjust for this, the photographer is likely to make the print lighter in the shadows, thereby not achieving a photographic black. Weak shadow details that show up in the print may well be too indistinct to appear in a halftone made from the print.

Fine reproductions sometimes can be made from photographs which appear too dark in their original form through contast control in the halftoning process.

Making the Halftone

A halftone negative or positive is produced with as many as three separate exposures: the main, the flash, and the bump. The main exposure, made through the halftone screen, forms the primary dot structure of the image.

A brief second exposure, called the flash, is made through the screen, but the photographic image is not used for exposure—only raw light. This adds substance to the weaker shadow dots which did not receive enough light from the main exposure. Flashing can be done from a lamp in the darkroom, from a lamp inside the camera, or by photographing a plain sheet of white or yellow paper. Flashing varies with the camera, methods, and materials used.

A third exposure, called a bump or highlighting exposure may be used to increase the size of highlight dots. This very brief exposure of the photograph, done with the halftone screen removed, is less commonly used.

Using the process camera, the operator may expand or brighten a somewhat muddy photo, lighten the shadows, or give weak highlights more substance. Poor image content in an original won't go away, but improved contrast can make a drab original seem more lively.

Many of the steps in shooting a line image, shown in Figure 6–10, are similar for halftones, except for flashing and use of the halftone screen.

Alterations for Effect

Many processes and techniques are available for modifying the effect of a picture. A photographer can control exposure of a photographic print by blocking light from certain areas (dodging) or allowing a controlled stream of light to strike selected areas (burning). Overall lightening can be done by bleaching the print.

Other print alterations include retouching or spotting the print to eliminate blemishes. Airbrushing may be done to eliminate distracting elements or to provide contrast and dramatic effect. Significant retouching is usually done on oversized prints so that the artist's work is concealed when the image is reduced.

A dropout (or highlight) halftone eliminates the dots in selected highlight areas so no ink is deposited on the press sheet and the highlights become pure white when printed. The highlight dots may be removed during exposure of the halftone negative or, preferably, by applying opaquing fluid to the negative to block highlight dots.

Artwork and photographs are sometimes outlined for reproduction. Outline art, sometimes confusingly called a silhouette, has had its background removed. Most commonly, a peel-apart masking film (such as Rubylith) is cut to cover only the wanted image area, leaving the background uncovered. The film mask is photographed to create a window into which the halftone later can be positioned. The background, which does not show through the window, is effectively blocked out. Masking is the best method here. Painting out the background of a negative or photograph is tedious and less effective.

A vignette is similar to an outline except that the edges of the image are not sharp—they fade gradually into the background. Vignettes emphasize the subject, but retain a hint of background, which may be pleasing. Vignettes are best done at the enlarger by blocking light around the edges of the photographic print during exposure. However, they also can be achieved by the airbrusher.

Photographers are likely to complain about misuse of a photograph when its rectangular form is altered. Outlines and vignettes may be accepted, grudgingly, as sometimes necessary and helpful, but mortises and notches are considered anathema. Geometric shapes such as ovals, triangles, and stars—or more innovative shapes such as trees, fruits, and letters—may cause wrathful outcries from the photographer. However, such alterations of shape may be useful to the designer with a sound sense of aesthetics, especially if made big and bold. An important key to effectiveness is that the image be appropriate to the message and not look like a gimmick.

A notched halftone allows the extension of type block or caption space into the picture's otherwise rectangular area. Pictures with closely related content may be overlapped, yet retain individual identity. Charts or graphs related to a photo's content may be inserted into a notch for more concise presentation.

A mortise is a hole cut into the interior of a picture. Another picture or type may then be placed into the mortise. When overprinting (surprinting) is desired but the photograph to be used is not suitable, a mortise may serve the purpose. This is especially true when a tint is added to the cutout area to avoid destroying the effect of the photograph. Using a mortise to insert a caption rarely is effective enough to justify its intrusion into the image. Indeed, mortises are rarely sucessful except when another illustration is inserted.

Nonrectangular shapes may function together as a collage to condense a large amount of pictorial content into a small space. Such assemblies often work well—especially in color—to attract attention and direct emphasis.

Cutting a photo into starbursts or palm trees may destroy the photograph, leaving only its texture. A shape which repeats a form in the photograph may help convert a poor photo into a successful element of the layout, but odd shapes must be used with care.

Creating Illustrations in Production

Dramatic or artistic effects may be achieved in many ways. Photoillustrators often plan on, even depend upon, techniques available to the printer to create special results.

A line conversion (Figure 6–13) changes any continuous tone original into a high-contrast image devoid of middle tones. With only black and white remaining, the result resembles a line drawing. The effect is easy to achieve. The original is photographed in the process camera without the use of a halftone screen. Exposures usually are adjusted by the camera operator to achieve good results. Originals with strong, dominant shapes, good contrast, and little fine detail work best for line conversions.

FIG. 6–13
A line conversion.
Reprinted from *Discovery* magazine by permission of Northwest Pipeline Corporation, one of The Williams Companies.

Reverses (white images in a dark field) also are popular with designers. Usually these are done with either line images or type. However, reversing a halftone creates a dramatic negative image in print. Most reverses are made by the camera operator shooting images on special reversal film or paper.

Posterization may be performed by the photographer or the printer, in black-and-white or in color. A color posterization is best produced by the photographer or his processing laboratory. Black-and-white posterization works better when done by the printer. Figure 6–14 shows a three-tone posterization.

The technique of posterization is straightforward, although not easy. Blacks and whites are retained, as

FIG. 6–14
Three-tone posterization.

Reprinted with permission from Clip Bits Magazine, published by Dynamic Graphics, Inc., © 1981 by Dynamic Graphics, Inc.

FIG. 6–15
Images made from special-effect halftone screens.
Caprock Developments, Inc.

ROUND DOT (100)

SQUARE DOT (100)

ELLIPTICAL DOT (100)

MEZZOTINT (75)

STRAIGHT LINE (62)

SUNBURST (100)

LINEN (50)

MEZZOTINT (150)

STEEL ENGRAVING (50)

WAVY LINE (60)

CONCENTRIC CIRCLE (60)

STEEL ETCH (50)

in a line conversion, while all intermediate tones are compressed into one or more middle tones. The result is an image in at least three tones—black, white, and middle gray. Sometimes one or two additional grays are produced.

Special-effect screens, including the standard halftone screen, may be used to reproduce middle tones. Both continuous tone and line images may be reproduced as parallel or wavy lines, concentric circles, mezzotint or steel-etched patterns, fabrics, and many other variations. All should be used with caution since they tend to become an important part of the illustration.

Figures 6–15 and 6–16 show the wide variety of images that can be made using screens available for producing special effects without halftone dots.

Special screens should be used sparingly. Since they tend to obliterate fine detail, the images they form carry more symbolism than information. The temptation to use special screens as an afterthought to make up for poor images should be resisted. Screens sometimes emphasize rather than hide imperfections.

Cropping and Segmentation

Pictures may be cropped to desired size and shape as the print is made or during the layout process to improve clarity and conciseness. Extraneous information or background is removed while sufficient context is retained to maintain the needed communicative effect. To put it most simply, crop away what you don't need, but be careful not to crop so much that the meaning is lost. Cropping may be used to focus attention on important content or to enlarge and emphasize detail.

Cropping is also used to fit a picture to space available. Many photographs may be improved by changing the proportions, perhaps from rectangular to square, or even from horizontal to vertical. On the other hand, many illustrations lose their effectiveness if they are cropped to a preconceived space rather than adjusting the space to make best use of the image. Note in Figure 6–1 how bold cropping of some of the Statue of Liberty images adds to their dramatic effect.

FIG. 6–16
Straight-line screen adds middle tone detail, but gives a striking effect not possible with standard halftoning. Compare this image to the screenless line conversion in 6–13.
Reprinted from *Big Yard*, July/August 1980, by permission of Pacific Gas Transmission Company. Cover design by Sidjakov, Berman & Gomez.

Segmentation—the extraction of small sections of a main image—can help carry a visual message through several pages (see Figure 6–17).

Tell the artist or photographer the probable sizes and shapes required so they can design their work to avoid excessive cropping.

Sizing

The size of a reproduced graphic image strongly influences its visual impact and ability to transmit information. A small reproduction may reduce the visibility of necessary detail. A picture that emphasizes mass rather than detail may retain its effectiveness in a small size. A large reproduction may show detail but lose overall impact because the viewer cannot see the entire image at a glance.

When several images are viewed at the same time, whether on a single-page or multipage layout, relative size becomes important. Individual pictures may lose their effect when size and proportion are repeated. Generally, a variety of sizes and proportions, with both vertical and horizontal emphasis, is preferred. Layouts of related pictures often are improved by the

FIG. 6–17a
Segmentation.
Courtesy *Panhandle Magazine,* a Panhandle
Eastman Corporation publication

selection of a dominant picture—the largest in the spread—around which others can be placed to maintain perspective and relative importance.

Perspective in layout is important for the reader. In a group of related pictures featuring a principal subject, the size of that subject should be about the same in each picture. A photograph depicting an overall view of the subject or its environment, accompanied by a close-up, would not need to conform to this rule.

Cropping and sizing helps maintain perspective. Sense of perspective also is affected by the way pictures

"read"; that is, how the eye follows images in a single picture. Some pictures read left-to-right, others top-to-bottom. The sizes and positions of illustrations, viewed as a whole, affect the total feeling of perspective in a layout.

Positioning

The location of an illustration within a layout space affects its emphasis. Although people in most western cultures begin reading at the upper left of a page, an

FIG. 6—17b
Altered segments.
Fluor Magazine

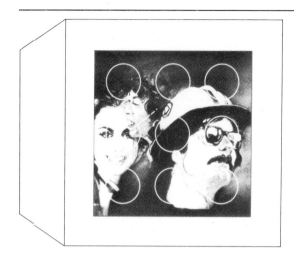

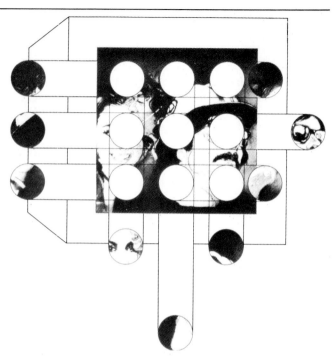

FIG. 6–17c
"Exploded" segments.
Reprinted by permission from New Jersey Bell Magazine,
Issue 1, 1979.

illustration position in the upper right corner often is more visually emphatic. Emphasis tends to decline as pictures are placed farther down a page. Illustrations near the bottom receive increased emphasis if they are reproduced in larger sizes.

The apparent direction of movement (sometimes called line of force) in an illustration helps determine its position. Pictures contribute more to overall layout effectiveness when they move the eye toward the center or through a logical sequence of information. Figure 6–18 indicates the effects different tonal values can have upon eye flow within a layout.

High-key and low-key images also are positioned relative to type blocks in a layout. Low-key (darker) pictures are often more effective when anchoring the bottom of a page. Type blocks tend to be light in tone, or high-key. Whether type or art, high-key areas generally work best higher on the page.

Low-key illustrations also may be effective when used as a frame for type blocks. Separating blocks of related type with illustrations, regardless of tone, may serve a purpose, but it more often reduces the emphasis on both type and pictures. This can result in a bland layout.

Bleeds and Margins

Bleeds are a popular device used to increase picture size or improve layout impact. Bleeding, or running the picture off the page, may change the balance of

composition unless the photographer knows in advance that a picture is intended to bleed. Small pictures rarely work well as bleeds.

Extra care in design is needed if some images in a layout will bleed and others will be kept within page margins. Such combinations are more effective when images are not the same size. A large picture which boldly dominates type blocks and other images may be especially effective as a bleed.

Bleeds may not be possible at all unless the publication will be printed on oversized paper for later trimming.

The designer of a printed piece selects margins and other white space as part of the overall publication format. This is discussed more fully in Chapter 9. The amount of white space around and between pictures tells readers a lot about the relationships of images. Related pictures, and the text which goes with them, should be separated by minimum space—about one to two picas. Unrelated images should be set apart by about twice that amount, unless the overall layout utilizes much open space. Panels of pictures of the same shape and size depicting a sequence of action need very little separation between images.

Scaling and Proportioning

When a picture or other graphic image to be reproduced is in hand, it must be scaled and

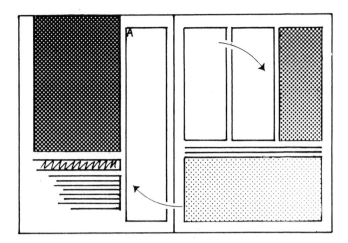

This double-page magazine spread shows a rotation of photographic tones from dark to light. Such rotation would encourage eye movement in roughly the directions indicated. Such horizontal/vertical/reversed horizontal often is seen as a desirable eye movement pattern for magazine spreads.

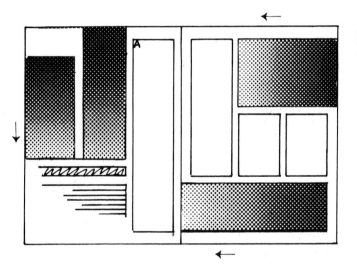

An alternative eye movement pattern, again suggested only by photographic tones, is one that points to the beginning of the story from all directions. Note that the dark-to-light tones are within *each* photograph in this illustration.

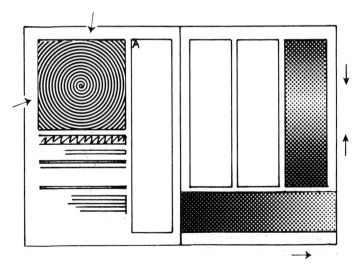

This layout indicates the disruptive effects of tonal direction. In the upper left is a photograph of monochromatic tone but which, through strong internal lines of force pulls the eye constantly toward it. Both photographs on the right contain tonal directions which draw the eye away from the beginning of the story.

FIG. 6–18
How placement of pictures with varying total values may affect eye movement.

proportioned to fit the layout. Production instructions also must be provided.

Scaling is easy with a little practice. Most professionals use a calculator or a device called a proportion wheel. The diagonal-line method of scaling also may be used, especially to double-check calculations.

First, let's label the dimensions of the image. Width is always given first, then depth. Four numbers are used: width and depth of the original image and width and depth of that image when reproduced (see Figure 6–19). Remember, measure only the part of the original

that is to be used, not the margin nor any section that is to be cropped out.

Usually, one of three kinds of problems must be resolved:

1. If the width of an original image is enlarged or reduced to a given reproduction width, what will the reproduction depth of that image become?

2. If the depth of an original image is enlarged or reduced to a given reproduction depth, what will the reproduction *width* of that image become?

3. Where must an image be *cropped* to make it fit into a given width-to-depth ratio?

FIG. 6–19
Labeling illustration dimensions.

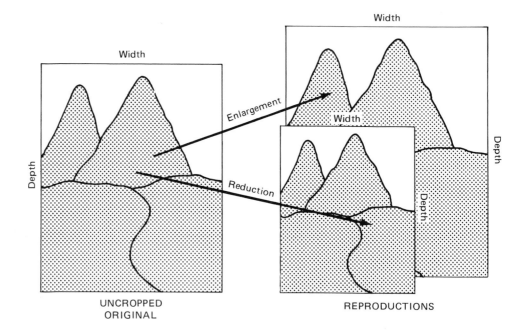

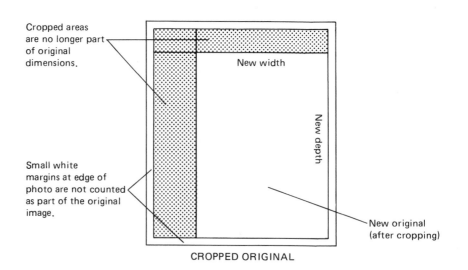

Example 1. *Finding the new depth.* We have an 8 × 10 inch image that must fit into a 4 inch width. What will the new depth become when the image is reduced to fit the new width?

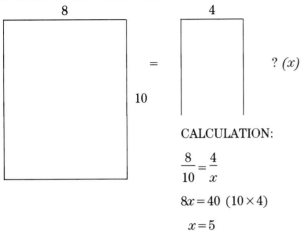

=

? (x)

CALCULATION:

$$\frac{8}{10} = \frac{4}{x}$$

$$8x = 40 \quad (10 \times 4)$$

$$x = 5$$

Answer: the new (reproduction) depth is 5 inches.

In this problem, *width was the critical factor,* so this determined what the *depth* must be.

Figure 6–20 shows how the *diagonal line method* can be used to check the answer. A line drawn from the upper left to lower right corner of the original (A to B) should always pass through a similar corner of the reproduction rectangle (A to C).

Figure 6–21 shows how we would work the problem on the *proportion wheel:*

1. The original width of 8 inches (on the inner wheel) is aligned with the reproduction width of 4 inches (on the outer wheel). Hold both wheels together firmly while you read the other figures.

2. We now see that the original depth of 10 inches (inner wheel) lines up with 5 inches (outer wheel).

3. We also notice that the number 50 appears in the small window that gives percentage of original size.

Mark the percentage of the original size on the picture—either on the back, on a tissue overlay, or on a small slip taped to the bottom. This lets the printer know what settings to use on the process camera when photographing the image. *Don't use the words "reduction" or "enlargement."*

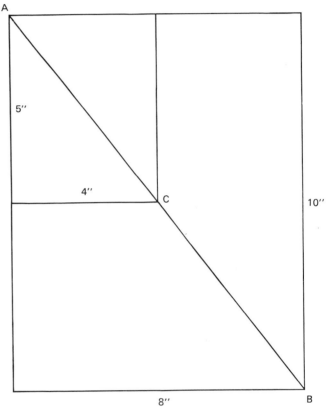

FIG. 6–20
Checking the answer to example 1 with the diagonal method.

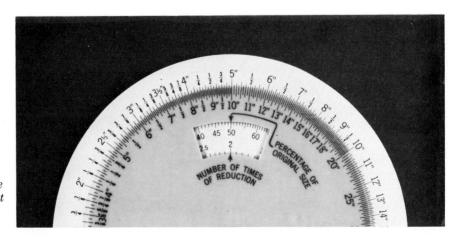

FIG. 6–21
The proportion wheel quickly gives the answer to example 1, plus it tells what percentage of the original size our final image will be.

Example 2. *Finding the new width.* Imagine we have a $5 \times 7\frac{1}{2}$ inch image that must fit into a 9 inch depth. What will the new width become when the image is reduced to fit the new depth?

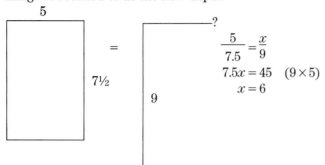

$$\frac{5}{7.5} = \frac{x}{9}$$
$$7.5x = 45 \quad (9 \times 5)$$
$$x = 6$$

FIG. 6–22
A diagonal check of example 2.

Answer: the reproduction width becomes 6 inches.

The diagonal-line method is used in Figure 6–22 to verify the answer.

Figure 6–23 shows how the proportion wheel is used to verify the answer and to establish the percentage of the original size—120 percent.

Sometimes a picture must fit a layout area for which both reproduction width and depth have already been determined. In this case, we must determine how much to crop the image so it will fit.

FIG. 6–23
Answer to example 2 on the proportion wheel.

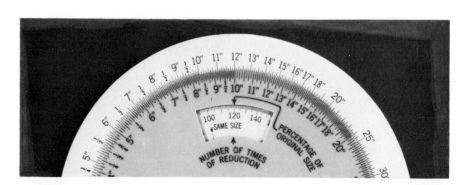

Example 3. *Finding a crop.* Suppose we have an 8 × 10 inch image that must be cropped to fit a layout size of 6 × 5¼ inches We can see that the original must be cropped to a horizontal shape to match the shape of the layout area. How much of the depth of the image must be cropped?

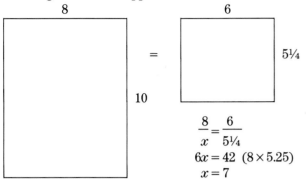

$$\frac{8}{x} = \frac{6}{5\frac{1}{4}}$$
$$6x = 42 \quad (8 \times 5.25)$$
$$x = 7$$

FIG. 6–24
Finding a crop for example 3 with a diagonal.

The diagram in Figure 6–24 shows how a diagonal line will help determine where the image is to be cropped. Again, the proportion wheel can be used to arrive at the answer more quickly.

In Figure 6–25, note that we work backwards on the wheel in this instance. Match the original width (8 inches) to the reproduction width (6 inches). Then find the reproduction depth (5¼ inches) on the outer wheel. The number on the inner wheel below 5¼ in. is the reproduction depth (in this case, 7 inches).

Since the depth of the original must be cropped from 10 to 7 inches, the answer to our problem follows: 3 inches must be cropped from the original to make the picture similar in proportion to the layout area. When the 8 × 7 inch image is reduced to 75 percent of its original size, it will fit just fine.

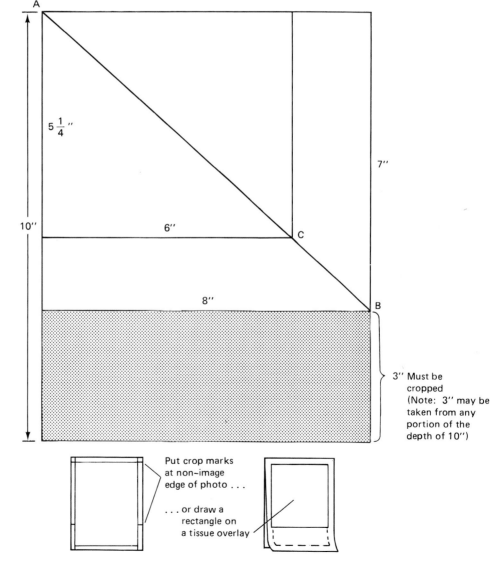

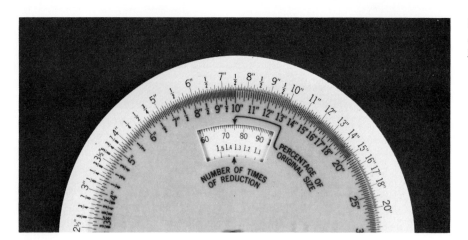

FIG. 6–25
Finding a crop for example 3 with the proportion wheel.

It is important to note here that 3 inches could have been taken from any part of the image depth—bottom, top, or both—depending on where the cropping would look best.

Reproducing Graphic Images

Those who work with graphic images understand the potentials and limitations of printing processes. They work within established requirements for reproduction in publications. When they cooperate fully with printers, the results are much more satisfactory to both the creators and the consumers.

Three main variables determine how an image will reproduce: the type of image, the printing process required, and the grade of paper to be used. The last two are discussed in detail in other chapters. They are mentioned here since they are vital to our understanding of image quality.

Line art is similar to type in the way it reproduces in print. A sharply defined, high-contrast image will reproduce well on a single side of almost any paper with any of the major printing processes, as long as the lines in the image are not too fine nor too closely spaced. Gravure printing requires that even fine line work be broken into small dots or cells. Therefore, originals for this process should not be made up of extremely fine lines.

Continuous tone images require halftone screening for both letterpress and offset printing. The cells created in gravure for printing photographs generally yield better simulation of the photographic emulsion. Subtle middle tones often can be retained.

The tremendous versatility of offset lithography has contributed to its dominance among printing processes. Large image sizes don't increase production expenses as much as in letterpress. The sharpness of line images and halftones, and the ease with which screen tints and type can be integrated with other images is well known to designers.

Sometimes the ease of offset image preparation encourages carelessness. Line art done in marker or pencil instead of ink, as well as poor-quality photographs, occasionally are submitted for printing. Unfortunately, offset usually reproduces the defects of these originals quite faithfully.

Solid areas of ink and large halftones sometimes do not fill in completely in letterpress printing. A blotchy appearance may result, spoiling the contrast of the original.

Recent developments in flexography have made this process more suitable for long, high-speed runs of both line and halftone matter.

Papers that appear to be the same may have quite different characteristics for reproducing illustrations. Variations in ink absorption, finish, texture, and opacity all may affect the finished image. Moisture content, press speed, relative position on the printing plate (which may affect how much ink the image receives), all should be considered.

Of all the variables that affect image quality, however, the submission of good originals is the most important and the most easily controlled by the illustrator.

7 Text and Image Processing

Capturing Words and Pictures Through Electronics

Print media specialists are fond of observing that newspapers—inexpensive, portable, packed with millions of bits of text and graphic data—are still some of the best information delivery systems available. As we examine the exciting technological advances used in communications graphics today, it is worth noting that no system for gathering nor medium for delivering information is more important than the quality of that information itself.

Introduction

A charming song from Rodgers and Hammerstein's stage play *Oklahoma* begins with the line, "Everything's up to date in Kansas City; they've gone about as fur as they can go."[1] Street lights and horseless carriages seemed as marvelous at the turn of the century as semiconductors, lasers, and robots seem today. Yet, only the pace of change has quickened, not the certainty of change itself. "New technology" is no longer new, and the electronic revolution has become an evolution, a way of life. An entire generation of Americans is growing up at ease with electronic mail, laser-scanned groceries, and video games—the artifacts of an electronic age.

McLuhan may have misjudged the attachment humans have to the printed word, but his prediction of a "global village," where instantaneous mass communications bind the world's societies together, rings more true each day. Long distance, high-speed transmission of data, text, and graphics for publication is now part of a worldwide information sharing network. And few of us would say, "we've gone about as fur as we can go," or predict where the next few steps will lead us.

It may be comforting to remember that, while the

people of the world now communicate in microseconds instead of months, the messages remain the same. Technology simply provides better and more exciting tools for human communication.

Electronic Information Processing

Communication professionals are far from alone in their use of electronic information processing. Many specialized applications found in computer-based media communication systems have been adapted from similar systems found in business, technical and military organizations. Word processing, teleconferencing, voice mail, electronic filing, micrographics, computer graphics, and document processing are familiar terms in corporate offices.

The Four Means of Sharing Information

Today's information processing addresses four communication areas—*text, image, data,* and *voice*. Distinctions between systems according to which of these they will process are blurring rapidly. In all probability, systems of the future will be able to handle all kinds of information with equal facility.

Electronic Processing and Sharing of Information

Although the focus of this chapter is on text and image processing, it is important to remember that all forms of information processing require the same activities. Information must be:

• *gathered, developed, and evaluated;*

• *coded and entered*—symbols or keystrokes captured and input to the system;

• *stored*—recorded with some degree of permanence within the system;

• *processed*—edited, organized, and filed appropriately in the system;

• *shared*—retrieved and output to other systems or directly to information consumers.

A daily newspaper is a familiar example of information processing in action. Reporters, photographers, and others gather stories, advertisements, and graphic images for the day's edition. Reporters evaluate and enter the text of their stories at video display terminals. Photographers and ad or graphics staff members scan or otherwise record their images electronically. All text images are stored by the newspaper's computer, where they are available to the editors. When all information has been edited and positioned electronically, it is transmitted to a type and image-setting device which outputs it in a form that can be printed in the newspaper.

Except for the electronics involved, a reporter on the 1890 *Kansas City Evening Star* might have readily understood the steps necessary to publish a modern newspaper.

Computer Components and Functions

Any discussion of text and image processing, as with all forms of electronic communication today, begins with computers. What is this device that *Time* magazine named Machine of the Year for 1983, passing over its warm-blooded competitors? A digital computer is a super-efficient adding machine that processes binary (yes and no) codes with breathtaking speed and accuracy.

The familiar personal or home microcomputer illustrates, on a small scale, the most important components and functions of a computer. A simple computer consists of:

1. The **microprocessor**—this is the "brain" contained on a tiny chip. Also called a CPU, or Central Processing Unit.

2. The two types of **memory**—temporary RAM (random access memory) and permanent ROM (read only memory). Information in RAM is entered through the keyboard or from a disk. This memory contains both the applications (program software) being used and the information being processed. Information in RAM is not permanently stored in the machine, but is usually stored on a disk. Any information that we enter, data or a program, is easily saved and printed out when needed.

ROM is made up of both a permanent machine language put in by the manufacturer to make the computer run, and an operating and assembling system to translate software programs into language the machine can understand. Two popular operating systems are CP/M™ (control program for microprocessors) developed by Digital Research Inc., and MS-DOS™, licensed by Microsoft.

3. The **monitor**—basically a TV display screen (a cathode ray tube or CRT).

4. The **keyboard**—this lets us communicate with the computer. Most keyboards have four sections—one for regular typing, one for editing, one for telling the computer what to do with what we typed, and one for formatting—programming certain keys to do particular functions.

5. A **disk drive**—a record/playback device into which magnetic floppy diskettes may be inserted. The computer can read information such as software programs from the disks, and information typed into the computer's memory can be transferred to the disks. Where a lot of data storage is needed, hard disks capable of accepting millions of characters are used.

Any piece of equipment that lets us communicate with the computer is called an input/output device. Such devices may be built into a console that houses the microprocessor and memory, or they may be *peripheral* units connected to a main console. Input/output devices include a monitor, a keyboard, a floppy or hard disk drive, a printer, and a telephone modem that permits the computer to telecommunicate with other systems.

Thoroughly disgusted with computerese by now? It's probably a good thing the common pencil was invented years ago; today, we would have named it a "data stream input system interfaced with an abrasive character-erase module.[2] Be patient—plain English will return.

A computer crunches its data in bits and bytes. A bit is a single binary digit, like a yes or no switch. Eight bits usually make up a character or numerical digit. Memory is described in approximately thousands (K) or millions (Mb) of bytes; for example, 256K RAM holds roughly a quarter-million characters; a ten-megabyte (Mb) hard disk stores more than 10 million characters.[3]

The more bits a computer can handle as a *unit* the more powerful it is. Today, both 16-bit and 32-bit personal and small office computers are commonplace.

Many students are already familiar with these computer basics. Bookstores are crammed with literature for dedicated computer buffs and the mildly curious alike.

Computers used in text and image processing for mass communications operate much like their familiar relatives in the home and office. However, because editors, printers, designers, and other communication professionals put unusually heavy demands on their systems—speed, accuracy, flexibility, and highest quality output—their equipment and software are highly specialized and generally expensive. The major task requirements include the following:

Text processing—an efficient, sophisticated means of capturing, editing, filing, and assembling many forms of information for publication, and preparing it to be output in printable form.

Word processing interfacing—receiving and translating information produced by business- and document-oriented systems so that it can be text processed and output in printable form.

Typesetting—converting the information from text and word processing systems into typographic form ready for printing. Some typesetters allow *direct entry;* that is, a keyboard can be used to send information straight to the typesetter to be set immediately.

Image setting—capturing, storing, and outputting line and halftone illustrations in printable form. Many typesetters today are actually image setters, since they generate both text and graphics from similar digital signals.

Hard proofing—generating copies of electronic information for approval or filing by using a dot-matrix, line, wheel, electrographic, or other printing device.

Soft proofing—previewing finished designs, pages, and layouts electronically (on the screen) before they are set in printable form.

Integrating text and images—electronically manipulating all elements of a single page (area composing) or many pages (paginating) so they can be output in assembled form ready to print.

Archiving—recording and managing information generated by the organization.

Data base sharing—interacting with systems containing large amounts of general purpose or specialized data to develop publishable information.

Because firms which require text and image processing vary widely in size and requirements, most computerized systems today are customized to fit a facility's needs. Until recently, much of the hardware and software produced by one manufacturer was incompatible with that produced by another. This forced purchasers either to 1) install a single, compatible system; 2) buy different systems to perform separate tasks; or 3) employ expensive consultants and technicians to *interface* units from different vendors in order to create a customized system. Today, most manufacturers' hardware can be interconnected readily. Also, standardized operating systems and flexible applications programs are available, although manufacturers naturally try to encourage use of their own proprietary software.

Because equipment interfacing is no longer difficult, and because most hardware today can be *upgraded* with improved software programs, computer systems are less likely to become obsolete as a company's needs change. For example, a typesetting company that wishes to offer clients the capacity to translate word processed material directly into typesetting codes can purchase an *interface* device for their typesetter.

Text Processing

Of the many functions required of computers in print-oriented communications, the transformation of words into type is most basic. Text processing systems have become indispensible to this function.

Any hardware/software combination that allows us to enter, store, edit, and output information can perform simple word processing. Text processing systems perform all these functions with the editing sophistication, storage, and file management capacity, output quality, and speed required for mass publication. Keyboards and software are designed so that all data can be *formatted* for typesetting with size, space, and copyfitting commands in place.

Most of today's systems are fully *integrated;* that is, they include all front-end steps that lead up to typesetting, and the back-end functions of type and image setting.

System Configurations

Almost any combination of equipment and applications is possible, but the following are most common:

Single station—also known as *direct input*, it is the smallest form of integrated system. While described and sold as a typesetter, the station actually performs text processing functions in addition to setting type. Often, an *off-line* (stand alone) terminal can be attached, which doubles inputting capacity. Direct-input stations usually perform functions simultaneously, such as typesetting in the *background* while the operator is keyboarding information in the *foreground*.

Modular cluster—a small, integrated system made up of terminals, peripherals, and a typesetter. Often providing up to eight workstations, such a system provides many of the same functions possible with large front-end systems but may be limited to a relatively small number of devices (typically eight) that can be attached.

Front end—A large, integrated, multiterminal system designed to drive one or more typesetters. An important distinction between such systems is the location of computing power. Some house most of the logic in a central controller, and *share* it among a large number of so-called *dumb* terminals. Others distribute logic between *smart* terminals or small *clusters* of terminals (linked to a *concentrator*). Both systems require central processing units and hard disk drives for storage of data.

Shared-logic front-end systems can utilize inexpensive terminals, including those designed for home and personal use. However, they require very dependable hardware since a failure of the CPU can incapacitate the entire system.

Front-end systems drive typesetters that operate as *slave* units—that is, the typesetting machines simply follow the commands of the system instead of applying their own built-in logic for hyphenating and justifying lines of type.

When a text processing system is composed of several linked clusters of equipment grouped according to processing function, department, or distance, it is called a *network*. All clusters in the network are able to telecommunicate with one another through a main central processing unit. Some networks operate primarily as a *local* area group serving a single company, organization, or general work environment. Newspaper and other media-oriented systems operate as both local and national or international information networks. Of course, through telecommunications and other transmission technologies, even the smallest network can share data over long distances.

Figure 7–1 (a–e) depicts several ways in which text processing systems can be organized. These are only general models, since every system must be tailored to fit special needs.

FIG. 7–1a
Direct input system—the Omnitech 2100 Typesetter.
Mergenthaler Linotype: Mycro-Tek Products Division

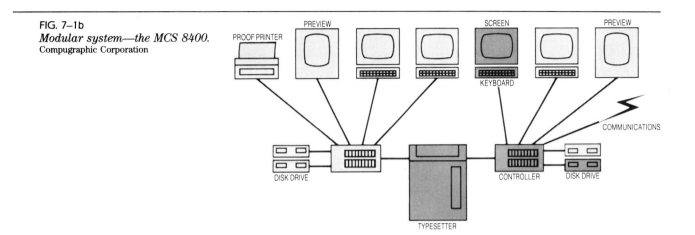

FIG. 7–1b
Modular system—the MCS 8400.
Compugraphic Corporation

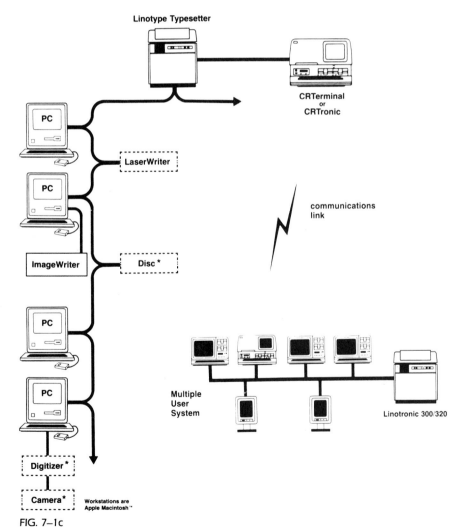

FIG. 7–1c
*Small front-end system showing two recent approaches to sharing resources.
The multiple user system, lower right, shows the networking of a number of
Linotype terminals and a typesetter. The other system depicts linkage of many
personal computers into a front-end configuration.*
Allied Linotype

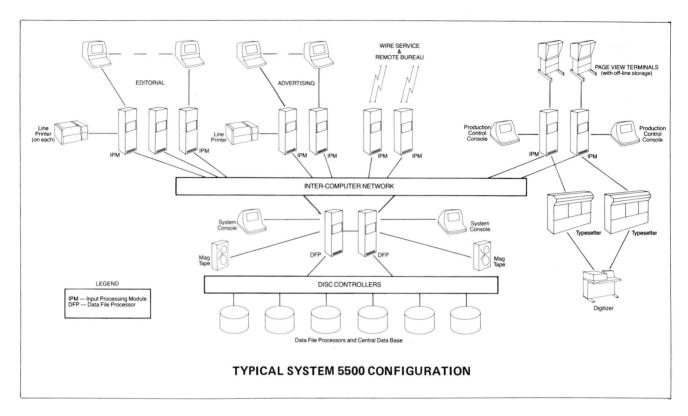

FIG. 7–1d
Large front-end/back-end network for newspaper.
Mergenthaler Linotype: Mycro-Tek Products

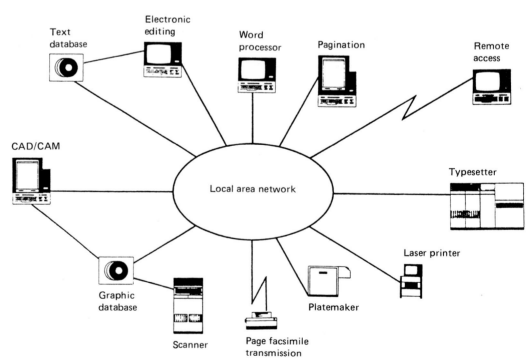

FIG. 7–1e
Local area network system capable of many different information-sharing applications.
Dunn Technology, Inc.

Copy Preparation on the Video Display Terminal

The basic *workstation* unit of any text processing unit is the VDT, video (or visual) display terminal. On it the operator can enter, store, and edit copy, and can create permanent files. Most terminals today have detachable keyboards, tiltable screens, variable brightness controls, and other features to make data entry easier.

Figure 7–2 (a) shows the keyboard of a popular newspaper terminal, the Linoscreen 400, made by Mergenthaler Linotype. It is divided into four areas, or *keypads*.

Area 1: Cursor control. Provides keys to allow movement of the cursor (a small video marker) that indicates the spot where copy is to be entered or edited.

Area 2: Typewriter. Located here is the main keyboard similar to an electric office typewriter.

Special commands such as point size and leading can be entered here.

Area 3: System functions. This pad contains function or format keys—keys that can be programmed to insert *strings* of command codes into the copy with a single keystroke. The pad contains *dedicated* keys (defined by the manufacturer) and *user-programmable* keys (those that can be changed by the operator).

Area 4: Editing. The final pad is for editing. It allows us to delete, insert, search for (and replace) characters, words, and paragraphs, and to *quad* (move) a line to left, right, or center.

The terminal shown operates in three *screen modes*—wide screen for normal copy, split-screen for viewing two different files at once (for editing or merging), and a half-screen mode for viewing a single section of copy in a two-column mode, as shown in Figure 7–2 (b).

FIG. 7–2a
Linoscreen 400 keyboard.
Mergenthaler Linotype: Micro-Tek Products

1. Wide Screen

2. Split Screen

FIG. 7–2b
Three screen modes of the Linoscreen 400.
Mergenthaler Linotype

3. Half Screen

Terminals usually have enough temporary *buffer* memory to store and process screenfuls of information without having to *dump* characters into the full screen memory. Simple keystrokes will *scroll* the copy up or down on the screen. Figure 7–3 shows how the screen provides a *window* that lets us see only a portion of the copy at a time as it scrolls by. Some systems provide huge buffer capacity for *virtual* scrolling. This allows unrestricted viewing or editing of files containing millions of characters.

The Linoscreen 400 displays 25 lines on the screen. The top line is a control line that shows commands entered and messages to the user from the computer. The other 24 lines are for entering copy. Each line can display the standard 80 characters.

Characters can be displayed in modes that simulate various type forms such as bold or italic, and can be shown in reverse or with a strike-through line. Characters can be hidden completely if necessary, especially when command codes might distract a person in the process of editing material on the screen.

Virtually all terminals allow the following operations:

Entry—All text to be processed requires initial *keyboarding;* that is, the words must be captured electronically at the terminal. While unsophisticated terminals may be used for this, data entry will be quicker and easier if the terminal is equipped with or has access to sophisticated editing programs.

Newspapers and other organizations that prepare large amounts of material for publication prefer simplified text entry procedures to speed operations and minimize training. While even complex text has become much easier to input, more intense training is required where typographic and design elements are exacting and widely varied.

For the majority of straight text entry, only a few simple commands such as paragraph indention, paragraph endings, and the editing functions listed below are needed. The operator doesn't even have to strike a return key at the end of a line—if the next word will not fit on the line, it is automatically *wrapped* (moved) to the line below.

Complex typesetting codes can be inserted into text through the use of the preprogrammed format or dedicated keys. A reporter can, for example, insert all the commands for setting a story—typeface, size, and spacing—with a single keystroke.

Editing—By moving the cursor to the desired location on the screen, the terminal operator can delete a character, word, line, or whole block of copy. Often it is easier to *replace* rather than to delete. This is like a strike-over on a typewriter, except that the new character completely deletes the former character.

In *insert mode,* new characters can be filled in to the left of the cursor. Copy to the right of the cursor is moved over to make room, and additional lines may be formed on the screen as the copy is repositioned.

Many terminals provide special editing aids such as selective brightening, underlining, or flashing of important sections of text.

Common Editing Enhancements

Most computer systems today provide enough memory to allow the following common editing capabilities:

Block moving—rapid repositioning of text. The beginning and end of the desired section is first defined by the cursor, then the move is executed. Sometimes blocks can be stored in a terminal's local memory, then reinserted where needed.

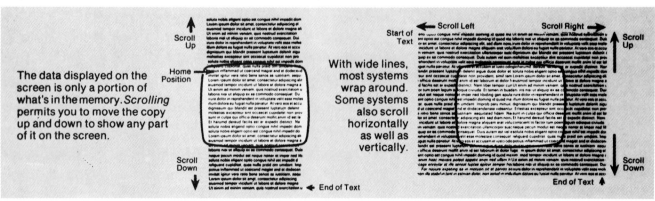

FIG. 7–3
The "window" of viewing area on a VDT screen.
Mergenthaler Linotype

Merging—split-screen terminals can display different text on both sides, allowing the operator to define certain sections on one side to be merged with the other. This is especially helpful when different versions of a news event, especially from wire services, must be rewritten to form a single story.

Search-and-replace—any character string can be defined by the operator, automatically located wherever it occurs in a file, and replaced.

Increased Logic Enhances Editing Power

Some editing functions require large amounts of computer logic, either at the terminal or shared from a central processing unit. Such functions include, but are not limited to, *global search-and-replace, hyphenation and justification,* and *soft-copy previewing.*

Global search-and-replace allows the operator to search out and replace many words or phrases in many files, even entire disks. This is often done in the *background,* so an operator can keyboard other material while the search is being performed by the computer.

Hyphenation and justification (H&J) is the important function of converting keystrokes into type by computing the number of characters that can fit onto each typeset line. This requires that the computer be programmed to remember the *exact unit width* of every character in every size and style of type to be set. As the units are counted and lines filled, the computer then hyphenates words where necessary. Although some counting systems display each line and require hyphenation decisions by the operator, most keyboarding is done as a continuous stream of characters that are later fitted automatically.

Today, nearly all systems use a combination of *logic rules* and *exception word dictionaries* to hyphenate. Logic rules consist of an internal program which would tell the machine such things as when to place hyphens before suffixes like *-ly* or how short a word can be and still be hyphenated. Since rules of logic produce correct hyphenations only about half the time, the computer first consults its exception dictionary (perhaps containing 300,000 words) for proper hyphenation. Nearly all the problematic words would be found there, but logic rules will be applied if the word is not in the dictionary.

Typesetting devices are usually fully capable of their own hyphenation and justification. On-screen H&J is a tremendous advantage, however, because it accurately predicts how the typesetting machine will set the story *before* actual setting occurs. The benefits for copyfitting stories and preventing waste of materials are discussed later in the chapter.

Soft-copy previewing (visual formatting) is also possible on many of today's systems. The operator's terminal may have a preview mode, or a separate preview screen may be used nearby. When a story is called up for previewing after keyboarding, it appears not only in hyphenated and justified form, but also in the desired size and position on the screen. All command codes are suppressed (hidden), letting us see the "page" as it will be typeset. This previewing procedure is *passive,* since we are unable to manipulate the images directly while in the preview mode. Some systems allow for *interactive* manipulation, whereby sizes and positions of characters can be changed while the image is being displayed on the screen. Both passive and interactive procedures are forms of *area composition.*

Information Control Via Front-End Systems

Most of the entry and editing functions described in the previous section are common to even small, single-station systems designed mostly for setting type, as well as many home and personal computers. Front-end systems bring additional editing, file management, and storage capabilities to text processing.

Those investing in a front-end system have to consider many things, including:

Size—small systems usually have from two to eight terminals, larger ones can have hundreds. Cost, of course, varies widely.

Design—some systems are better for excellent quality commercial typography, others offer excellent full-page and multiple-page typesetting; still others are designed primarily for handling masses of information for rapid publication, as in newspapers. The latter may be known as an *editorial system.*

"Friendliness"—many systems today are easy to use without the necessity of learning computer language or memorizing complicated codes.

Growth potential—most systems now allow computing power and memory to be added in modules as the need arises. Limits on the number of terminals that can be added to controllers and the compatibility of software and hardware between manufacturers is important.

Systems for Newspapers and Magazines

The Atex 5000, a popular system for daily newspapers, with up to 32 terminals, operates as a powerful front-end system designed for both editorial and advertising information control. Specifically designed to be easy to use, it helps illustrate some major features of newspaper-oriented systems (although exact capabilities do vary between systems). Atex also offers newspapers and magazine systems that can handle up to 500 terminals, but the general capabilities are similar to the system shown here.

The Atex 5000 uses a "distributed database" approach, whereby each file is duplicated automatically as it is created. Backup systems such as this are common for newspapers where sudden loss of data from system failure could spell disaster. Terminals share the logic from the two 260K central processing units and storage on two 80 megabyte hard disks.

Atex, like many other companies, uses a hierarchical system of file management. All information processed in the system is organized by *groups*, *queues*, and *files*, all of which can be called up on the screen as directories. Figure 7–4 shows how these three levels of information correspond to the departments, desks, and individual stories in a typical newspaper organization. Security of all information is protected by the use of passwords.

Typically, a reporter beginning a story would *log on* (request access) to the system by entering his or her name and password. When access is approved, the reporter is guided (or prompted) through interactions with the system by a system of *menus*, plain English lists of instructions. For example, one item on the menu states: "To start a new story, name it:———. Then press EXECUTE."

Before beginning to write, the reporter would use a few keystrokes to call up any messages left for him in the system (by the editors, for example). An electronic response could be sent to anyone with access to the terminal. Important messages can be made to flash when the operator logs on to the system.

The reporter can begin keyboarding the story using common entry and editing functions. Scrolling is unlimited. Formatted keys provide typesetting codes or specific information preprogrammed by the reporter (such as a byline). When the story is completed, the reporter can request an estimate of story length. Should any procedure be forgotten, pressing the "HELP" key brings immediate assistance.

Editors can call up the story on their terminals at any time and easily insert, move, or delete words, paragraphs, or blocks of copy, or merge the story with another. After any necessary headlines are keyboarded, a special program is run which relates typefaces to the sizes of the newspaper; this allows a rapid checking of proper fit for the headline. Finally, story hyphenation and justification (H&J) is performed. The computer rapidly determines how many characters will fit on each line when typeset,

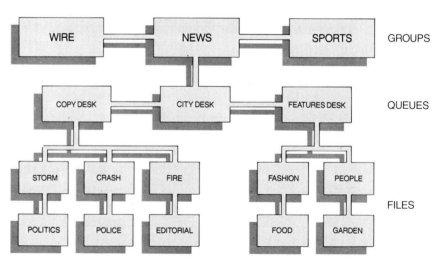

FIG. 7–4
File management system for Atex publishing system.
Atex, Inc., a Kodak company

and displays the story line-for-line, including hyphens. This way, an exact length for the story is determined. Hyphenation can then be adjusted by the editor if needed.

Editors can check the status of all stories by writer, by subject, or by page, simply by defining the directory needed. Wire service copy in the system is routed automatically to the appropriate queue (desk) for processing with local material. Once the material is ready for printing, page makeup is the next step. As we shall soon see, page makeup and graphics terminals are also becoming an integral part of publication systems.

Commercial and In-Plant Front-End Systems

While the system described above is ideal for editorial applications, many users of front-ends require systems designed more specifically for handling many kinds of typeset output. Such users include commercial typesetters, book, magazine, and periodical publishers, advertising and public relations agencies, as well as thousands of in-house or corporate printers and publishers.

Although some of the same capabilities listed below are also found in newspaper editorial systems, the following are becoming nearly universal in commercial front-ends:

• highly refined typographic software programs that include automatically paired or multisector kerning, ragged setting, wrapping, contouring, and vertical justification
• ability to interface with almost any input/output device
• sophisticated hyphenation/justification packages
• passive or interactive preview and full-screen composition
• large, expandable directories and dictionaries
• modular expansion of overall logic and memory
• pagination (automatic paging of long documents and publications)
• simple, mnemonic coding
• software orientation to prevent equipment obsolescence
• standardized formatting and programs for keyboarding and setting complex jobs
• digital storage and manipulation of line and halftone images

Preliminary & Composition		End Format	EF	Change Ragged Range	CR
Job Command	JC	Re-use Format	RF	Cancel Ragged Margin	XR
Position Fonts	PF	Delay Lines	DL	Quad Left	QL
Position Disk	PD	Merge Copy	MC	Quad Right	QR
Change Column Width	CC	Use Global Format	UG	Quad Center	QC
Change Face	CF	Point Mark	PM	Justify Line	JU
Change Point Size	CP	Return Mark	RM	End of Paragraph	EP
Change Leading	CL	Point Maximum	PX	End of Take	ET
Zero Leading	ZL	Return Maximum	RX	Track 1	T1
Increment Auto Leading	IA	Define Vertical Rule	DV	Track 2	T2
Extra Leading	EL	Set Vertical Rule	SV	Track 3	T3
Reverse Leading	RV	Tabs & Indents		Upper Rail	UR
Break Galley	BG	Tab Set	TS	Lower Rail	LR
Hang Punctuation	HP	Tab Number	TN	Additional Composition	
Cancel Hang Punctuation	XP	Tab Proportion	TP	With Leaders	WL
Quad from Middle	QM	Tab Text	TT	With Characters	WX
Font Italic	FI	Mark Tab	MA	No Escapement	NE
Font Roman	FR	Jump Tab	JT	No Flash	NF
Hyphenation & Justification		Quad Tab Line	QT	Cancel No Flash	XF
Allow Hyphenation	AH	Cancel Tabs	XT	With Rule	WR
Cancel Hyphenation	XH	Indent Take	IT	Underscore	US
Discretionary Hyphen	HH	Indent First Line	IF	Small Caps	SC
Allow Letterspace	AL	Indent Hang	IH	Cancel Small Caps	XC
Cancel Letterspace	XL	Indent Paragraph	IP	Automatic Fractions	AF
Change Interword Space	CS	Indent Text	IX	Cancel Fractions	XA
Set White Space	SW	Indent on Left	IL	Change Character Width	CW
Kern Character	KC	Indent on Right	IR	Continue Tape	CT
Allow Kerning	AK	Skew Left	SL	Secondary Dictionary	SD
Allow Ligatures	AG	Skew Right	SR	Enter Dictionary	ED
Cancel Ligatures	XG	Ragged & End of Line		Kern Numeric One	KO
Formats & Make-up		Ragged Right	RR	Cancel Kern on One	XO
Store Format	SF	Ragged Left	RL	Superior Character	SU
Use Format	UF	Ragged Center	RL	Inferior Character	IN

FIG. 7–5
Computer command codes.

One great advantage of many of these systems is cost; some are in the range of tens of thousands of dollars versus hundreds of thousands, even millions, for large newspaper publication systems.

A Word on Coding

Every system that drives typesetting equipment must be able to insert codes or commands that tell the typesetter what to do. These commands include: *typeface, point size, line width, leading (interlinear spacing), letter tightening (tracking), positioning (left, right, center), mode (justified, ragged, contoured), indenting, tabbing,* and many others. Easy to remember mnemonic codes made of two-letter initials preceded by a command keystroke are common. Figure 7–5 shows a variety of commands and their meanings on one popular system.

FIG. 7–7
Pagination by Penta Systems International.

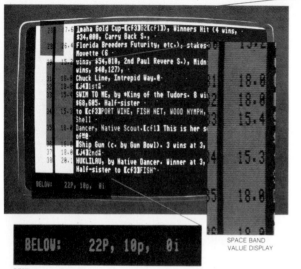

SPACE BAND
VALUE DISPLAY

DEPTH VALUE MEASUREMENT DISPLAY

FIG. 7–6
*Hyphenation & Justification pass on
sophisticated front-end.*
Penta Systems International

AA & PE DISPLAY
(VERIQUICK)

INDENT LEFT VALUE DISPLAY

A Few Highlights of Commercial Front-Ends

The Computer Composition International CCI400, a shared-logic, interactive system, offers five levels of H&J: a 4,000 word root dictionary, a set of logic rules, an infinitely expandable exception word dictionary, up to 99 additional specialized dictionaries, and operator controlled discretionary hyphenation. Each job file can have access to 1,000 universal and 500 specialized typographic formats.

Every job is recorded on oné or more of the system's seven directories, depending upon its status. A *master queue* records all jobs, and there is a directory for jobs ready to *proofread, compose* (insert commands and hyphenate), *typeset,* or *output* to a printer. In addition, jobs coming from *outside sources* and those *already set* also have their own directories. A system operator can call up a job from any directory at any terminal.

Sophisticated software allows "flashback H&J," whereby any job can be displayed on a screen fully hyphenated and justified, with kerning, letter tightening, loose line indicators, and numerical values

showing minimum and maximum word spacing ranges for each line. Figure 7–6 shows a file after an H&J "pass" has been made.

CCI also offers a program called InfoMix that allows complex data from many sources to be sorted, merged, and coded for typesetting.

Pagination

An important tool for today's publications is automatic paging of publications and documents. Most front-end systems offer *pagination* programs for magazines, books, catalogues, and similar materials. Some offer full-size page makeup programs for newspapers, as well. According to Penta Systems International Inc., its Penta Page pagination program, will make up a 300-page book (normally a 20–30 hour job) in four hours. Figure 7–7 shows four pages and the automatic paging functions that the program performs. Even simple paging programs usually include automatic, multicolumn page breaking, insertion of running head and foot lines, folios (page numbers), top and bottom alignments, and rules.

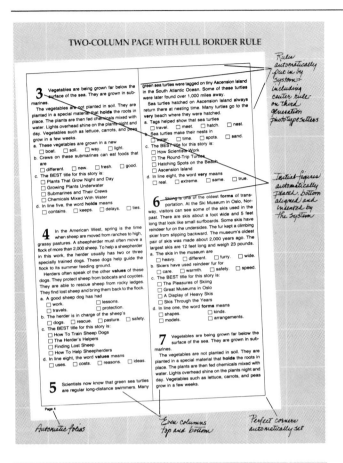

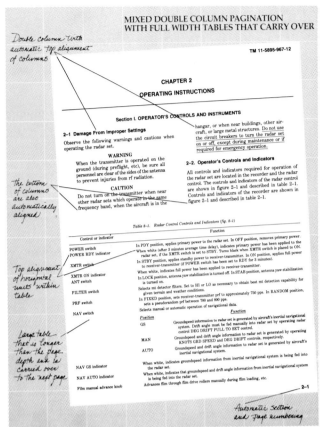

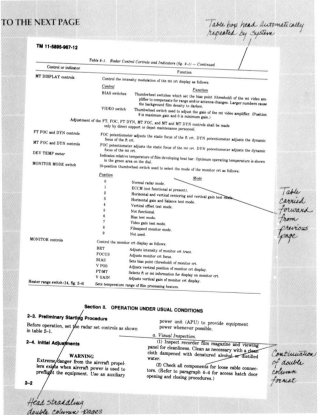

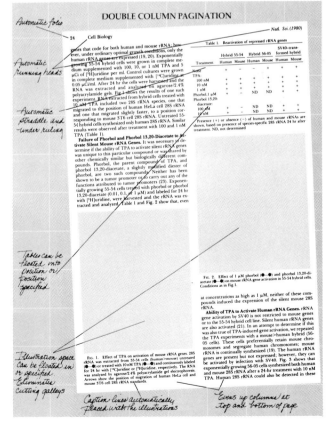

FIG. 7–8
Specimen of automatic typeset modes.
Computer Composition International

NIGHT AND DAY

Many of the new daytime outfits for spring could be stamped with the guarantee "Good for sixteen hours." Suits with bolero jackets, dresses with camisole tops, soft pantsuits and pajamas can all go from the office to a night out.

University presses haven't indulged, traditionally, in the publication of creative work. They have remained faithful to their original purpose: the production of scholarly monographs and studies too erudite for the general reader. True, there have been exceptions. About 15 years ago some university presses relented a bit and began to issue slim volumes of poetry. And last year the University of Chicago Press published *Carnival*, Isak Dinesen's posthumous stories. And last year Harvard University Press announced that it will begin publishing short novels in an attempt to revive the art. Louisiana State University Press began publishing fiction in the spring with Madison Jones's *Passage Through Gehenna*. The editors declared then that "fiction has become comparatively less profitable and much

A Day in the Life   Interview by William Brashler

Bill Veeck

Moment to moment off-season with the owner of the Chicago White Sox

Sleep is a habit, you know, and not a particularly good one. I never liked going to bed for fear I'd miss something. I got up this morning at 6 a.m. I'll sleep sometimes to 6:15. Four or five hours is usually enough. In winter it's still dark when I awake; but in the summertime, on those days made for a ballgame, the view of Lake Michigan and Chicago from our apartment on S. Chicago Drive is just beautiful.

My address is in the book if you're curious. Mary Frances and I have always lived in the cities of our ball clubs. I've always felt if you're going to work in a city, you ought to live there. We've had big places, enough to accommodate the nine kids I've had in two marriages—four or five of them are usually around the house. Best apartment we ever had was right in Sportsman's Park, in St. Louis, in 1952, when we owned the Browns. It worked out wonderfully.

Mary Frances gets up with me and spoils me terribly, you understand. She makes coffee for me, then breakfst for the kids, who are headed for school or work, and I prop my wooden leg under my stump and hobble off to the bathroom, a place I go to each morning to soak what's left of my right leg. After a short, undistinguished Marine career in the Pacific during World War II, I became an albatross around Uncle Sam's neck and spent twenty-two months of my hitch in hospitals trying to overcome the infection in a leg injury. The infection

William Brashler is a writer-novelist currently living in Chicago.

Full Page Make-up performed by the CCI-400 Front-end Typesetting System.

Chicago magazine, a hardback book or two, a pack of salem longs, the telephone, my horn-rimmed bifocals, and coffee. I still smoke—some years back doctors were certain the habit was killing me, but I wriggled out of that one—and I'll drink five or six cups of black coffee.

The phone calls come in and go out, Mary Frances comes in and chats, but mostly I read. A low-grade infection (the doctors call it that, but I'd hate to see the high-grade stuff) plagues my inner ears, and I enjoy no hearing in my left ear and only 30 percent in my right. I could turn down my hearing aid, but I concentrate better with an irritant like noise, and I like to keep an ear open for things. I'm not

me a call. I'm waiting for an artificial left knee. Mine is shot from overwork. I want the new knee to be a good one; I'd love to play tennis again.

The wooden leg is not much of a problem. It's hinged at the knee and locks when I walk. I used to go through two or three a year, but now they last longer. Or sometimes they do. At some inauspicious times they've collapsed and sent me sprawling. Bionic man I'm not. I try to find a pair of pliers and bolt things together until I can get to the repair shop.

Out of the tub, I shave, but don't use colognes or after shaves. I strap on the wooden leg, which is hollow and has a hole drilled in it just right for an ashtray, put on a pair of dark, casual slacks, an orange, short sleeve shirt or a turtleneck sweater, and my one black suede shoe. I'm not much of a dresser, and I don't own a tie. On the most important day of my life—when I married Mary Francis—I didn't wear a tie. They make my neck break out. Mary Francis says that when I get dressed up I look like I've forgotten my pants.

Without breakfast, something I seldom have, I call a cab for Cominsky Park. I don't drive anymore—besides, cabbies have always been a great source of knowledge for me, and I spend the ten-minute ride to Thirty-fifth and Shields and the ball park gabbing. By 8:30 I'm in the door, and the rest of my day is organized disorganization.

Waiting for me are Sandy Gruark, my secretary; Rudie Schaffer, my partner; Roland Hemond, who handles player person-

Sophisticated pagination programs control *widows* (one-line paragraph endings that appear at the top of a column) and *orphans* (one-line paragraph beginning at the bottom of a column). They also provide placement of rules, justify columns vertically, and position graphics within the page.

Included in many automatic paging programs are automatic calculation and insertion of initial letters, run-arounds, notches, skews, contours, and other shaping of body type. Many of these modes are provided automatically with very little calculation on the part of the operator. Figure 7–8 shows examples of these special typesetting modes.

The Modular Composition System (MCS) by Compugraphic Corporation represents a growing trend of manufacturers to blend the power of a front-end with the convenience of a direct-entry station. Its advanced software allows editing of the last two lines of copy during H&J calculations, three separate tabbing functions with up to 20 tabs, 10-line skip scrolling, and horizontal and vertical ruling. It also has the popular CP/M-86™ control program that allows the operator to take advantage of Star Program software

such as Wordstar™, Spellstar™, and MailMerge™. This gives the system word processing, business, financial, and mailing list capabilities similar to personal computers.

Typesetters and Image Setters

Only about a hundred years have passed since the advent of machine-assisted typesetting. In only the last thirty years, we have progressed from the Linotype, capable of setting an "amazing" 12 to 14 lines per minute, through four stages of typesetting automation. We now stand squarely in the "digital era," with machines capable of setting several thousand lines per minute.

Practical phototypesetting began with the Intertype Fotosetter, a modified linecaster that substituted photographic film for hot metal. Not long after, phototypesetters were introduced which looked and worked nothing at all like the earlier hot metal casters. In the late seventies, the age of photocomposition reached its peak, and recent trends have been toward typesetting devices that link digital storage of type and other images to laser and other nonoptic output.

Figure 7–9 depicts four successive classes of typesetting devices that together have completely replaced hot metal linecasters. Some people refer to classes of machines as *generations*, although not all agree on what constitutes a new generation.

Edward Gottschall describes the four classes of typesetters according to how characters are *stored* and how characters are *generated*.[4]

1. Photo/optic—master characters are stored on film strips, disks, drums, or grids. A beam of light is flashed through a character contained on the rapidly spinning master and optically projected onto film or paper. Lenses within the machine alter character size.

2. Photo/scan—uses master photographic fonts similar to the photo/optic class, but characters are scanned and converted to tiny lines (or dots) on a cathode ray tube. The digital image is then projected onto paper or film.

3. Digital/CRT/scan—uses no photographic character masters. Instead, it stores digital, prescanned records of each character. The digital image appears on the CRT as tiny lines similar to photo/scan devices.

4. Digital/laser—completely bypassing both photographic masters and CRT scanning, this class of typesetter stores digital records of characters which

are used to control an exposure laser. The laser beam produces ultrafine *raster* lines directly onto photosensitive paper.

Typesetters which store digital type can also easily store and set any other graphic image, including line and halftone illustrations and logo symbols. The term *image setting* has now become popular to describe such diverse applications.

While photo/optic typesetters have been eclipsed somewhat by technology, many are still in use today. Improvements in their speed and text-processing capabilities, especially for direct-entry (single-station) models, have allowed them to remain competitive with the generally more expensive digital models. Also, used machines and typeface masters, which exist in large numbers in the marketplace, can often be upgraded for less cost than new digital equipment. Figure 7–10 shows a diagram of how a typical photo/optic system operates.

One disadvantage of photo/optic systems is that only the small number of typefaces *on-line* at a given time can be used. Typically, two to four master fonts can be mounted, each containing four typestyles. This means that only 8 to 16 faces can be used before the operator has to replace a master font. While 16 typefaces may seem like a lot, by the time regular, italic, bold, and bold italic versions of each face are mounted, the actual number of distinct families may be reduced to four.

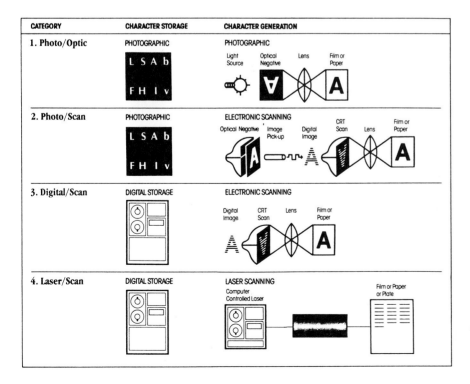

FIG. 7–9
Four general classes of typesetting devices.
Edward M. Gottschall, *Graphic Communication '80s*, p. 143.

Another problem is speed. Many such typesetting machines operate at speeds as low as 30 lines per minute, although more advanced models rate up to 150. If numerous lens changes (for point sizes) or scan-position changes (for type styles) are required, speed may be reduced even more.

Type sizes usually range from about 5 to 74 points, with a zoom lens providing all sizes in between in one-half point increments. Leading is usually available in one-quarter or one-half point increments.

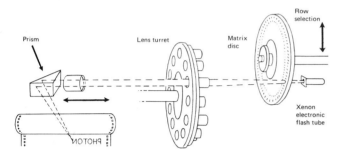

FIG. 7–10
How a photo optic typesetter operates.
Reprinted with permission from *The Printing Industry* by Victor Strauss,
© 1967 Printing Industries of America, Inc.

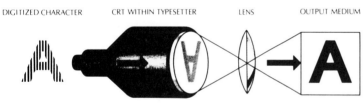

DIGITIZED CHARACTER   CRT WITHIN TYPESETTER   LENS   OUTPUT MEDIUM

FIG. 7–11
CRT scanning.
Autologic, Incorporated

FIG. 7–12
Digital images are "painted," stroke-by-stroke, on a cathode ray tube. On digital machines such as Lintoype's 202, characters are stored as an outline which defines an area to be filled with tiny scan lines during the imaging process.
Mergenthaler Linotype

Shaping up. **Digital technology "paints" characters, stroke by stroke, in blue-white light on the CRT tube. By calling a character's outline from font memory, the imaging system knows where to begin and end each stroke. Digitized fonts, pioneered by Mergenthaler, have eliminated the need for font replacements.**

Digital Models

As the cost for digital technology plummets, the much greater speed and flexibility of such systems may well send the photo/optic systems down the same obsolete path as the hot metal linecaster.

How does a digital typesetter work? Figure 7–11 shows one basic principle of CRT-scanning devices. Characters are stored as digital data on a magnetic disk. Each character is made up of hundreds or thousands of magnetic *start* and *stop* signals that control how the letter is "painted," one tiny line at a time, across a cathode ray tube (CRT). These overlapping vertical (and sometimes horizontal) strokes—as many as 5,700 per inch—form letters on the CRT screen. A mirror casts the image through a lens onto the photographic paper. Figure 7–12 indicates how a letter is made up of scanning strokes. The strokes shown here are greatly exaggerated; they are actually much smaller relative to the letter size. Also, the figure depicts a full letter made up of many strokes; actually, the strokes are reflected *one at a time* from the CRT screen to the photographic paper at an extremely high speed. This means that a full letter never appears on the screen at any given time.

Figure 7–13 depicts how an image (actually made up of thousands of scan lines) is reflected from the CRT to the paper or film.

Digitized typesetting offers tremendous advantages, including:

Easy font storage—on disks, not character masters. With expanded memory, several hundred fonts can be stored. Fonts can even be sent by telecommunications from one location to another. Also, since more characters can be stored per font (typically 128), more ligatures and pi characters (boxes, bullets, etc.) can be stored.

High speed—typically ranging from 175 to several thousand lines per minute.

Excellent image quality—while early models produced characteristic "jaggies" (where scanned lines attempted to reproduce delicate curves), the increased numbers of lines and careful overlapping of strokes now makes digital type superior to phototype in many cases.

Flexibility of image formation—sizes ranging from 5 to well over 96 points are common. Characters can be italicized, expanded, or condensed electronically.

Figure 7–14 shows a number of letter modifications and image generations possible on the Omnitech 2000.

The precision of digital output speeds up the preparation of forms that require rules, boxes, borders, and reverses. Photo/optic typesetters are able to produce horizontal rules easily, but are usually unable to form precise connections of images to provide fine vertical rules. Many digital typesetters control the size and placement of scan lines precisely, so even hairline vertical rules can be produced.

In order to *build* a box or multicolumn page, a typesetter must have *reverse leading* capability; that is, it must be able to back up the typesetting paper after setting one vertical rule or column down the left side and then set another one beside the first. Most typesetting devices have from about 16 to 24 inches of reverse leading capability. Many digital typesetters, or their parent front-ends, offer *software reverse leading*. In effect, the machine "looks ahead" to see exactly which words, rules, or portions of other images must appear *across the page*. It then produces the images necessary in that area, building columns or rules piece by piece. For example, on one horizontal sweep, the typesetter may produce the first line of column one, the first line of column two, and a portion of a box in column three. During the next sweep, it produces the second line of column one, the second line of column two, and another portion of the

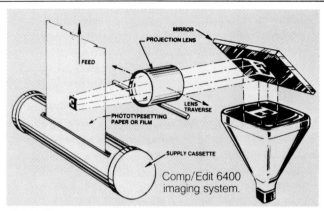

FIG. 7–13
CRT-scan imaging.
Varityper

box in column three. This way, the typesetter does not need to back up the typesetting paper and produce each column separately. Many photo/optic typesetters can accommodate software reverse leading.

The Omnitech 2000 laser typesetter operates entirely on the ability to look ahead to know where to expose images. This laser/scanner uses digital signals to control an exposure laser. The laser paints tiny horizontal scan lines—723 per inch—across a specially treated, silverless paper to form images. The Omnitech

FIG. 7–14
Capacities and letter modifications possible on the Omnitech 2000— sizes up to 127 1/2 points, reverse video, forward or backward slant, vertical and horizontal rules, and type expansion/reduction.
Mergenthaler: Mycro-Tek Production Division

2000 has only four moving parts and requires no mechanical backing up of the paper during exposure. It scans a constant 34 square inches per minute, regardless of image content. This is equivalent to an 11 × 17 inch page in a little over five minutes.

It is worth noting that the Omnitech is made by Mergenthaler Linotype. It is a strange irony that the machine to first demonstrate the potential of fourth generation digital/laser technology in an affordable direct-entry typesetter should so thoroughly render obsolete the same company's earlier machine—the Linotype—which had itself so completely revolutionized the way type would be set.

Another well-known machine by the same company, the Linotron 202, stores some 560 digitized type fonts on-line when fitted with a 10-megabyte disk drive. The 202 uses no lenses at all, even though it is not a laser typesetter. Instead, the paper moves across a special face plate on the CRT. The plate, made of thousands of tiny fiber-optic tubes, transmits the scan lines from the CRT directly onto the paper above. This machine uses *vector storage;* that is, it stores the *outline* of each letter as a digital record. This record is used to tell the CRT where to create its thousands of tiny letter strokes.

The 202 can produce 136 point sizes directly on the CRT, whereas many other digital typesetters depend upon lenses to enlarge or reduce the CRT scan lines as they are painted on the screen. Figure 7–15 shows the Linotron 202 and a sample of a capital H produced in all sizes of Poster Bodoni in 32 seconds, without the aid of lenses.

Because of the simplicity of digital typesetters, some have been miniaturized and produced as desk-top models. The CRTronic 200 is an example, shown in Figure 7–16. This direct-entry model sports 24 typefaces on-line, setting from 4 to 128 point sizes in one-tenth increments at a speed of 125 lines per minute. It can be fitted for traditional photographic output, or for dry-output, silverless paper.

Proofing Hard Copy

Typesetting systems today not only offer *soft proofing* by displaying the finished type form on the screen— they also present more sophisticated *hard copy proofing.* Several types of printers are used in text processing and typesetting operations, including:

dot matrix printers—for rapid verification of wording, spelling, and overall accuracy of text;

FIG. 7–15a
Linotron 202, digital CRT typesetter.
Mergenthaler: Mycro-Tek

FIG. 7–15b
Sample of 136 point sizes set on the Linotron 202 in 32 seconds.

correspondence quality—or letter quality printers for more readable proofing of text, or where actual typeset output may not be necessary for reproduction; **reproduction quality printers**—using laser, electrographic, or other means to create plain- or coated-paper copies.

Figure 7–17 shows a sophisticated high-speed reproduction quality proofing device called the "bit blaster."

Area Composition and Page Makeup Systems

Full on-screen graphic control of composed pages is rapidly becoming universal where text/processing must include display type and graphics. The term *area composition* refers to typesetting all elements of a page *in position* to eliminate piece-by-piece pasteup.

Advertisements and pages smaller than newspaper sizes are the main products of area composition terminals. *Page makeup* generally has come to mean the electronic assembly of full newspaper-size pages, including body type, headlines, illustrations, and other graphic elements.

As with other areas of graphics development, advances in technology have blurred the differences between area composition and page makeup systems.

FIG. 7–16
The CRTronic 200 desk-top digital typesetter.
Mergenthaler

FIG. 7–17
The Autologic "Bit-Blaster" proofer.
Autologic, Incorporated

Virtually all such systems are designed with the motto "what you see is what you get" in mind—regardless of the information being processed.

One thing many people require from an area composition or page building terminal is *interactivity*—the ability to allow the operator to make copy changes on a keyboard or a specially designed graphics tablet and see the results immediately on the screen. Less expensive systems usually require that information be entered first, then "previewed" on the same screen or at an adjacent terminal.

Experts disagree as to whether or not the value of area or page terminals warrants their expense, which is often five to ten times that of regular terminals. Some feel that the effort required to call out each section of copy, position it, size it, and fit it with other page elements often requires as much if not more time than traditional keyboarding of commands—especially if the operator has a soft-preview screen to verify page output after keyboarding. Others are concerned that too much design freedom is granted terminal operators, although in most cases a master layout is used as a guide. Naturally, those who manufacture area and page terminals don't agree. It is clear that

FIG. 7–18
The Comp/Set 4800
Varityper

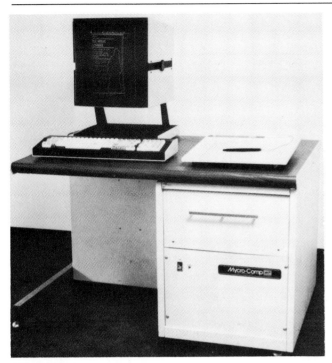

FIG. 7–19
The AdComp workstation.
Mergenthaler

production people want them, and that they have proven themselves in a variety of applications.

The Comp/Set 4800 by Varityper is an interactive area composition terminal designed to produce its effects entirely at the keyboard. Its large 13-inch screen can be used to call up a keyboarded job, where it can be sized and positioned. A single keystroke will increase the size of a letter, word, or line, or move one to any desired position. Appropriate software allows boxes with vertical rules or multicolumn forms to be composed quickly, as long as the typesetter producing the final page can produce them. Figure 7–18 shows the Comp/Set 4800.

Designed more as a display ad composition station, the AdComp workstation uses a large screen, a keyboard, and a graphics tablet activated by an attached electronic pen. As with most such ad makeup stations, copy for the ad to be produced is keyboarded on a simpler input/editing terminal first, then transferred by disk to the makeup terminal. The keyboard is used mostly for short corrections or insertions of copy.

To create an ad, the operator simply touches the electronic pen to the appropriate spot on the clearly marked graphics table. Copy can be sized, repositioned, contoured, or skewed to fit unusual

shapes. Boxes or rules can be drawn rapidly. When the ad is complete, the machine converts all information to typesetter codes automatically. The AdComp workstation is shown in Figure 7–19.

Newspaper Page Makeup

The APS-Raycomp III by Autologic is a display ad and page makeup system designed for maximum control of all content on full newspaper pages. Its 144-inch screen can display up to 20,000 characters. As type is called up, positioned, and sized, the operator can call up a half-size or double-size image of the page or zoom in on a single area to refine the layout or typography. A command tablet and digitizing pen may be used for all functions, or anyone more oriented to typing can perform them on the accompanying keyboard.

Figure 7–20 shows the comprehensive APS-Raycomp III Command Tablet activated by the electronic pen.

Figure 7–21 shows some of the steps involved in assembling a full page on the APS–Raycomp III screen.

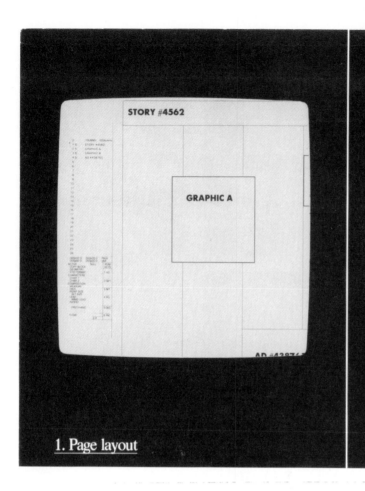

STORY #4562

GRAPHIC A

AD #43874

1. Page layout

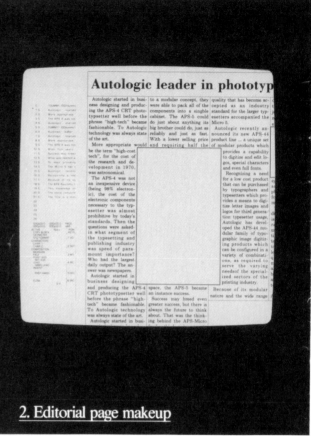

Autologic leader in phototyp

2. Editorial page makeup

FIG. 7–21

Electronic page layout on the APS-Raycomp III—(1) Areas for ads, graphics and captions are defined. (2) Text is placed into the layout, then edited and vertically justified on the screen. (3) Pre-scanned graphics are called from electronic storage, sized and cropped. (4) Display ads— fully composed as one piece through the use of hundreds of pre-defined typesetting formats—are positioned. (5) Any special page elements previously stored in a huge database (adv./editorial morgue) are retrieved and merged with the page. (6) The final page is completed ready for outputing as a single piece in proof or reproduction quality.

Autologic, Incorporated

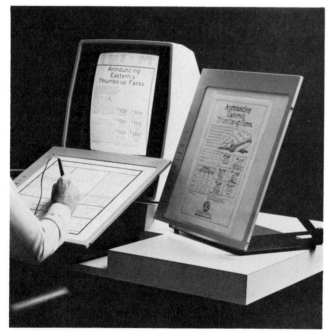

FIG. 7–20

The APS-Raycomp III Command Tablet.

Autologic, Incorporated

126

Autologic leader in phototyp

3. Graphics

4. Display ad composition

5. Database management

6. Completed page

Computer Control of Illustrative Graphics

While many levels of sophistication exist today in area and page layout systems, the major goal for manufacturers (and users) is to acquire the ability to control both text and graphics in the pretypesetting and prepress stages. Recent advances toward this goal seem to ensure the demise of systems that either leave *holes* in a layout for pasting in graphics or do not allow full on-screen image editing.

In order for line or continuous tone images to be captured and manipulated as characters are, they, too, must be converted into digital information. In the most widely used process, images are scanned (digitized) by a device that breaks them into thousands of individual *picture elements* (also called *pixels* or simply *pels*). These tiny image signals correspond to positions on the *raster scan* lines of a video screen.

Images can be created on a CRT in two ways— through raster scanning, such as in a regular television set, and through the use of vectors. Figure 7–22 shows a diagram of both methods. Raster methods are by far the most popular today for graphic display, even though they require much more memory.

Raster scan lines are nothing more than rows of picture elements. Some 500 scan lines appear on a standard TV screen (CRT), and each is divided into 500 fixed points to form pixels. An electron beam within the tube *refreshes* the illumination of pixels 60 times a second.

Since highest quality display and output are important in working with print graphics, illustrations are

digitized at high levels of resolution. The more pixels per inch, the better the image at high levels of resolution. An array of electronic sensors, or charge-coupled devices (CCDs), in the illustration scanner "read" the image and convert it to lines and tones at a resolution of 1,000 or more pixels per inch, horizontally and vertically. The digital record of pixels for each image is stored in the computer and called to a high-resolution viewing screen as needed.

In systems that accept line work only, the pixels are simply represented as black or white, yes or no. In continuous-tone systems, each pixel can be represented by any one of 256 tones of gray. In color systems, to be discussed later, pixels can be assigned any one of thousands of possible colors.

After the page or area composition and makeup has been completed, it is output as a single piece, including graphics. At this stage, the pixelized signals forming continuous-tone images must be converted into halftone dots for conventional printing systems. If nonimpact proofing or printing is to be done, the stored data is converted to whatever output medium is required.

Figures 7–23 (a and b) show the versatility of the ECRM Autocon II laser graphics system in digitizing and manipulating the dimensions and tonal reproduction of a simple photograph.

A powerful line-art digitizing station, the APS-44 Modular Digitizing System, is depicted in Figure 7–24. Allowing complete editing of logos, special characters, and entire type fonts, this set of devices includes a scanner, a control terminal, a graphics display terminal and a graphics editing tablet, along with disk drives and other electronic components.

Both typographic and line images can be "read into" the system from a disk or from the scanner, then sized and refined in any way and stored for later image setting. A graphics tablet, controlled by a hand-held, cable-connected *mouse,* allows easy control of images on the graphics display screen.

Image Setting and Nonimpact Printing Methods

A long-term revolution in image reproduction is underway and is nowhere more evident than in systems where full integration of text and graphics takes place. The technology is called *nonimpact,* since no printing platen or other striking mechanism need touch the paper to transfer an image.

FIG. 7–22
Edward M. Gottschall, *Graphic Communication '80s*, p. 33.

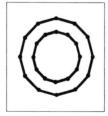

1. **Raster Scan**   2. **Random Scan**

FIG. 7–23a
The ECRM Autocon II, graphics digitizer.
ECRM

FIG. 7–24
The APS-44 modular digitizing system.
Autologic, Incorporated

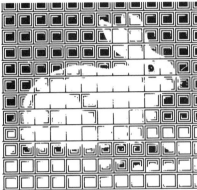

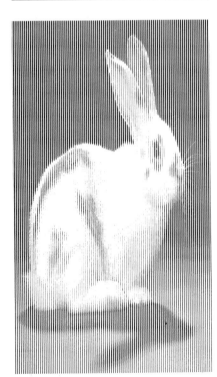

FIG. 7–23b
*Special effects possible with ECRM
Autocon II.*

A leading expert in graphics technology applications, S. Thomas Dunn, president of California-based Dunn Technology, describes nonimpact printers (NIPs) as intelligent copiers. He says that such devices differ from office copiers in that they produce images/signals on a photoconductive surface by means of a computer-guided laser. Dry toner adhering to the image impulses is transferred to plain paper and fused by heat.

Intelligent copiers don't work from a typed or printed original—they read electronic data in roughly the same fashion as a typesetting machine. For this reason, they are becoming popular proofing devices for quality hard copy as a prelude to typesetting. Figure 7–25 shows the operations of a nonimpact device, the Canon LBP-10, a semiconductor laser beam printer. It is capable of printing 10 letter-size pages per minute with a printing resolution of 240 dots per inch. It requires a user-supplied *raster image processor* (RIP) to convert electronic data into laser-controlling signals.

Demand Printing—A Growing Market

The need for high-volume production of multipage documents has created a rapidly expanding market for printing *on demand.* Demand printing is that which is done by nontraditional methods that do not require image-carrying plates. Sophisticated electronic printing systems—nonimpact printers, coupled with an electronic text and graphics front-end—offer high-speed, plateless demand printing and threaten quick-printing commercial markets.

FIG. 7–25
The Canon LBP-10 laserbeam printer.
Canon USA Inc.

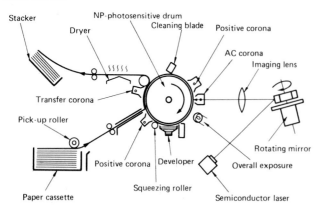

Figure 7–26 shows the Xerox 9700, a well-known electronic printing system capable of producing 8½ × 11 inch pages at the rate of 120 pages per minute. A graphic input station allows digital scanning and processing of any information—logos, graphs, signatures—to merge with type up to 24 point. A less expensive model, the 8700, produces 70 pages per minute. No negative, plates, nor other traditional prepress materials are required. While reproduction quality is not as good as traditional printing, for many business applications it is ideal, especially where large mainframe computers already exist.

Electronic Prepress Systems

Near total control of all aspects of creation, assembly, and prepress preparation of images has now been accomplished through the development of *electronic prepress systems.* Such systems allow highly interactive makeup of pages, both black-and-white and color. Not only can all information, text, illustrations, even color separations be entered electronically into the system, but many special effects can be generated electronically as original images.

The Scitex Response-300 system is one of the most complete systems in use today. The system accepts digital data from standard color separation scanners, along with electronic text and other output from graphics scanners or archival storage. Operators at color and page assembly consoles call up, correct, modify, and assemble all type and graphic images interactively. When pages are completely assembled, they are stored in disk memory for later exposure to film or plates, or used to control a laser head for gravure cylinder etching.

The amount of control such systems give over color images can be amazing. Electronic masking can eliminate an image background, allowing it to be replaced with another scene or changed to another color. All colors in the original can be changed, as can shadows, highlights, flesh tones. Electronic airbrushing is done with a hand-held stylus. The area to be retouched is greatly enlarged on the screen; then, the operator simply touches a paint on the screen which contains the desired color, then begins the detailed electronic brushing-in of color on the graphics tablet. The original image can always be called up if a mistake is made, and electronic masking protects adjacent images from being affected. The stylus can be used for detailed drawing as well.

After all color images have been processed, including

cropping and sizing, type can be called up from storage, merged with the images as a reverse, surprint, or any other special effect. Digital records of pages are created automatically as the operator works, so that appropriate data is stored and ready to output in the desired form.

The Response-300 is a *color* electronic prepress system (CEPS), and, at a price of roughly one million dollars, it is found only in large publishing or printing organizations that can justify the enormous up-front expense. A less expensive *monotone* (black and white) system (MEPS), called the VISTA is also available from Scitex. It is designed for use by art directors and graphic designers for sophisticated page assembly, and can be linked to the Response-300 system and a front-end system to form a complete text/graphics processing network. Figure 7–27 shows how the VISTA system fits into a complete electronic prepress network.

Electronic Publishing—The Way of the Future

It is not difficult to imagine that, along with the wondrous control of text and image communications we exercise today, we also have found ways to manage them as part of larger information networks. This, indeed, is the case.

Time, Inc. recently signed a $10 million contract with Crosfield electronics for installation of an electronic front-end prepress system for their magazines, including *Life, Time, People,* and *Sports Illustrated.* The pact concentrates total production of pages with the editorial staff so that flexibility and efficiency might be increased. Time, Inc.'s system, one of the largest to date, joins many others as a model of full integration of graphics technologies for mass communication.

FIG. 7–26
The Xerox 9700 electronic printing system.
Xerox Corp.

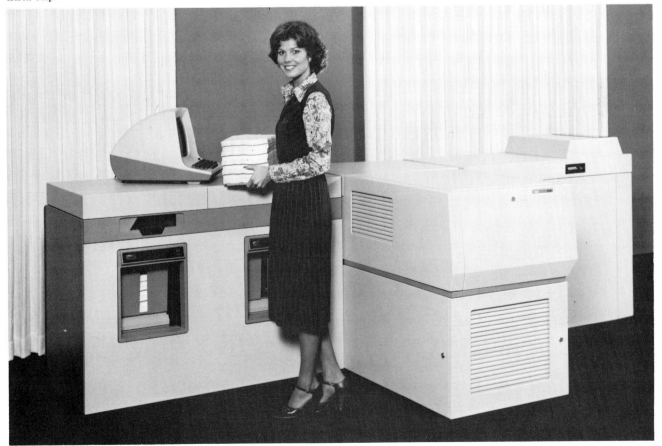

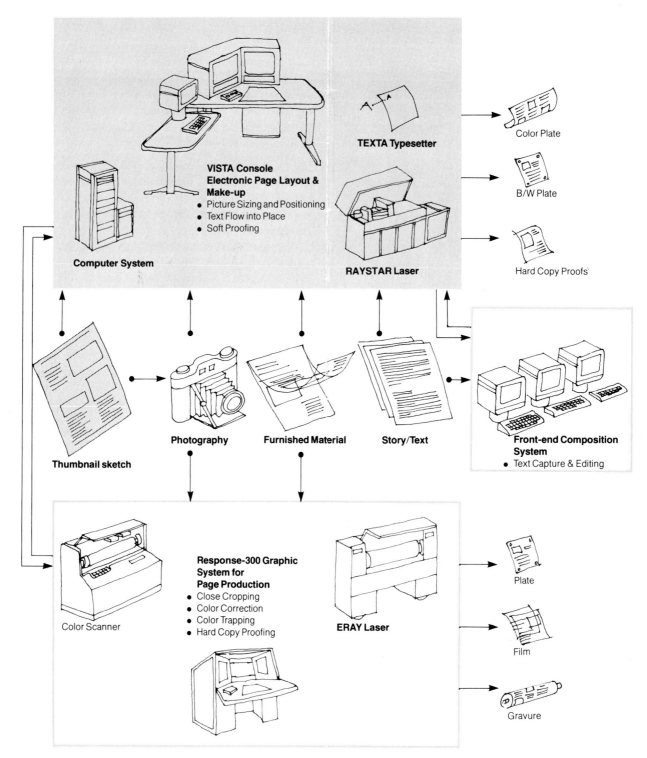

FIG. 7–27
How the Vista system fits into a text/graphics network.
Scitex America Corp.

FUTURE INFORMATION SYSTEM

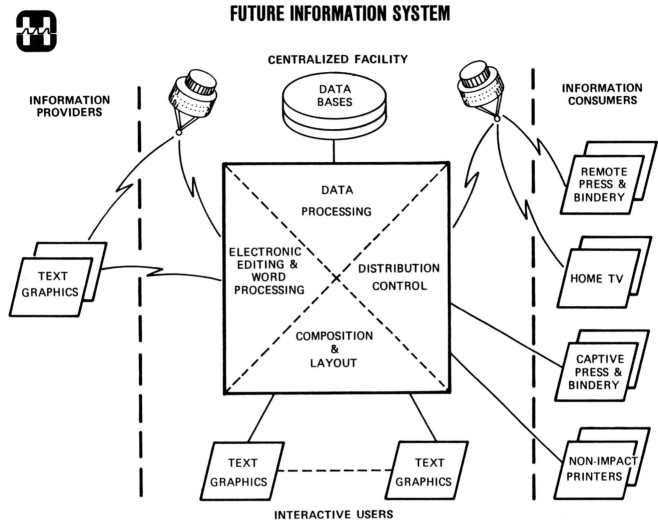

FIG. 7–28
Model of an electronic publishing system in the near future.
Harris Corporation

Figure 7–28 shows a generalized model of a future information system, proposed by Walter Fredrickson of Harris Graphics. Fredrickson foresees the increasing linkage of all information-sharing publics as technological advances continue.[5] Communication graphics will remain a major link in this chain of human communication.

Notes

[1]Richard Rodgers and Oscar Hammerstein II, "Kansas City," from the stageplay *Oklahoma* (New York: Williamson Music Co., 1943).

[2]This electronic-era tag for the common pencil was coined by Jack Catran in "Flushing Out Space Fallout—A Smiling Brush With Jargon," *Los Angeles Times*, August 9, 1982.
[3]Adapted from Peter Johnston's "Solving the Puzzle of Low-Cost Business Computers," *Graphic Arts Monthly*, 54, No. 5 (May 1982), pp. 142–149.
[4]Edward M. Gottschall, *Graphic Communication '80s* (Englewood Cliffs, N.J.: Prentice-Hall, Inc., 1981), pp. 141–142.
[5]Presentation by Walter Fredrickson of Harris Graphics at a meeting of the Web Offset Section of the Printing Industries of America, Inc., May 13, 1981, in Chicago.

8 Modern Design Problems

From Identity Symbols
to Effective
Page Structures

By now we are probably more aware that messages in print don't just happen. They are the products of planning and the creative use of design tools and techniques. Let's take a closer look at the design process as it relates to the creation of some familiar kinds of messages we see every day.

Introduction

In Chapter 2 we discussed some important elements of the design process and how they relate to communication graphics. Here we see how those elements directly influence printed design. Design principles are especially helpful in solving two common problems in printed communication: the development of *identity symbols* and the layout of *printed advertisements*. By evaluating these, we can see the importance of design to other messages in print.

Organizational Identity: Determining the Message Sponsor

Throughout recorded history, identity symbols have been linked to the fates and fancies of mankind. Their purveyors have used them to rule, to teach, to intimidate, even to grasp at immortality. Royal marks embellished the tombs of Eygptian kings. Fearsome dragons leered from the bows of Viking ships, later to

be replaced by images like the Union Jack and the Jolly Roger. The Christian world spanned centuries of faith clinging to a simple cross. In winning a war of survival, the free world learned to despise another crosslike image once regarded as a good luck symbol—the swastika.

Trade and service marks, the devices that identify commercial products and services, came into prominent use by craft associations and guilds in medieval Europe. These marks helped distinguish the work of members from competitors and assigned status to merchants and buyers alike. More elaborate symbols of heraldry such as family coats of arms also became prominent then. Today's trade symbols, used throughout the world, are descendants of these early marks. Indeed, the power, utility, and mysticism of symbols is legend.

Now, the positive power of symbols is harnessed by designers for individuals, commercial businesses, noncommercial organizations, governments, and other entities that require unmistakable identity. The purpose of these symbols is to convey identity messages clearly and effectively.

Capturing the Essence

Like expensive perfume, where the smallest drop of fragrance is sufficient, an effective symbol is the essence of identity. A very little says it all.

Organizations, companies, institutions, and events are infinitely more complex than the images they use to identify themselves. Identity symbols distill complexity down to easily understood graphic statements: this is what we are, this is what we stand for, this is what we produce, this is what we provide.

Affective qualities can also be suggested through illustration technique, color, type choice, and creative interpretation: we are reliable, up-to-date, trustworthy, businesslike, down-to-earth.

Design firms often perform extensive research to determine what questions and concerns a symbol should address. However, some leading designers such as Paul Rand, who developed the IBM logo, do not always feel the need for such elaborate study.[1]

If a firm is redesigning its symbols, it often performs what Jay R. Headly calls an *inventory of identity assets*.[2] An asset inventory helps an organization determine which past or present identity images it should retain in planning new designs. Such assets might include a signature, a friendly looking figure, or a catchy slogan for which the public has demonstrated wide recognition and positive appeal. Well-known examples of symbols that survived as important identity assets in updated designs are the multicolored NBC peacock, the curious RCA dog listening to "his master's voice," and the stylized signature, Coca Cola.

Important Characteristics of Identity Symbols

Simplicity

Because symbols carry a heavy load of identity information, they are often reduced to the bare essentials of shape, line, tone, and other design structure. Usually all parts of the final image contribute significantly to message clarity or help set the proper message tone. Redundant or too obvious information is eliminated.

A classic case of an evolution toward simplicity is the John Deere trademark. Figure 8–1 shows the original 1876 mark, then the version designed nearly a hundred years later in 1969. The earlier mark is highly ornate: widely spaced and arched hand-lettering, a detailed

FIG. 8–1
Courtesy of Deere & Company

rendering of a deer leaping over a group of indistinct objects. A city name, Moline, is as prominent as other elements in the design.

The modern John Deere mark is clean: a stylized profile of a proud buck with its head held high, antlers and legs shown in profile as abstract sets of images; no need to show all four legs or all the points on the antlers. The tail is proudly aloft, not drooping. A leaping deer is enough; no confusing clutter is shown. The type is clear and easy to read, yet businesslike. The city name, once important in identifying the manufacturer, has been deleted.

Distinction

A good symbol sets its owner apart from others, especially its competitors. Resources spent on establishing and protecting identity are wasted if they unintentionally promote others.

It is especially difficult to create a unique symbol with a single initial letter. Since many organizations are

General Mills

FIG. 8–2
Courtesy of General Mills, Inc.

likely to have a similar initial, a designer may be forced to create an original letter rather than use one from a common alphabet. The Westinghouse logo contains a unique *W* that says much about the electronic nature of the company's activities.

Symbols don't have to be highly unusual to be effective. Rand's IBM logo is a plain, slightly modified square serif typeface containing distinctive horizontal lines. General Mills' simple *G* is unconsciously handwritten each day by millions, a claim few companies can make about their own symbols (Figure 8–2).

The logo recently adopted by Sun Oil Company shows how a familiar image—the sun—having been used symbolically for thousands of years, still provides a highly original identity for its user. The unique sunburst forms a striking new identity for this modern energy company.

Figure 8–3 shows the Sun Oil company's logo, created by the design firm Anspach Grossman Portugal. Figure 8–4 depicts a page from the corporate identity manual established for Sun Oil to ensure that the design is used properly and consistently.

There are many kinds of distinctive symbols:[3]

the trademark, a name, symbol, or other visual device, in any combination, used to identify a product or distinguish it from other products;

the service mark—similar to the trademark, but used for service-related enterprises;

emblems, insignias, and seals—used by nonprofit organizations;

collective marks—used by trade associations; and

certification marks—used for quality assurance (Good Housekeeping or Underwriter Laboratories).

The term *logo* also is used to describe distinctive symbols or marks.

Function

Besides being simple and distinctive, a good symbol must be functional in two important ways:

1. The symbol can be duplicated easily in a variety of ways using many printing methods. The image is bold enough to be reduced without loss of detail in lines and tones. Little or no special modification of the image is required if more than one printing process or another communications medium (such as television) is employed.

2. The symbol is flexible enough to look good in a variety of formats: full-page advertisements, letterheads, signs, report covers. If normally done in color, the symbol retains its integrity in black-and-white (the Mobil symbol, for instance, looks fine when the red *o* is replaced by one formed by a white outline). Atlantic Richfield employs a multicolor scheme to distinguish its many divisions, using a separate color for each. Their elaborate corporate color identity survives intact because of exceptionally clean typography.

Types of Identity Symbols

Elinor and Joe Selame, in their excellent book, *Developing a Corporate Identity: How to Stand Out in the Crowd,* discuss seven major types of identity symbols: the seal, the monoseal, the monogram, the signature, the abstract mark, the glyph, and the

FIG. 8–3
Courtesy of Sun Company, Inc.

Do not use broken or imperfect logotype

Do not letterspace the logotype

Do not overexpose the logotype

Do not fill in the rays of the logotype

Do not place the logotype on heavily textured background

Do not combine with other type

Do not reverse the logotype

Do not shape typography around logotype

The use of text copy and the logotype should be as simple as possible for maximum impact. Copy should not be fitted around logotype as this demonstration indicates.

Do not place the logotype in unauthorized shapes

FIG. 8–4
A page from the Sun Company's Identification Standards Manual, depicting what not to do with the symbol in print.

alphaglyph.[4] Figure 8–5 shows a well-known example of each type of symbol. Here are their general characteristics:

seal—a symbol with a generally geometric shape, usually made up of a dark background with reverse lettering in a rather standard typeface. Full words, not initials, are used.

monoseal—similar to the seal, but with initials, often in stylized lettering.

monogram—distinctive initials presented without a border or background.

signature—a fully spelled name with no background in a highly distinctive type.

abstract mark—a symbol which depicts overall essence, philosophy, or feeling (such as quality-oriented, progressive, futuristic).

glyph—a stylized symbol that suggests function, product, service, or raw materials.

alphaglyph—similar to a glyph, but formed completely or partially by the initials of the organization.

Figure 8–6 depicts the several stages required by this author to design a simple signature logo. The problem in the design, produced for a continuing series of publications on the future of humanities education, was to create a forward-looking, single-color signature emblem that could be used in a variety of sizes and formats. In this case, neither drastic change in size nor the switch from positive to negative form seemed to alter the overall sense of the design, nor affect its reproduction.

Identity symbols epitomize good design. They depend upon several principles we have mentioned before to be effective. They must be simple, easy to see, and

FIG. 8–5
Classification of identity symbols.

Seal
Ford Motor Company

Monoseal
Westinghouse Electric Corporation

Monogram
International Business Machines

Signature
Eaton Corporation, Cleveland

Abstract mark
Chrysler Corporation

Glyph
United Way of America

Alphaglyph
Goodwill Industries of America

FIG. 8–6
*Steps in designing a typographic logo
include: (a) Rough sketch of a key
element; many drawings or tracings
may be necessary. (b) Setting type
elements. (c) Another rough, this time
combining the two type elements. (d)
The final logo is shown in both positive
and reverse form. Note that the
Ouline type has been filled in with a
graduated pattern of horizontal lines.*

a.

b.

c.

d.

LITERACY FOR THE
TWENTY-FIRST CENTURY

BY WILLIS E. McNELLY, Ph.D.
DEPARTMENT OF ENGLISH, CSUF

In 1946, a time much less hectic than the era we face three and a half decades later, the late great poet Wystan Hugh Auden wrote and delivered the Phi Beta Kappa poem at Harvard. Entitled "Under Which Lyre," it retells in wise, witty language the ancient battle between what Auden calls "precocious Hermes" and "pompous Apollo." I can visualize him now—rumpled tweeds extending beneath his bright scarlet Oxford gown, its lapels dusted with cigaret ash, his craggy face belying his insouciant eyes and smile—plumping firmly on the side of Hermes, the heart. Hermes' sons "love to play," he said, "and only do their best when they are told they oughtn't." "Apollo's children," he continues, "never shrink from boring jobs but have to think their work important." Later in the poem he presents the "Hermetic Decalogue," a few lines of which seem to speak to us with special import as we approach 1984.

(Continued on the next page)

The Political Meaning of the
General Education Program

BY DAVID DEPEW, Ph.D.
DEPARTMENT OF PHILOSOPHY, CSUF

The liberals who are the uncles, even if not the fathers, of our political culture did not believe that the virtues of democracy came without some vices. On the contrary, John Stuart Mill, and later in America John Dewey and Walter Lippmann, could agree with intelligent conservatives like DeToqueville that popular sovereignty tends to bring in train cultural vulgarization, anarchic competition of private interests, periods of mass politization followed by sullen withdrawl, and inconstancy in the conduct of public policy. Have we escaped these evils?

To the extent that we have it is largely because of two considerations.

First, we have left the conduct of affairs to a marked degree to an elite, though one which recruits from a fairly wide base. Second, we have heeded the advise of men like Mill and used public education to extend to the entire citizenry an ability to participate in public affairs.

Only the latter tactic, of course, is fully compatible with democratic institutions. We would like then to think of it as dominant over the other. Is this, however, the case?

For his part, Mill envisioned democracy as an unlimited exercise in public discussion, and hence as the rebirth of the classical notion of a public realm. Yet it has become

easy to understand. Abstract symbols require more abstract visual processing by the audience. Therefore, they may require more verbal explanation and more reinforcement through repeated use.

Good identity symbols capture a main idea or concept efficiently and present it in a unique, interesting, and attractive way. Figure 8–6a shows several recently designed symbols that present particularly creative solutions to design problems.

Advertising Layout and Design: The Sponsor's Message

We have seen how message sponsors identify themselves through symbols. Now we can examine how sponsors construct the messages that accompany those symbols.

All forms of print layout employ at least some design principles. Advertising layout generally requires a design that works well in a limited area—a partial page, a whole page, or two facing pages in a publication. Therefore, the application of design principles is easy to see, and we will concentrate on this form of layout for now.

Basic layout principles help tie together a total design package—a creative concept, well-written copy springing from a platform of solid objectives, careful typography, and appropriate illustrations. Without the other elements, even a good layout is a silk ribbon on an empty package.

Basic Principles of Display Ad Layout

A number of general principles guide advertising layout practice today. Art directors and designers often use them subconsciously or call them by different names, but these principles aid creativity and ensure that messages can and will be read.

Good layout is like music from a fine orchestra. It is harmonious, balanced, rhythmic, emphatic. It repeats themes and uses devices such as dynamic contrast to surprise and stimulate the audience.

The visual composer seeks effectiveness similarly through the following:

unity—the overall sense that the message is an integrated whole, that its elements are related and state a single, unified theme.

harmony—the matching of elements such as tone, line weight, and shape, so they blend well and are pleasing to the eye.

rhythm—the sense of purposive repetition and even image flow. This might take the form of a duplicated image or pattern, reinforced by a consistency of tone, alignment, or spacing.

balance—a sense of equilibrium or stability in the distribution of elements. Balance may be a formal, symmetrical placement of elements on a centered axis. It may also be a more informal counterbalancing of varying weights and sizes along a centered or other axis.

contrast—differences in size, shape, tonality, or other important elements that help attract interest and promote easy reading.

direction—the selection and positioning of elements to promote eye movement in the layout. These include illustrations that point to other layout units, italic typefaces, and images arranged along a straight or flowing axis. All suggest action or movement.

sequence—an arrangement that helps ensure that reading flows along a definite path. Illustrations grouped into a story panel are often similar in size, tone, and alignment so they may be easily read as a sequence from top to bottom or left to right.

emphasis—a technique whereby an element or a group of elements is accented by size, content, color, or position to make it stand out.

FIG. 8–6a
Recent logo designs.
Reprinted with permission of the Port of Los Angeles and Stanley Blacker.

proportion—the matching of element shapes and sizes so that they blend or contrast effectively with one another in a layout; for example, similarly rectangular or square photographs used in varying sizes in a design. Proportion refers also to a sense that the size and emphasis of an element is in keeping with its real importance and that its scale matches other objects shown.

Diagrams in Figure 8–7 show examples of each basic layout principle.

Layout Charactersitics

Advertising layouts are sometimes grouped into a few general classifications for description purposes. Most ad layouts, however, are made up of a combination of many design characteristics related to the layout principles mentioned earlier. Layout characteristics include:

Structure—the type of balance, alignment strategy,

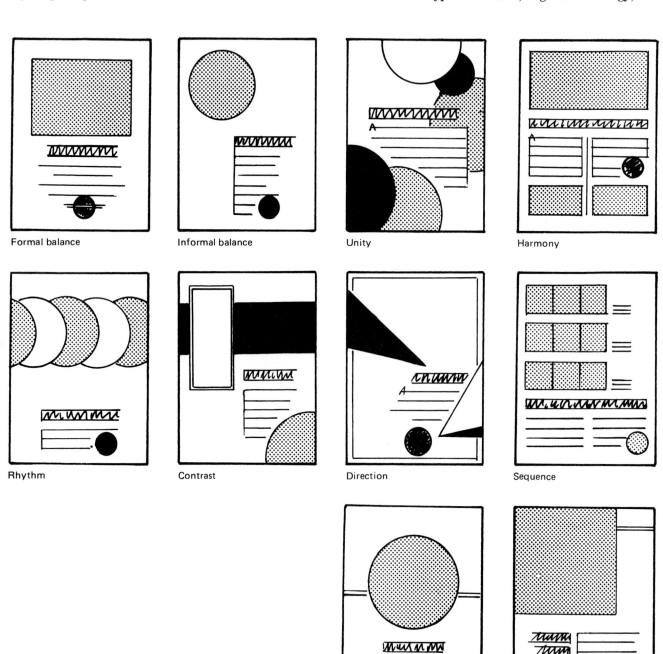

Formal balance   Informal balance   Unity   Harmony

Rhythm   Contrast   Direction   Sequence

FIG. 8–7
Basic layout principles.   Emphasis   Proportion

repetition of patterns or grids, or unusual tilt of the visual plane;

Primary emphasis—whether stress is placed more upon illustration, type, graphic devices, or a combination of these elements;

Secondary emphasis—the space, background, color, borders, bleeds, and other elements that complement the primary elements;

Affective tone—visual treatment that denotes action, seriousness, softness, humor, or other mood-setting qualities.

Structure

All ad layouts have some unifying structure around which the message is built. It may include:

Balance—in ad layout, balance may be formal or informal. Sometimes ads are purposely unbalanced for dramatic effect. Every element, including white space, contributes to layout balance.

Alignments—pictures, type, and other important elements are often placed along invisible but easily identified lines—up and down, sideways, at an angle, or curved.

Patterns—grid blocks, rectangular panels, columns, and other repeated patterns are used to organize alignment points, sizes, spaces, and sequences of elements more readily.

Grids and other repetitive patterns form individual blocks, groups of blocks, intersections, and white space that help us organize the positioning of photographs, type, and other layout elements. They help establish consistent separation and balanced use of space.

Occasionally, an eclectic layout style is seen, in which pictures, type, and other elements of various shapes and sizes are scattered over the page with little apparent relationship to one another.

Primary Emphasis

Illustrations. Photographs and other illustrative images are familiar vehicles for presenting advertising messages. When used for main emphasis, they are usually large and often bleed off the page. They may be rectangular, or made up only of a subject's silhouette against a white or toned background. They may fill the page completely and contain type superimposed in a pleasing area. Layouts that use large, dominant illustrations are sometimes called *picture window*, *picture dominant*, or *poster ads*.

Emphasis also can be established by fitting together picture rectangles of different shapes and sizes to form a single, harmonious unit. This approach is

sometimes called *Mondrian* layout because it is loosely patterned after the technique of the famous Dutch painter, Piet Mondrian. Picture panels, usually of similar shapes and sizes, also may be used for primary emphasis, but often simply supply a sequence of information that is secondary to the main picture or headline.

Some layouts merge small illustrations with body type to help break up gray masses and to illustrate points in the copy. Whether or not this integrated picture-type area can be considered an ad's primary emphasis depends upon its size and treatment in relation to other elements.

Type. Both headlines and body type may be emphasized more than illustrations. A particularly creative headline may be the most captivating part of an ad, and should be set off by larger size, bold or unusual type, or extra space. In so-called *reader* or copy-heavy ads, a large block of body type dominates the layout, so it should look easy to read.

Graphic devices. Multiple borders or rules, screen tints, unusual identity symbols, and other elements that are neither type nor illustration can be the main focal point of a layout.

Secondary Emphasis

Reinforcing, secondary elements can be almost as important as a primary element. White space helps frame a layout and provides separation of units. Relationships between elements can be shown easily by how much white space separates them.

Negative space, the background tone, shading, or other images of lesser importance in a layout serve the same function as white space. They create a pleasing setting or frame for the design. Special color treatment, unusual borders, and bleeds sometimes can be displayed as prominently as photographs and type, but usually they accent or complement those more common primary elements.

Affective Tone

Professionals try to establish a certain mood for readers through creative design. Affective or mood characteristics of an ad are often subtle and may be related to other format characteristics. For example, a formal layout structure often suggests elegance or exclusivity.

Affective qualities resulting from the basic structure of an ad can be modified by elements within the ad. For example, the formal layout sometimes preferred by a conservative institution such as a bank suggests a businesslike, but perhaps stiff, nature. Friendly looking type and informal illustration may loosen up a no-nonsense layout by suggesting a warm, human dimension to customer services.

Designers know that the same oldstyle roman typeface that perfectly matches the pen and ink illustration of a steam locomotive will fall flat in describing a painting of a space station. Heavy borders on an ad for expensive lingerie don't fit. Neither do ornate borders on an ad for construction equipment. The selection of a blue background to accent a model's eyes and the matching of shadows in borders and type with the shadows cast in a photograph are less obvious examples of efforts to establish the right visual tone.

The layout principles of unity and harmony are closely related to the concept of affective tone.

One of the most common mood-setting devices is the use of bold, italic typestyles placed at an angle to show action.

Typical Layout Principles

In this section we look at examples of several contemporary newspaper and magazine layouts, and examine their characteristics and the main principles they embody. Not every possible combination can be shown, but many popular layout styles are represented.

PROMONTORY POINT
APARTMENTS

Because you believe your
apartment should be as
distinctive as your lifestyle.

Luxury 1 and 2 bedroom adult villas from $655
Newport Beach (714) 675-8000

FIG. 8–8   Promontory Point Apartments
This ad for an exclusive beach community complex is an
example of formal, symmetrical balance. Both sides are
essentially mirror images on a centered axis. The weight
of the headline type matches the weight of the tile to
provide harmony. The shape of the title matches the shape
of the ad, providing a sense of unity and proportion.
Primary emphasis is on type, with white space providing
strong secondary accent. The affective tone is one of style,
elegance, and understatement.
Irvine Pacific

FIG. 8–9   PDP Word Processing
Here, formal symmetry is used not so much for elegance but to provide a simple layout structure that will best display a complex illustration. A closed circle is one of the most pleasing visual shapes, and the designers have put our natural desire for shape continuity to good use by suggesting that their product is a unified package. Primary emphasis is on the illustration. Subtle rhythm is provided in the similarly shaped segments around the circle.
Data Processing Design, Inc.

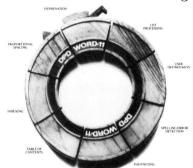

The Whole Truth About PDP-11 Word Processing.

When you explore word processing systems, you'll find a number of systems that offer part of the package. And frankly, if you're only looking for text editing, almost any system will do.

But if you're looking for full word processing capabilities, you should insist on a system that can handle all of the tough tasks you'll encounter.

For starters, a good system will have list processing, the vital function that generates correspondence, reports, even statistical analyses. It should have user defined keys to simplify repetitive operations.

And a good system can handle the details. Like spelling error detection, automatic renumbering of footnotes, tables of contents and indexes. And proportional spacing and hyphenation for profes-

sional looking documents.

A good system can be shared. It should be able to support up to fifty terminals. And it must be backed up by successful installations and a strong service team.

After you've examined the options, we think you'll select WORD-11, the only system with all the sophisticated features you could want.

For the details, please call Data Processing Design Inc., Corporate Office; 181 W. Orangethorpe Ave., Suite F, Placentia, CA 92670. (714) 993-4160. Telex 182-278. New York Office: 420 Lexington, Suite 647, New York, NY 10170. (212) 687-0104.

Data Processing Design, Inc.

Vincent van Dumb.

FIG. 8–10   Vincent van Dumb
A formal layout accents the primary photograph at the top. Large picture layout is one of the most popular styles of all. As with many such layouts, a catchy headline provides strong secondary emphasis, as does a small, related photograph. Contrast in dark-to-light tone frames the image and provides a sense of downward motion to the type below. Unity is aided by a small photograph which repeats the theme of the larger one. A bold, gothic headline harmonizes nicely with the dark image above.
Lear Siegler, Inc.

The Dumb Terminal® video display terminal has done it again.

For around $2000, you can have all the alphanumeric capabilities of the renowned ADM-3A Dumb Terminal, plus the full vector drawing and point plotting capabilities of a sophisticated graphics terminal. All in one neat package. That's less than half the cost of other comparably equipped graphics terminals.

The ADM-3A with Retro-Graphics™ gives you complete flexibility to develop bar charts, pie diagrams, histograms, even function plots. What's more, it's completely Tektronix® Plot 10™ software-compatible.

The package consists of an ADM-3A Dumb Terminal plus a single plug-in card engineered to fit neatly inside the ADM-3A without soldering, special tools, or a service call.

Retro-Graphics is a product of Digital Engineering, Inc., and is sold separately or installed in the ADM-3A by local Lear Siegler distributors. For the distributor nearest you, contact any Lear Siegler sales

office or Digital Engineering, Inc., 1787-K Tribute Road, Sacramento, CA 95815, 916/920-5600.

The Retro-Graphics-equipped Dumb Terminal. What does it mean to you? Draw your own conclusions.

Lear Siegler, Inc./Data Products Division, 714 North Brookhurst Street, Anaheim, CA 92803, 800/854-3805. In California 714/774-1010. TWX: 910-591-1157. Telex: 65-5444. Regional Sales Offices: San Francisco 408/263-0506, Los Angeles 213/454-9941, Chicago 312/279-5250, Houston 713/780-2585, Philadelphia 215/245-1520, New York 212/594-6762, Boston 617/423-1510, Washington, D.C. 301/459-1826, England (04867) 80666

DISTRIBUTORS: San Francisco, Consolidated Data Terminals, 415/533-8125. Dallas, Data Applications Corp., 214/231-4846. San Diego, Data Systems Marketing, 714/560-9222. Bedford, Continental Resources, 617/275-0850. Falls Church, Marva Data Services, 703/893-1544. Cleveland, W.C. Koepf Associates, 216/247-5129.

DUMB TERMINAL. SMART BUY.

LEAR SIEGLER, INC. DATA PRODUCTS DIVISION

Dumb Terminal® is a registered trademark of Lear Siegler, Inc.   Retro-Graphics™ is a trademark of Digital Engineering, Inc.   Tektronix® and Plot 10™ are trademarks of Tektronix, Inc.

FIG. 8–11   The Future of Front Wheel Drive
This ad is similar to the previous example. Formal balance is established and primary emphasis is again placed on a large photograph. Positioning of the photograph on the "floor" of the ad provides a strong sense of balance, as does the fact that the image bleeds off the page on all sides. A large reverse headline and pleasing graduation of background tones provide secondary emphasis. Black-and-white contrast in the type complements similar tones found in the wheels and accents the play on the word "bright." The ad's affective tone is a sense of harmony and quality.
American Exhaust/Div. of Mr. Gasket Company

Give your father something he can wear with any shirt.

Chances are your father already has a tie or two. Probably the ones you gave him last year. So this year, why not surprise your father with a Sheaffer fountain pen, ball point, rolling ball or gift set? Each one is cleverly designed to match his pencil stripes as well as his tattersall checks. And because Sheaffer pens are sensibly priced from $10, you can still buy him a tie, if you want to. Of course, he'll never be able to write with it.

SHEAFFER.
SHEAFFER EATON TEXTRON

FIG. 8–12   Sheaffer Pens
This intriguing ad is a combination of informally balanced pens and type against the formally balanced shirt. The rhythmic lines in the shirt give exceptionally strong secondary emphasis to the primary image of the pens. Here again, the full-bleed background lines lend stability and balance. Subtle directional lines, the tips of the collar, point us to both the body type and the right-hand pen. The arrow shape of the larger pen sets up an unmistakable reading sequence ending at the smaller pen below.
Sheaffer Eaton

Be prepared. Before you make plans for X.25 protocol, join the Pro/Testers™ all over the world. Pro/Tester, the X.25 protocol simulator from Applied Data Communications.

With the Pro/Tester, you can create:
• An economic system for programmer training.
• A tool for determining X.25 compatibility of your product.
• Easy standards updates with floppy diskettes.
Or test and debug systems. Even monitor lines, and validate X.25 protocol.

SOFTWARE UPDATE SERVICE
To keep you current with CCITT changes and future amendments, Pro/Tester comes with a software update service. Provided twice a year with complete documentation.
Use it for frame manual tests, frame automatic tests, packet level tests and interactive terminal interface testing.

Use it locally, or remotely via synchronous modems. It's CCITT designated and supports HDLC and BSC X.25 protocol.
The Pro/Tester is an independent stand-alone unit, complete with carrying case. Or it's available in a cabinet for permanent tabletop installation. All for only $11,900.
So unite with Pro/Testers everywhere. You'll discover why they're so committed: Field-proven reliability; designed by Applied Data Communications, a proven supplier to the computer industry for nearly a decade; and service you can trust—every time.

TOTAL FLEXIBILITY
Whether you manage a network, build terminals or computers or plan to integrate your equipment into a data communications network, it's time you knew the facts. For more of them, contact Applied Data Communications, (714) 731-9000. Or write 14272 Chambers Road, Tustin, CA 92680, or the Eastern Regional Office, (617) 273-4844, 50 Mall Road, Suite 209, Burlington, MA 01803.
Sales and service nationwide.

ADC APPLIED DATA COMMUNICATIONS
It's as simple as ADC.

PROTEST X.25

FIG. 8–13   Protest X.25
The surprise nature of this layout is highlighted by the sign's tilted plane. Balance is informal. White space lends strong secondary accent to the primary sign image. The sign and the small picture set up an unusual upper-right-to-lower-left diagonal balance. This is a subtle accent to the protest theme—a search for alternatives. The affective mood is one of controlled tension.
Applied Data Communications

FIG. 8–15   Computer Automation
Like the Sheaffer ad, this layout combines formal and informal balance. The primary image—puzzle pieces grouped in a continuous sequence from upper left to lower center—is informally balanced against the slogan in the lower right corner. An overall puzzle pattern in the background, however, provides a sense of formal unity. Used like white space, this patterned negative space frames the main puzzle images. The designers seem to hope that in our natural desire for balance we will solve the puzzle by associating the shape of the missing puzzle piece with that of the empty space. What better way to involve the reader directly with ad content?
Computer Automation, Inc.

THE MACNEIL-LEHRER REPORT. UNERRINGLY ON TARGET.

Weeknights on PBS
with Charlayne Hunter-Gault

by WNET/New York and WETA/Washington, D.C., and made possible by grants from the Corporation for
Broadcasting, Exxon Corporation, AT&T and the Bell System Companies and member stations of PBS.

FIG. 8–14   MacNeil-Lehrer Report
Informal balance, a repeated image for rhythm, and primary emphasis on both type and illustration make this ad hard to ignore. Image direction from upper left to lower right is typical of many informally designed ads, but here it is also dramatic. Note how the light-to-dark tonal variation in the target images pulls us in the direction of the headline.
Exxon Corporation

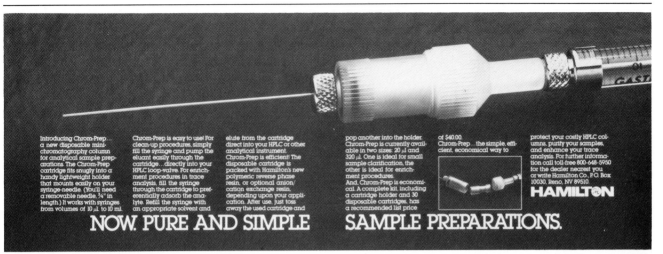

FIG. 8–16   Hamilton
Informal balance and a very strong directional pointer characterize this example. We can hardly imagine any reader failing to see where to begin reading this ad. Since the primary illustration is so dramatic, it serves the purpose of a headline. Secondary emphasis is provided by the headline below. A nice sense of proportion can be felt in the matching horizontal shapes of the primary image and the ad itself. The disproportionate size of the needle almost guarantees that we will notice the ad.
Hamilton Company

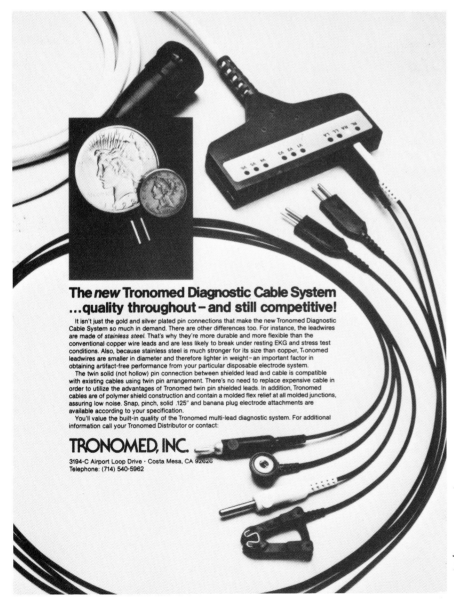

FIG. 8–17   Tronomed
Rarely will an ad have such emphatic directional lines. This layout contrasts a flush-left image group (the type and the secondary photograph of coins) against a circular primary frame. The coiled cables provide a pleasing, rhythmic pattern and strong lines leading to the logo.
Tronomed, Inc.

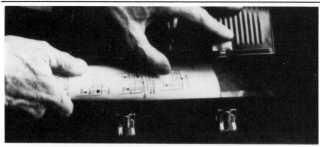

"But who could've gotten access to that information?"

"We couldn't figure it out. Confidential information was being leaked to our competitors. Then we realized that just about every employee had access to our files. So we installed an RES CARDENTRY® system and now we know who gets what information. And when. We should've done this a long time ago."

Yes, you should have.

The fact is, without an RES CARDENTRY system, you don't have the accountability you need — and are more vulnerable to white collar crimes. Such as the leaking of confidential material like your company's financial statements, customer lists, business plans, new product information, or any other type of information that could cost you money if it got into the wrong hands.

With an RES CARDENTRY system in your facilities, you can restrict personnel access to anyplace you think appropriate. So only authorized people can get into high-security areas. You also have printed records of all unauthorized attempts at entry, and records of who's been where, and when. You can even tell your system what time during the day certain people are allowed into restricted areas. Small wonder we're the

world leader in access control.

When we install a system, we issue each employee a RUSCARD™ with a personal code imbedded in the card so the code can't be tampered with or decipher[...] in a card reader at an [...] is transmitted back to [...] reads the code and eit[...] access to the cardhol[...]

To deny access to [...] your system which exi[...] longer valid. So it's ea[...] cards without disrupti[...] never a need to collec[...] ming for specific indiv[...] Your unwanted "silent [...] forever.

With an RES CARI[...] bring accountability t[...]   stop [...]   leaks [...]   ends [...]   troni [...]   mati [...]   Glen [...]   1-80[...]   In Ar[...]   Ext. [...]   know[...]   the i[...]

RES RUSCO ELECTRONIC SYSTEMS
A DIVISION OF **ATO**

FIG. 8–18   Rusco Electronic Systems
This ad uses the photograph as a shocker. Tight cropping of the image gives the impression that we are peering through a hidden peephole to catch a thief red-handed. Primary emphasis is shared between the headline's question and the answer provided in the photo: "Who could have gotten access . . .? Anyone." The obvious mood established here is one of concern for security. Formal layout helps reinforce the seriousness of the message.
Rusco Electronic Systems

FIG. 8–19   Penn
This ad demonstrates how little is required to transform a layout from static to dynamic. The round shape of a ball is a highly stable, perfectly balanced image. But add a blur for speed and a shadow to show impending rebound, and the image immediately becomes exciting. Anticipation of action completely dominates the emphasis here, so a large, bold headline is necessary to compete with such an image. A formal, centered axis is complemented by informal balance between shadow and logo.
Penn Athletic Products Company

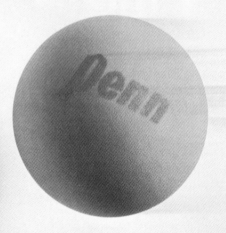

The one part of your game that never has an off day.

Rely on it. When all your shots are working. And when they're not. Because even on those days your game's more off than on, the Penn®Ultra-blue racquetball performs. Shot after shot. Game after game. Whether your style is power or finesse.

This ball is so lively, consistent, and durable, we back it with the guarantee no one else offers: If any Penn ball should fail before the label wears off, return it to the place of purchase or to Penn for two new balls.

Ask anyone who plays it. Once you try Penn, you won't go back to anything else. Because even if you're off, it's on.

Penn Athletic Products Company, 200 Mall Boulevard, Monroeville, PA 15146.

Now You Can Cruise The Oceans, Guided By Satellite.

Imagine the thrill of setting out on a long voyage, knowing you're in for a thoroughly enjoyable time. You're going to relax and sail with confidence, secure in the knowledge that you're being guided by the most proven, most accurate worldwide navigation system available.

It's the famed Transit satellite system, designed to guide the U.S. Navy's Polaris submarines. And we're your direct link to its many benefits.

Steer Any Vessel by Manmade Stars.

We introduced the world to commercial satellite navigators in 1968. And now, we're offering you our many years of experience in the remarkable new MX 2102 Satellite Navigation System, the first we've designed for pleasure craft owners. Small size, low power dc operation and lower cost make the MX 2102 a welcome addition to power and

118 20.26W

sail craft that cruise or race offshore.

With the MX 2102, you can take full advantage of the Navy's investment in accurate navigation. Transit blankets the earth and is unaffected by local weather conditions. You'll always know where you are, anywhere, anytime, in any weather. Position fix accuracy is a nominal 0.1 nautical mile.

The Facts Are Always In Easy Reach.

Although the MX 2102 is highly sophisticated, it's about as simple to use as a pocket calculator. Press one or two keys, then view the facts you need on a crisp, clear digital display. You can select up to nine waypoints, and your MX 2102 calculates and displays course and distance to each. It works automatically, signals when you reach your first waypoint, then displays course and distance to the next.

You can relax and enjoy the voyage, secure in the knowledge that you have the most accurate navigation information available. What's more you save sailing time and fuel.

A World Of Navigation Information.

Information is displayed constantly and automatically. The MX 2102 gives your current position in latitude and longitude, updated by each satellite position fix...and GMT, accurate to one second. On your command, the MX 2102 displays...

☐ Speed and heading
☐ Dead reckoning since last position update
☐ Set and drift

☐ Great Circle or Rhumb Line distance and course to waypoint
☐ Total distance run
☐ Course and speed of advance

☐ Heading to steer
☐ GMT of next and future useable satellite passes
☐ Last position fix data

Enjoy Satnav On your Boat.

Both sail and power craft can benefit from the MX 2102, thanks to simple installation and dc operation. Made of light, yet sturdy cast aluminum, the compact MX 2102 easily mounts overhead, on a bulkhead, on a table, or flush-mounts in a panel.

The slender, lightweight antenna can be installed in any relatively unobstructed location. It connects to the MX 2102 with a single cable. Installation is surprisingly low in cost. And so is the entire system, for that matter.

Find out more. Let the company that introduced the world to satnav introduce you to a totally new experience in navigation ease and cruising safety. We'll help guide you through the waters of the world!

Write today for a free full-color brochure and the name of your nearest Magnavox marine dealer. Magnavox Advanced Products and Systems Company, 2829 Maricopa St., Torrance, California 90503 U.S.A. Telephone (213) 328-0770, or call toll free (800) 421-5864.

Magnavox Magnavox Advanced Products and Systems Company

Navigate with an Experienced Hand.

FIG. 8–21   West Coast Air Charter
This spread depicts a striking emphasis on the dominant image of the plane, yet the strong alignments of the type help lead the reader through the message. Note that the flush right mode of the headline and the placement of the logo keep extraneous white space from detracting from the open space around the plane. Observe how the sweeping wing span at the bottom provides a solid "floor" to balance the ad.
West Coast Air Charter

FIG. 8–20   Magnavox
This double-page spread is a good example of merged emphasis. Because the body type and photographs have been blended together artfully under a banner headline, all elements claim a fair share of emphasis. The generally informal balance is strengthened by top and bottom rules and attractive subheadings. Image weight is roughly equal between pages, and a strong logo unit and slogan are positioned for sure reading. Indeed, the bottom rule guides eye flow eventually to the hand in the lower right. Sometimes this style is called storybook or rebus layout. Care must be taken not to allow illustrations to intrude so much that they break type lines into pieces. This forces the audience to jump over them in search of the remaining words on the line, and disrupts reading. Natural, flowing type contours are fine, however. Often rules between columns strengthen such layouts.

Harmony of tone and general illustration style is important to such layouts.
Reprinted courtesy of Magnavox Marine and Survey Systems Division, Torrance, California

Smith International, The Natural Resource.

Drilco Industrial is the leading manufacturer of blast hole and raise drilling string tools for the mining industry. A complete "swivel" joint permits drilling operations in elevated, inclined, or vertical positions, with vibration dampeners. Drilco Industrial products include drill strings for construction and mine shaft applications to 20 feet in diameter.

Dyna-Drill pioneered the development of the positive displacement, fluid activated drilling motor. This tool can be used in minerals exploration, methane gas drainage from coal seams, rescue operations and pipeline construction in surface-to-surface operations.

Servco markets tungsten carbide hardfacing material in several different forms for a wide variety of industrial applications.

Smith-Gruner supplies steel-tooth and tungsten carbide insert drill bits to the mining, water well, mineral exploration and construction industries. These bits are designed in a wide range of sizes for a wide range of formation types.

Mining Tools manufactures carbide cutting tools for underground coal mining, carbide drill bits for roof supports in underground mining and carbide cutter bits for road scarifying and trenching. Mining Tools developed the Dust Hog™ roof bit for mine roof support work.

Tungsten Carbide Mfg. supplies durable, wear resistant tungsten carbide inserts which provide the cutting and digging action on Smith Tool's rock bits as well as on other oilfield and mining tools.

Willis is a world leading manufacturer of precision flow control valves for a variety of energy related applications including coal gasification/liquefaction processes.

Smith International's group of operating divisions is a valuable resource to keep in mind. Each division lets you tap into a different combination of tools, products and services.

Behind that variety you'll find a consistency of quality and a high degree of practical knowledge that's hard to match anywhere else.

For information about our divisions products and services, write to Smith International, Inc., 4343 Von Karman Avenue, Newport Beach, CA 92660. Telephone (714) 752-9000.

Sii SMITH INTERNATIONAL, INC.

Smith International, The Natural Resource.

Drilco is a leading supplier of drill collars, kellys, stabilizers, tool joints and drill pipe assembly, reamers, Shock Sub vibration dampeners, Mud-Chek kelly valves, degassers and hydraulic catheads.

Dyna-Drill pioneered the use of the positive displacement motor in directional drilling operations. Still the industry leader, Dyna-Drill has modified the tool for straight hole operations.

Emco provides the free world, both land and offshore, with directional drilling and well bore surveying services. Trained directional consultants and survey technicians are on 24 hour call to meet the needs of companies requiring directional services.

Servco supplies the drilling world with stabilizers, under-reamers, hole openers, milling tools and well abandonment services. Servco is the word for downhole tools and services.

Smith Energy Services uses the latest in hydraulic fracturing and acidizing techniques to increase well productivity.

Smith Tool is a leading manufacturer of a wide range of rock bits in many types. The rock bit line includes milled tooth and tungsten carbide insert (TCI) bits backed by reliable, proven bearing systems.

Tungsten Carbide Manufacturing (TCM) is an industry leader in metallurgy techniques. TCM manufactures wear resistant tungsten-carbide inserts which provide the cutting and digging action on Smith Tool's rock bits as well as on other oilfield and mining tools.

Willis controls the flow of fluids in operations all over the world with their line of multiple orifice valves, chokes, actuators, gate valves and pig ball valves.

Smith International's group of operating divisions is a valuable resource to keep in mind. Each division lets you tap into a different combination of tools, products and services.

Behind that variety, you'll find a consistency of quality and a high degree of practical knowledge that's hard to match anywhere else.

For information about our divisions products and services, write to Smith International, Inc. 4343 Von Karman Avenue, Newport Beach, CA 92660 Telephone (714) 752-9000.

Sii SMITH INTERNATIONAL, INC.

FIG. 8–22   Smith International, Inc.

This design illustrates the application of layout integration principles discussed in Chapter 2. The intent was to present the individual contributions of Smith International's major divisions to natural resource development, while also demonstrating the divisions' overall relationship within a single corporation.

The solution here requires attention to proximity, alignment, and consistency of style: the photos are small and close together; they are aligned top and bottom, allowing the captions to begin in alignment; finally, they are consistent in shape, tone, and cropping. This layout treatment provides a sense of community with equal emphasis on all divisions.

The supporting cast in both productions includes banner headlines and logos, general copy blocks, and background illustrations. All help to reinforce overall message unity. Note the general copy blocks to the right of each row. The block in the lower row is less likely to be mistaken for a division copy block because it has been set wider.

Ad reprinted by permission of Smith International, Inc.

Layout Procedures

Advertisements for print pass through many creative and production stages. Sometimes a layout designer plans the ad around structured visual ideas already selected to match written copy and existing illustrations. At other times, an ad concept may evolve freely as the designer experiments with creative layout ideas that generate ideas for illustrations and copy. Either way, it is the quality of the final design that counts.

Everyone charged with the creation and production of an ad should be involved in the earliest stages of planning. This helps avoid wasted effort, focuses creative talent, and ensures that the final layout isn't a surprise either to the client or to those whose job it is to please that client.

Ideally, an ad should not go into final stages of finished production until it has been approved. Copy should always be edited carefully before typesetting and type should be proofed before it is pasted. Most importantly, though, the client should approve at least a rough layout of the ad before a single line is set or illustrations are produced.

Layout Stages

Four basic stages may be required in layout:

Thumbnails are several miniature sketches showing different ideas for size, placement, etc.

A full-size rough is the final idea drawn out in pencil or gray marker. Illustrations are not detailed, but rapidly outlined and shaded. Headline type is traced into position, body type is designated by horizontal lines.

A comprehensive (comp) is a much more detailed version of the ad. Headline and body type are more carefully drawn. Illustrations are rendered in more detail and liquid markers are used to indicate all color.

A finished comprehensive is a close simulation of the printed page. It may contain typeset or transfer-letter headlines, self-adhesive color film, transfer body type, and reproductions of finished art pasted into place.

Many ads require only the first two stages. A fairly neat rough layout can be done in a much shorter time than that required of comprehensives. A great amount of time and effort can be wasted in production of comps for clients or production personnel who are sophisticated enough to make decisions or paste-up from a good rough. Comprehensives are necessary when the color, typography, or complexity of an idea makes it difficult to judge unless presented in detail.

Figure 8–23 shows the stages in the production of a newspaper advertisement and the simple tools required.

Notes

[1]"Heraldry for the Industrial Age," *Time*, October 18, 1982, pp. 84–85.
[2]Jay R. Headly, "It's What's Behind the Logo That Counts," *Journal of Organizational Communication*, 7, No. 4, 1978, pp. 5–7.
[3]These symbol classifications are suggested by Jan Michelson in "LOGOS: Imagination, Inc.," *CLIP BITS*, April 1982, pp. 10–13.
[4]Elinor Selame and Joe Selame, *Developing a Corporate Identity: How to Stand Out in the Crowd* (New York: Chain Store Publishing Corporation, 1975), pp. 42–46.

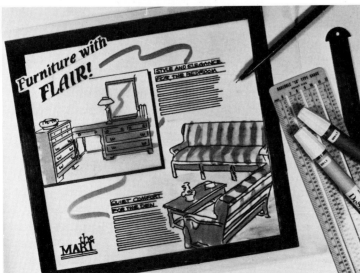

FIG. 8–23
Two stages in the preparation of a newspaper ad. Above, miniature "thumbnails" depict two general plans. Below, a marker "comp" somewhat more detailed than a pencil rough is shown actual size.

9 Publications Design
Planning a Creative, Cohesive Visual Package

The design of publications is but a special application of some of the basic principles we have just examined. Every publication has its own special purpose and format, but the common objective of all publications is to provide information and entertainment as effectively as possible to a selected audience. Lasting design facilitates reading, excites without resorting to gimmickry, and showcases what really matters—the publication's content.

Introduction

In the previous chapter we saw how design principles could solve problems in the construction of identity symbols and advertising messages. In this chapter we'll use these same principles and others to examine problems in publication design. We will see how a structured plan can help design *continuity* when content must be displayed on more than one page.

An amazing variety of printed publications exists today, with new forms being established all the time. While some are one-time, special purpose efforts, most are intended to be issued on a scheduled frequency— daily, weekly, monthly, quarterly, annually, or otherwise. Newspapers, magazines, and newsletters form the bulk of regularly issued publications.

These usually follow an established format to simplify their design, since complete redesign of each issue would be both impractical and confusing to readers. Infrequent publications such as brochures and annual reports allow ample time for totally fresh design. However, many of the same design principles that help organize the content and streamline the production of regular publications can also be applied to irregularly scheduled or one-time publications.

Common Elements of Publications

Given the marked differences between publications in subject matter, audience, and intent, it might be surprising how much they have in common in terms of design. A good way to compare designs of publications is by format. The term *format* means different things to different people, but generally is assumed to include the elements discussed in this section.

Overall Configuration

Newspapers—broadsheet (full size), tabloid (half size), magapaper (quarter size), or other.

Magazines—usually designated by the trimmed page size (such as 8½ × 11 inch), and may also include descriptive terms such as self-cover, saddle-stitched, web offset, four color.

Booklets and Brochures—descriptions similar to magazines, plus any special features (oblong, embossed cover, etc.).

Folders, described in Chapter 14, are specified according to the type of fold (for example, French fold), number of folds, and number of panels per side.

Specific Page Organization

This is a detailed plan for sizing and positioning all content, including:

• Type page size—the working area within the page margins

• Number and width of columns

• Width of margins and spaces between columns

• Any grid pattern or other design plan, other than columns and margins, that aids in positioning content

Typographic specifications and details about photographs, illustrations, and white space are sometimes regarded as part of the format, as well.

As with the printed advertisements discussed in Chapter 8, layouts containing editorial matter also can be viewed as having *structure* (balance, alignment, patterns); *primary emphasis* (on illustrations, type, or graphic devices); *secondary emphasis* (from white space, background tones, etc.); and *affective tone* (an overall mood).

Regardless of a publication's format, certain guidelines seem to apply for effective presentation of content to readers. The following items are adapted from a checklist presented in the *Newspaper Design Notebook*, the official publication of the Society of Newspaper Design.[1] This offers a common sense basis for evaluating the display of not only newspaper page content, but that of most other publications as well.

Checklist for Functionally Integrated Design

Organization

• Are readers guided smoothly and naturally through the page?

• Do all elements have a reason for being?

• Are all intended relationships between elements readily apparent?

• Are packages (modules or blocks of content) clearly defined?

• Does the design call attention to itself instead of the content?

• Does the page appear "cluttered"?

• Do any type or art elements appear to be lost or floating on the page?

Readability

• Do any elements interrupt reading or cause confusion?

• Is the line width of any text too narrow or too wide for easy reading?

• Is text set to follow the contour of adjacent art easy to read?

• Do any headlines (or other display type) compete excessively with those in adjacent columns?

• Are the starting points for all stories easily determined?

Accuracy and Clarity

• Does the layout accurately communicate the relative *importance* of the stories contained on the page?

• Do the art elements accurately convey the *tone* and *message* of the stories?

• Are logos consistent and differentiated from headlines?

• Are the devices (such as borders) used in a layout appropriate for the content of the page?

Proportioning and Sizing

• Are all elements sized relative to their *importance*?

• Does the page have a *dominant element* or package of elements?

• Does the shape of an element appear contrived or forced?

• Do any logos or headlines seem out of proportion to the size of the story or column?

Efficiency and Consistency

• Do all areas of white space appear as if they were planned?

• Is spacing between elements controlled and consistent?

• Are areas of white space balanced on the page?

• Is all body type set at the most efficient line width for the information presented?

• Is the size of the spaces between columns (*alleys*) constant?

• Does the number of elements and/or devices used in a package seem excessive?

Much of what makes any publication format effective can be found in the above list. Keep these points in mind as we analyze the designs in this chapter.

Several good books listed in the bibliography cover publication layout in detail. We won't attempt here to show all possible variations of formats, layouts, and types of pages. Instead, we will try to distill the best ideas from a few representative samples in order to present some solid, useful criteria for publication format and design.

Newspaper Formats

While newspapers obviously vary in size and content, for purposes of *design*, we can group most newspaper publications together. They each usually present many items which vary in content and length (news, features, opinion, etc.) together in a relatively large page format. Most display on the front page the most important news, content directories, features, photographs, illustrations, or other items. Most also have special pages or sections inside that organize and highlight special subject matter.

Whether the items are intended for the general public, employee, special interest, or other audiences, they must compete with all other items on the page and within the issue. To some degree, they must also compete for reader time and interest.

There are three common newspaper formats: the well-known broadsheet, the tabloid, and a recent variation, the magapaper. A broadsheet paper is a full-size newspaper, roughly $14 \times 22\frac{3}{4}$ inches. It can contain as many as eight or nine columns, but today's papers tend to use fewer—most commonly the so-called "optimum" format of six columns.[2] A *tabloid* is about the size of a full-size page folded in half.

The front-page size of a *magapaper* is approximately $8\frac{1}{2} \times 11$ inches, the result of folding an oversized tabloid a final time to create a magazine page size. The magapaper sometimes contains all three formats in one—a magazine-style cover, a tabloid on the opening spread, and full-size newspaper when completely unfolded.

The size of the image printed on the paper page is of more practical importance than the paper size itself. A uniform system of advertising space units, developed in 1981 by the American Newspaper Publishers Association, has helped standardize type page sizes of newspapers. Adopted by the American Association of Advertising Agencies and many other industry organizations, this Standard Advertising Unit System recommends a 13-inch type page width, and a $22\frac{3}{4}$-inch paper page depth for broadsheets. It also lists two ideal tabloid type page widths—$9\frac{7}{8}$ inch and $11\frac{1}{8}$ inch. Tabloid paper page depths range from roughly $13\frac{1}{2}$ to 15 inches. The 25 standard ad unit sizes are designed to ensure that ads planned for a variety of newspapers can be fitted to any of their formats with the least amount of resizing.

Until the design-conscious 1970s, the appearance of newspapers remained surprisingly resistant to change. From the turn of the century, when the use of engravings began to brighten the usually gray masses of type, through the mid-century scramble to convert to offset printing, newspapers seemed immune to the application of truly functional design principles. The *makeup* of a newspaper (a term implying a mechanical "construction" project rather than the formation of a thoughtful plan) usually resulted in too many narrow columns, awkward wrapping of stories around other elements, unesthetic use of art and photographs, and generally uninspired typography, layout, and design. All of this has changed, due to the impact of color in print and television media, increasingly youthful readers with more active lifestyles, and new communication technologies.

Figure 9–1 shows more than a century of design evolution at *The New York Times*.[3] The front page in Civil War days was devoid of illustration, display type, or anything more than single-column story. By the middle of this century, the *Times* displayed its famous symmetrical front-page makeup—a completely balanced page, where elements on either side of a central axis formed the mirror image of those on the opposite side.[4] By the beginning of the 1980s, the *Times* exhibited a much more esthetic, less "forced" appearance. Note the return, after more than a century, to the original six-column format.

Basic Elements of Front Page Design

Figure 9–2 depicts the main newspaper design elements on the front page. The example shown is the Los Angeles *Herald Examiner*, a major daily.

The front page elements include the following:

Nameplate.   This is the title of the publication, displayed in large, often stylized type. The day or date, frequency of publication, and volume and issue number, price, and other information may appear, as well as special borders or illustrations. The trend is toward less ornamentation, modern typefaces, especially in noncommercial newspapers, and more white space. Some still call this element the flag, the masthead, or the banner, but design professionals generally prefer the term nameplate.

Standing heads.   Also called logos or constants, these include the index, calendar, weather, and other regular headings. Section heads and other constants also occur inside the newspaper. These standing elements are usually similar in design and may be set apart graphically from other headlines by rules, boxes, tints, and other devices.

FIG. 9–1
The New York Times *today and in 1864.*

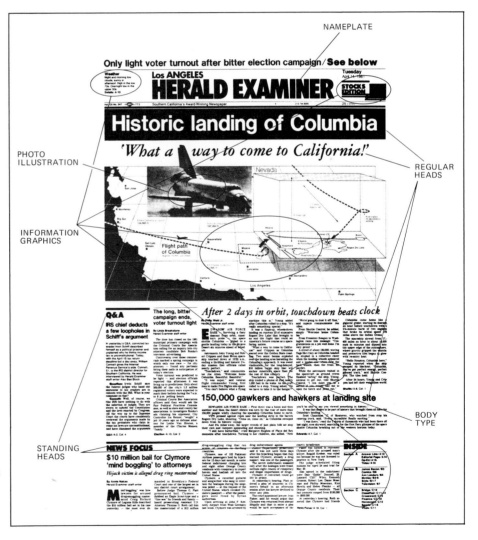

FIG. 9–2
Front-page design elements, shown on a major daily newspaper with a prize-winning approach to modern design.
Copyright 1981 Los Angeles Herald Examiner.

Regular heads.   All display typography 14 point and above is considered a headline. This includes *kickers*, small display lines above a larger headline; and *decks*, multiline headings of a smaller size placed directly below a larger, related heading. *By-lines*, identifying the writer of an item, and special typographic treatments such as an oversized lead paragraph may also be set display. Newspapers rarely use sizes above 60 or 72 point, unless special graphic impact is desired. Since the size of a headline has traditionally been associated with the importance of a story, editors tend to be wary of size "overkill" simply for visual effect.

Typefaces formerly found only in advertising and special sections now can be found throughout the newspaper in addition to the always popular romans and sans serifs. Designers have not abandoned the useful role of limiting the number of type families to two—perhaps three—and mixing different weights, stances, and forms of those families freely. The mixing of roman and sans serif headlines, once taboo, is now considered acceptable; but strange, disjointed effects can result if the faces are not selected carefully for compatibility. Usually, very bold sans serifs—in caps—mix well with romans and other serif styles.

Arbitrary rules, such as banning certain headline sizes from appearing above the fold, have given way to the placement of type according to overall page impact.

Body type.   Today's formats lean toward six-, five-, or four-column displays, and many have variations that create an attractive mix of column widths. Point sizes of 8, 8½, and 9, with ½ to 2 points of leading, are popular. Column widths of around 15 picas provide a pleasing number of characters per line. Roman types remain almost universal for commercial newspapers, because of their longstanding popularity with editors for readability. More innovations, run-arounds, contours, ragged right, etc., are found today. Caption styles vary widely, and often imitate magazine styles.

Information graphics.   Usually meaning charts, graphs, tables, or statistics boxes, these elements have undergone radical face-lifting in recent years. Exhibiting colorful, easy to understand illustrative styles similar to that used by Nigel Holmes of *Time* magazine, these graphics increasingly stand alone as simple modules containing a complete story or theme. When supplementing a story, they employ graphic elements that match its theme. These information-packed graphics add interest and often color to newspaper pages. In *USA Today*, the colorful "nation's newspaper" published by the Gannett newspaper

group, information graphics were elevated to the status of stand-alone items in an attempt to allow readers accustomed to television to absorb news and features rapidly.

Illustrations and photographs.   Nowhere has the move away from tradition been more complete than in the creation and use of both photographic and nonphotographic illustration. Commercial newspapers rely heavily on staff artists and illustrators to produce captivating line and continuous tone graphics, previously found only in the editorial/opinion sections of a few major dailies. Pictorial coverage of news and features has become highly sophisticated as more and more photojournalists are trained to think in terms of overall visual communication. They produce images that lend themselves to unusual, dramatic treatments as well as to the transfer of information. Much more sensitive conversion of continuous tone images into symbolic "art" occurs today. Commercial newspapers' greater reliance on visual statements demands a *graphics editor* with a solid news sense. That editor is the link between the art and photography staffs and the more "word oriented" editorial staff. Noncommercial newspapers—those published for companies and other organizations—depend no less upon the graphic sense of editors and designers and photographers to convey their information effectively.

Using the Design Ingredients Effectively

In Figure 9–2 we see the front page of a modern commercial newspaper and its basic design elements. This attractive front page, highlighting the landing of the first space shuttle, contains many features now considered fundamental to effective newspaper design:

Modular layout—all stories are rectangular, not irregular, to prevent reader confusion.

Wider columns—in this case, the popular six-column format.

Aesthetic headline typography—an especially nice mix of sans serif (in reverse) and roman lends emphasis to the quotation above the art. The sans/roman treatment is repeated in the hijacking story. The page employs size variations of only two type families. All heads are flush left to give a uniform sense of alignment to the columns. Bold initial letters, a trademark of many newspapers today, also brighten the page.

Coordinated contents—those items used each day—Q&A, Inside, and New Focus—match the design of the modern nameplate in weight and all-caps treatment.

Exciting informational graphics—a well thought-out linkage of the shuttle photo to both a flat map and a perspective view of the landing site. Note how two simple touches, the shadow beneath the box and the little arrow leading from the flat to the perspective views, make the illustration "friendlier" to read.

Guiding the Reader Through the Page

Over the years, many notions have evolved about how people read newspaper pages. Editors were taught a few basic page structures to use to give their front pages some variety and to help readers make sense of the day's news. Creative editors quickly learned that no "ideal" layout exists, just as no day's news is identical to the next. Here are six of the choices they might consider in putting together their page:

formal balance—divide the important elements equally, perhaps even symmetrically, on both sides of a vertical axis down the page.

informal balance—distribute the visual weight of important elements equally on both sides of a vertical axis.

upper half prominence—place all or most important items above the fold.

upper corner prominence—place the most important item in the upper left (or right) corner.

quadrant distribution—distribute items so that every quadrant contains something important.

horizontal distribution—run all or most stories as wide rectangles across the page, with headlines over the full width of each story.

Figure 9–3 shows these six general front-page layout patterns. Many design editors believe that the visual

FIG. 9–3
Six front-page layout models.

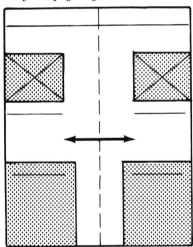

1. Formal balance

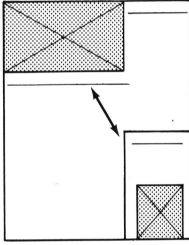

2. Informal balance

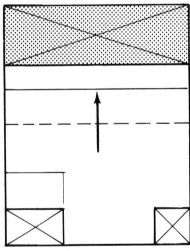

3. Upper half prominence

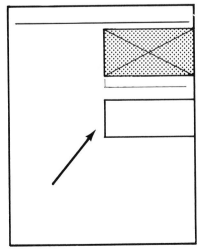

4. Upper corner prominence

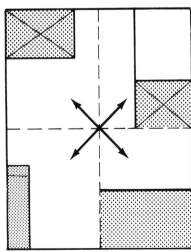

5. Quadrant distribution

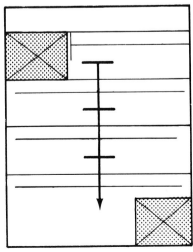

6. Horizontal distribution

content of stories is more important than their "correct" positioning according to traditional patterns. They pay great attention to creating an eye flow on the page that is best suited to the content, not simply one that fits a preconceived idea about natural eye movements. This doesn't mean that all the old rules (and a few basics) are completely ignored. Many editors still prefer top-half display of most important news stories, and often favor right or left as a focal point. They usually want visual page balance, and prefer not to have large gray masses of body type. They still lean heavily toward serif body types and use a variety of sizes and weights for headlines. Strict rules for story placement, however, are fading.

In short, there is more confidence on the part of designers that readers can be guided through the page by the use of visual cues and sensible, uncluttered organization.

In his recent book on newspaper design, Mario R. Garcia stresses the importance of a *center of visual impact* (CVI) on the newspaper page.[5] He maintains that the designer must first establish a focal point for the reader—a large photo or illustration, or an attractive headline or box—in order to initiate a structured reading flow. On the front page of the *Herald Examiner* (Figure 9–2), the large photo/illustration creates a powerful CVI, from which the reader can be guided into related content. Note that the lack of rules (lines) between the illustration and the two related stories below help the reader sense, visually, that the items belong with the illustration.

Alternative Structures

Garcia explains that a *structural approach* to newspaper layout utilizes orderly progressions down the page—large to small, small to large, vertical to horizontal, horizontal to vertical.[6] He emphasizes that elements should be squared-off, rectangular, and uniform. Figures 9–4 and 9–5 show a few examples of both size and shape contrasts that allow orderly structure.[7]

Newspapers and the Grid—Extending the Visual Plan

The use of *grids*—columns and rows made up of uniform rectangles to establish proportion and order on a page—is commonly seen in newspaper layout. Their predecessors, dummy sheets divided into columns, have been in use for decades. The

refinement of dummy sheets into useful planning grids for newspapers is a relatively recent development that began with Peter Palazzo's redesign of the *New York Herald Tribune* in the early 1960s.[8] Figure 9–6 shows the grid pattern of another New York newspaper, the Sunday *Herald*, which utilized the advantages of a 6-column, 17-row, 102-unit grid to organize pages into dramatic, horizontal divisions of space.

A Sampler of Designs

The following examples show how prizewinning commercial and noncommercial newspapers use modern layout techniques effectively to inform and entertain their readers. Both front pages and inside section pages are included.

USA TODAY—The Grand Experiment

Just about everyone interested in newspapers, commercial or otherwise, has been fascinated with a brash newcomer to daily publishing, *USA TODAY*. Launched in September, 1982, by Gannett, Inc., as "the nation's newspaper," this colorful daily is produced for 16 major markets through the use of an elaborate satellite transmission network and local printing facilities. It became the most visible indication to date of what future newspapers could become. Featuring full-color, exciting informational graphics, brief items reminiscent of television newscasts, and easy to read layout, this paper seemed to represent the wave of the future in newspaper design.

Chided for stimulating instead of communicating and failing to provide the depth of information usually considered a key advantage newspapers enjoy over television newscasts, the publishers held their ground. They maintained that most readers want short, national and local items that are colorful, informative, relevant, and written in a down-to-earth manner.

Given the findings of a recent study conducted by Yankelovich, Skelly & White, Inc., for the American Society of Newspaper Editors, *USA TODAY* may not be far off the mark. The study found that significant numbers of readers find newspapers dull, hard to read, and confusing, saying they contain misleading headlines and pretentious language.[9]

The future of *USA TODAY* is somewhat uncertain at this writing. Despite disappointing first-year advertising and circulation revenue losses, Gannett, Inc. has said that support for the project will continue.

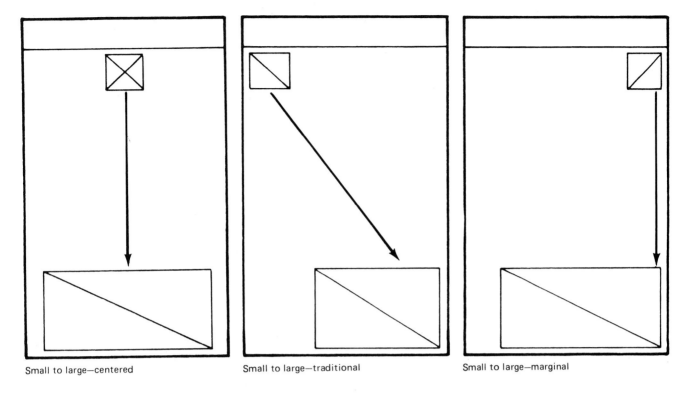

Small to large—centered   Small to large—traditional   Small to large—marginal

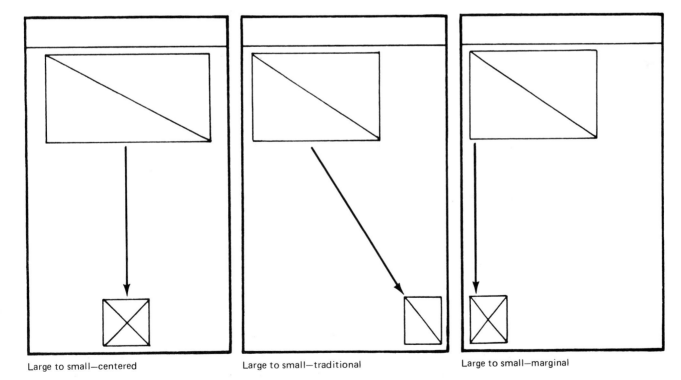

Large to small—centered   Large to small—traditional   Large to small—marginal

FIG. 9–4
Orderly progression of page elements, six patterns.
Mario R. Garcia, *Contemporary Newspaper Design: A Structural Approach.*
© 1981 Prentice-Hall, Inc., pp. 45–46. Used by permission.

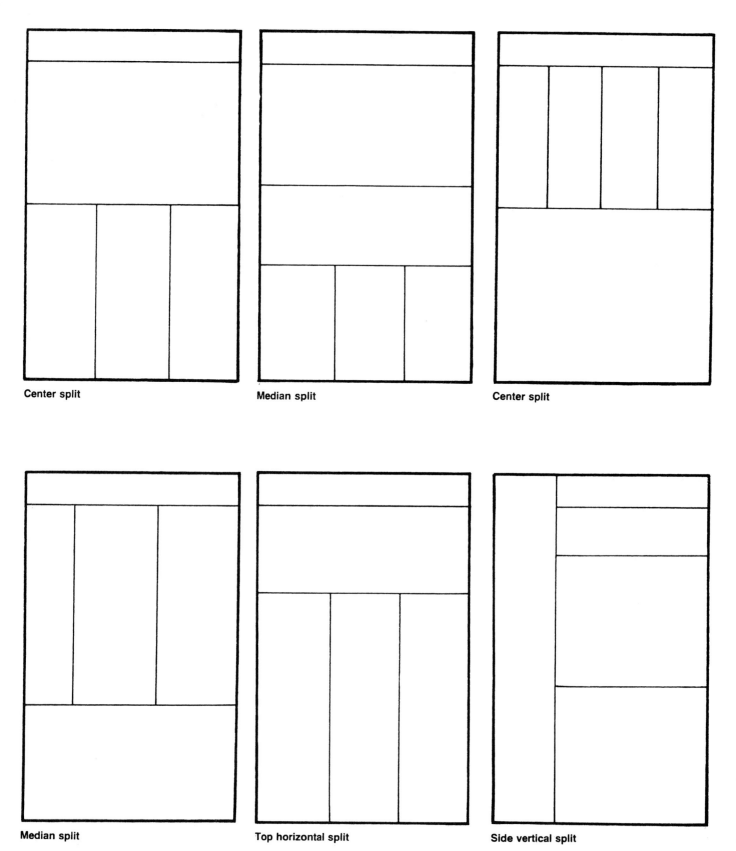

Center split

Median split

Center split

Median split

Top horizontal split

Side vertical split

FIG. 9–5
Basic page structures—eight models.
Mario R. Garcia, *Contemporary Newspaper Design.*
© 1981 Prentice-Hall, Inc., pp. 47–48. Used by permission.

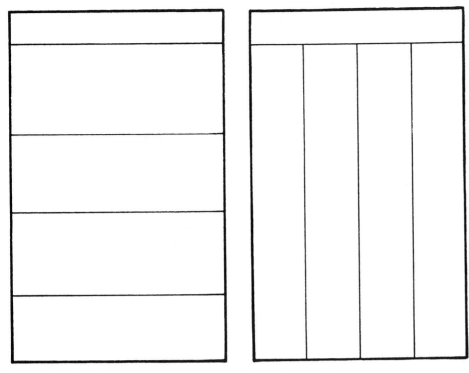

Complete horizontal   **Complete vertical**

FIG. 9–5 (continued)

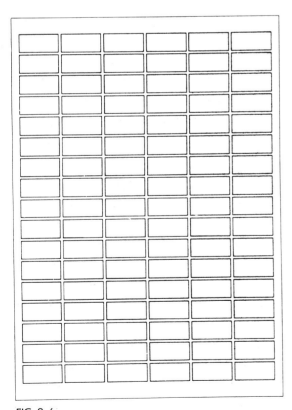

FIG. 9–6
Grid pattern of The Herald.
Allen Hurlburt, *The Grid: A Modular System for the Design and Production of Newspapers, Magazines, and Books.* © 1978 Van Nostrand Reinhold Co., p. 36.

FIG. 9–7

The Columbia Daily Tribune. *Long one of the most respected smaller dailies in terms of design, the Tribune's front page shows why. This page is an ideal example of "clean" design. The delightfully legible directory at the top accents a tasteful nameplate and three easy-to-read story modules. Note how the feature on the left of the page is set ragged right, with a screen-tinted background and column rules. The headline employs a reverse kicker. The section pages continue the understated elegance of the front page.*

FIG. 9–8

The New York Times. *Excitement is generated on these inside pages of The Living Section. The accent is on art, freely flowing through the columns of type, but without sacrificing readability or aesthetics. Liberal use is made of initial letters, boxed modules, and thin horizontal rules enclosing story by-lines. Headline sizing is very consistent. Each page has its own directory at the top to guide readers to the inside pages of the section.*

FIG. 9–9

The Orange County Register. *These pages, originally in full color, show the importance of good photography to overall design. These dramatic images, carefully positioned for graphic impact within a clean, uncluttered format, helped win a Pulitzer Prize in Photojournalism for this paper.* The Register *is a consistent innovator in contemporary design and editorial color.*

Regardless of what happens, *USA TODAY* publishers can claim one of the most significant attempts in this half-century to deliver a graphically innovative daily publication to a general audience. Figure 9–10 shows a sample of this remarkable milestone in newspaper design.

Noncommercial Newspapers

The approximately 1,700 daily and 8,000 weekly newspapers in the country make up only part of the newspaper design picture. At least as many newspapers are not published for the general public. They are produced for special audiences—primarily company employees and members of organizations.

The three examples that follow show some of the best principles of contemporary newspaper design as they apply to the tabloid format so common to these kinds of publications.

PRSSA Forum

Figure 9–11, the publication of the Public Relations Student Society of America, is an outstanding example of a thoroughly readable design. The front page is a four-column ragged right, sans serif format beautifully accented with column rules and stylish graphics. The contemporary nameplate, a fashionable marquee display of shadowed script against parallel rules, ties in with the outside vertical rules to create a functional page frame. All rules lead to the primary graphic emphasis, the city skyline. Headline typography is exceptional—careful word and character spacing, flush left, bold but not gaudy. Stories are essentially rectangular, but not boringly so. Note that only the first column extends the full length of the page, then the next columns are displayed horizontally to allow the second story to start in an advantageous spot for reading. When front-page content can be limited to a single theme, this total package concept is ideal.

On inside pages, note that while many subjects other than the New York conference are introduced, the clean format of the newspaper remains intact. The page frame is retained, as are the rules and the ragged right body type. All stories but one in the lower half of page three are modular. Note the use of heavier horizontal rules at the bottom of pages and at the ends of boxes. The masthead, containing the flag (a reduced version of the nameplate) and information about the editorial office is set off with a light screen tint. This nicely balances the presidential address on the opposite page. Another plus is the liberal use of indexing graphics, the small black bullets on the executive profiles story and large numbers in the personality traits story.

INTERCHANGE of Old National Bancorporation

Figure 9–12 shows the attractive monthly produced by the Corporate Communications Department of a major financial institution. It sports a completely different

FIG. 9–10
USA Today, *an inside section page.*

FIG. 9–11
Front and inside pages of the PRSSA Forum *illustrate attention to design continuity.*

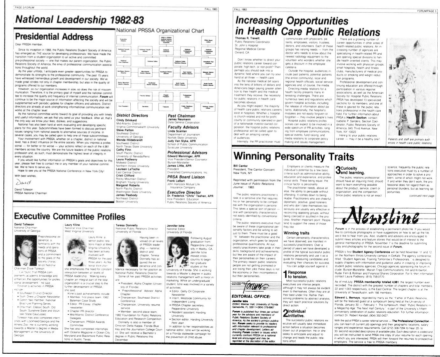

FIG. 9–12
Courtesy of Old National Bancorporation

approach to the 4-column front-page format. Here the elegant Olive Antique typeface used for the nameplate is carried through to all heads, body type, and captions, in several weights and sizes, resulting in complete typographic coordination. Here, too, column rules and page-framing rules are used, creating two vertical stacks of story modules. Note the unusual headline treatment of the "Branching Out" story. Pictures are grouped as a horizontal module at the bottom, and captions contain directional arrowheads. This technique prevents photographs from becoming scattered and allows great flexibility in caption writing.

BETWEEN BRANCHES of AVCO Financial Services

Figure 9–13 is an inside spread from an eye-catching publication that shows how "friendly" white space can be used to advantage instead of filled to capacity. The "promotions page" contains more than fifty "mug shots" (head and shoulders photos), as many captions,

FIG. 9–13
Courtesy of AVCO Financial Services

and three lists of additional names—a layout nightmare. By applying the fundamental rules of layout integration: proximity, alignment, and consistency of style, discussed in Chapter 2, whole units are formed that are a pleasure to read, not a pain.

What makes it work?

• all photographs are the same shape and size, and the head sizes are similar;

• each group of photographs is aligned in rows and pleasingly spaced apart from other groups;

• white space is to the right side, framing, not separating, the elements;

• captions are set flush left with each photo for a cleaner alignment;

• horizontal and vertical spacing between photos within each group is similar;

• the lists at the bottom of the page each have one short column that provides a bit of white space to separate the units.

Faces, not coats and ties, are important here. The editors and designers of *Between Branches* have done their job very well indeed.

Magazine Formats

As with newspapers, magazines come in a variety of styles to serve a wide range of objectives. Consumer, farm, business, association, industry, and special interest magazines in the United States number around 5,000.[10] In addition, public relations magazines are estimated to number in the tens of thousands.

Fortunately, standardization of formats and page sizes makes it easier to discuss design elements for magazines. The vast majority of magazines today have a paper page size of approximately 8½ × 11 inches. Magazines printed by sheet-fed methods may measure exactly that, unless trimmed for bleed. Most magazines printed on web-fed presses (usually with a circulation of more than 10,000) conform to the 8⅛ × 10⅞ inch standard recommended by the American Business Press. Of course, other formats such as digest, about 5½ × 7½ inch, and the once-popular 9 × 12 inch still are used by many publications.

The type page format recommended by the American Association of Advertising Agencies and the ABP is 7 × 10 inches, or a size in direct proportion to these dimensions.

Basic Components of Magazine Design

Although magazines contain a cover, table of contents, and special pages, we will focus primarily on the elements common to *spreads*—the horizontal rectangles formed by two facing pages. Figure 9–14 shows the main elements used in magazine page design. They are:

Paper page area—this is the size of the full magazine page after it has been printed and trimmed. Since magazine layouts are often thought of as *spreads* instead of single pages, the practical paper page area for layout is twice the single paper page.

Type page area—also called the live area—this is the size of the image areas on the page minus the margins. On a two-page spread, the *gutter* (the space between the pages) may also be considered part of the live area.

Margins—these include all space between the type page and paper page areas. They are designated as *head* (top), *foot* (bottom), and *side* (left or right). The two interior margins of facing pages form the page *gutter*.

Grid—this is the pattern of rectangles—rows and columns—which provide for orderly placement of elements. Magazines tend to use both *vertical* and *horizontal* grid lines, which define not only the number of columns (like a newspaper), but also a number of important space divisions down the page. Spaces between columns or grid units are called *alleys*.

Typographic elements—these include *titles* (headlines), *body type*, *captions*, *lead-in paragraphs*, *windows* (pull-outs) for quotations, biographical *blurbs*, *running heads* (to identify a department or section of the magazine), and *folios* (name of the publication, date, and page number).

Illustrations and photographs—all nontypographic images: line art or halftone, color or black-and-white.

Use of the Grid in Magazines

With traditionally more design freedom than newspapers, magazines have adopted countless strategies for displaying content on their pages. Because of the practical limits on page width, most formats have tended to be either 2- or 3- or 4-column. Historically, typography has been much less restrictive for magazines, and therefore sans serif body type, ragged right columns, extralarge titles and subtitles,

Paper page area

Kicker — *Business' Bill Buehler takes*

A Fresh Look at "Corporate Culture"

Main title

by Sidney Hollister

Initial

Lead-in paragraph

"Just as tribal cultures have totems and taboos that dictate how each member will act toward fellow members and outsiders, so does a corporation's culture influence employees' actions toward customers, competitors, suppliers and one another. Sometimes the rules are written out. More often they are tacit."

Business Week, October 27, 1980.

Illustration

Halftone

rule

text (columns)

Paper page area

Type page areas

Head

Window (pull-out)

"We have an organization now that's as good as any in the Bell System. Of course there's risk, but wherever there's risk, there's opportunity."

Gutter

Side

Blurb (bio)

"We cannot be all things to all people. We don't want to be all things to all people."

Foot

FIG. 9–14
Main elements in magazine page design.

169

and stylized typefaces have found easy acceptance for many years. With relative design freedom, magazines have often found that the structure of a good grid system helps prevent design chaos.

Figure 9–15 shows a very simple three-column grid employed by a major oil company to structure one of its publications. The most basic of grids, it is similar to what many magazines use—a simple breakdown of type page space into margins, columns, and alleys. In this case, the two horizontal lines at the top show where the upper limit for the title placement occurs and where a horizontal rule appears on each spread. While the three columns on each page could be divided vertically into grid units, the content of the layouts is easy to place without it. As with every tool in design, a grid should not be more complicated than necessary.

Figures 9–16 and 9–17 show the page without the grid present, and an overlay to indicate how the grid relates to the content.

Magazine grids should be functional, presenting options for layout that tie content together and allow variety and spontaneity. Simple grids may not always be the most functional. For example, in Figure 9–18

we see a two-column, 6-unit grid, the ultimate in simplicity. It's obvious that decisions about where to place type and pictures would be easy: they can be either one or two columns wide and can fill the page to the depth of one, two, or three units (or portions of a unit). After producing about the fifth layout on such a grid, we would find ourselves looking for something different. Of course, two pages together as a spread would allow four units across, but even then our layouts might soon begin to have a repetitive, checkerboard look.

Consider the possibilities for variety as grid columns and units are added. The two grids in Figure 9–19 suggested by Allen Hurlburt in his outstanding book on grids, indicate ways in which magazine pages can be divided into pleasing proportional units without sacrificing flexibility and ease of use.[11]

The first one, the 24-unit *Domus* grid, divides the page into twice as many horizontal and vertical units as our original grid in Figure 9–18. The second grid, a popular 12-column, 72-unit plan, shows how each unit of the Domus grid actually can be split into three more vertical slices. The black rectangles on the grid show how either 2-, 3-, or 4-column elements can be placed on the grid with ease.

FIG. 9–15
Basic magazine grid.

Happy and happier

Last fall Lois Frankel took a couple of days of vacation to ponder her career path. She just didn't feel her job at Russell Anaconda in Miami, Fla., was right for her. Now that's changed. Lois has changed the direction of her career, and she did it through the Atlantic Richfield Job Placement System (APS).

APS is essentially a "help wanted" program which has been active throughout Atlantic Richfield since 1974. Since that time, it has published over 2900 jobs and has received over 16,000 candidate applications. An average of five employees apply for every position, although the number has ranged from zero to 70 (the 70 figure was for an executive secretary position in corporate headquarters). The APS program expanded its distribution and listings to include Anaconda employees in January 1979.

"We weren't getting APS last fall," says Frankel, "but we all knew that it was just a matter of time. That's why I decided to sit tight for a couple of months and see what Atlantic Richfield had to offer."

The 28-year-old has a bachelor's in psychology and a master's in counseling. She worked for Russell Anaconda, a subsidiary of Anaconda Industries' Aluminum Division, as an employee relations representative. "I was told I had a future with Anaconda Aluminum, but I didn't feel I was on the right career track for me," she says.

In the very first issue of APS, Lois found a job she was interested in — as an employee relations representative for Anaconda Copper in Denver. After reaching the interview stage, she withdrew her name for consideration because the job seemed too much like her own. The next issue of APS was out and an even better career

Happiness is APS for Lois Frankel and Don Lewis, who found jobs through the program.

opportunity confronted her. Now she's a college relations representative at corporate headquarters, and one of the first Anaconda employees to cross the Atlantic Richfield/Anaconda job bridge through APS. For the woman who was once an assistant director at a university residence hall, she's happy to be back working with college students.

Another successful Anaconda applicant, Don Lewis, was happy in his job, but as he puts it, "APS enabled me to become happier." He had never looked at APS, because he just wasn't in the market for a new job. But when the corporate telecommunications department put a call out for a telecommunications analyst, he became interested. A friend brought the job to Don's attention. And the senior product engineer for Anaconda Telecommunications in Garden Grove, Calif., found an opportunity he couldn't pass up — a chance to go "corporate."

"I wanted to advance, but the opportunity to advance within that unit wasn't as great as it is here," says the 31-year-old. "The salary was certainly

a consideration, as well as the opportunity to come in contact with a broader spectrum of people." So Don crossed the manufacturing lines over to the side of the end user. And now the salesmen are trying to satisfy him.

"I'm on the first rise of the learning curve here at ARCO and it's a tremendous feeling," he says of his first month with ARCO. "I'm with a company that is a leader in the field of telecommunications and has one of the largest company-owned telecommunications programs."

Don says he wants to learn all he can — to become better prepared perhaps for another job in APS.

It's worth noting that the traffic on the Anaconda/Atlantic Richfield job bridge is not one-way. In the past few months, an average of two Atlantic Richfield employees have applied for each of more than 40 Anaconda jobs posted.

Meeting our commitments

It's 3 o'clock in the afternoon. Archie BigMedicine and his crew are sitting around a table in the lunchroom of Thunder Basin Coal Co. headquarters having coffee and soft drinks. The four men are on break from their janitorial duties, waiting for employees to leave for the day so they can vacuum, empty wastebaskets and dust furniture.

Andy Edgerton looks up, grins and says, "Hi," to a passing employee he recognizes. The young woman greets him in return. Across the table, Ward Cooly recalls the time he worked for a uranium mill in Jeffrey City, Wyo., and had his own apartment. Another member of the crew listens. BigMedicine, a tall, striking American Indian, looks on and smiles. He's proud that these men are learning to cope with the challenge of day-to-day living and that they have full-time jobs.

The men are part of a program run by Dignity Inc., a private, non-profit corporation based in Riverton, Wyo., that helps the state's developmentally handicapped find jobs. Thunder Basin Coal has a contract with Dignity for janitorial services in a unique "hire the handicapped" program.

BigMedicine, a Dignity supervisor and trainer, and his crew have been familiar faces around Thunder Basin for about a year now. They work Monday through Friday, from 1 p.m. until 9 or 9:30, cleaning the change house, along with all the mine offices. Archie turns in progress reports on the men regularly. The crew members all live together in Wright, in a trailer, where they share household duties

Developmentally handicapped make up janitorial crew at Black Thunder Mine.

and are supervised by a Dignity "life skills specialist."

Although the three men on this Dignity crew have all held jobs prior to now, Dignity originally drew people from institutions and individual homes throughout the state. BigMedicine says Dignity gives these people the satisfaction of knowing they can work and pay their own way.

"We decided to get involved with this program because we felt it was a chance to utilize a special service that's allowing handicapped people to develop pride and dignity in their own abilities," says Ron Norton, administrative manager, Black Thunder Mine. "We've been pleased with the results and are planning to involve the men in on-the-job training in cleaning with chemicals."

Bill Miller, employee relations manager here, points out that ARCO Coal is the only mining company in the Powder River Basin to hire a Dignity crew so far. Like Norton, he's pleased with the crew's services and applauds the company's involvement in this program.

Back in Los Angeles, Equal Employment Opportunity Manager Jerry Davis says that hiring the handicapped is one area which ARCO will be stressing even more in the future — along with employment of veterans and internal auditing of hiring and promotion practices throughout the entire organization.

"On the whole, in 1978, we successfully met our EEO goals."

The overall percentage of minorities employed within ARCO at the end of the year was 16.2, versus 13 percent in 1974. Exempt minority employees totaled 7.7 percent versus 5.5 percent four years ago. Total women employed at ARCO at year's end in 1978 was 17.6 percent versus 16.5 percent in 1974. Total exempt women in 1978 was 9.5 percent versus 6.0 percent four years ago.

Regarding the handicapped, Davis comments, "We were one of the first 300 companies selected in 1978 for a review of our Affirmative Action program for the handicapped. Our review was successfully completed."

He adds, "Increasingly, I find as I meet my counterparts from other corporations and representatives from governmental agencies, the name ARCO is well known not only for our solid EEO statistics, but for our special, outside programs like Dignity. These programs will continue to be an integral part of Atlantic Richfield's Affirmative Action program."

FIG. 9–16
Page layout based on grid of Fig. 9–15.
Courtesy of Atlantic-Richfield Company

FIG. 9–17
Layout with grid overlay.
Courtesy of Atlantic Richfield Company

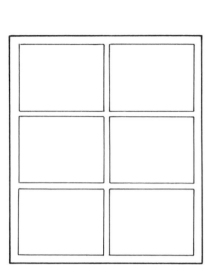

FIG. 9–18
A 2-column, 6-unit grid.

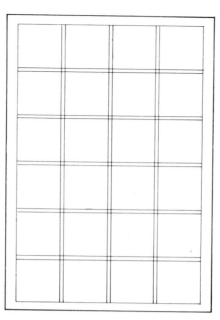

FIG. 9–19a
A 4-column, 24-unit, the Domus grid.

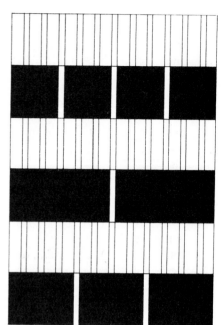

FIG. 9–19b
A 12-column, 72-unit grid
(note 2-, 3-, and 4-column options).

Two things should be noted here. First, both the Domus and the complex 72-unit grids may be thought of as extensions of the simple pattern we began with the 6-unit grid. Second, note that the alleys between the grid units do not necessarily stay the same as the number of grid units changes. The Domus alleys are narrower than either of the others. *Space between units is as important a part of the design grid as the number and size of units.* This space may vary, but rarely is more than about one-fourth of an inch.

Thinking in Spreads

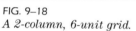

A useful grid plan is only the beginning in magazine layout. We must also consider how the reader will be able to decipher the layouts we put on those grids. A few guidelines can help. Since magazine pages are much smaller than those of newspapers, our eyes can take in the entire area of two facing pages as we turn the page—at least briefly. In his book *Techniques of Magazine Layout and Design*, Donald E. Hill states that readers tend to approach a spread in a roughly clockwise fashion.[12] They first enter the upper half of the left page, move across to and down the right page, then swing back into the layout again. Hill describes this pattern as a series of page viewing *dynamics—* primary, secondary, and supplementary.

Figure 9–20 diagrams the dynamic movement suggested by Donald Hill. Note that this eye-travel pattern is not unlike the upper left to lower right movement we have discussed in other types of layout. In this case, however, it involves a horizontal area. A major component of the notion of dynamics is that readers will attempt to cross the gutter between pages not once but twice at least. It is essential, then, that the gutter be minimized to avoid obstructing eye travel.

FIG. 9–20
Layout dynamics.

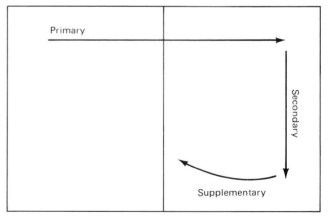

Diminishing the gutter can be accomplished by placing titles or photographs across it, by aligning pictures or type columns across the pages, and by allowing pictures to *bleed* into the fold. The layout in Figure 9–21 makes good use of the principles of layout dynamics.

Other design guidelines, in addition to thinking in spreads and minimizing the gutter, include the following, some of which we have encountered before in connection with other design problems.

1. Use white space effectively, mostly around the outside for framing.

2. Use the principles of layout integration: if elements are related, align them, draw them together, and make them similar to one another.

3. Keep spaces between related elements consistent.

4. Use a dominant graphic to get the reader's attention (somewhat like the CVI for newspapers); but don't let it completely dominate everything else.

5. Attempt to provide a visual path for the reader, so that the elements on the page may be read easily in their most logical sequence.

6. Above all, don't overwhelm the reader with "cute" graphic effects at the expense of losing the meaning of the material being presented. Simplicity is one of the most durable attributes of design.

Effective Application of Magazine Design

The spreads shown in Figures 9–22 to 9–28 illustrate a number of ways in which the principles of magazine layout and design can make content attractive and readable. In every example the major objective is to communicate. Many of these publications accomplish this with style.

FIG. 9–21
A page layout.
Source: AVCO Financial Services, Money Tree.

Avco's United Kingdom "Superleague" travel award competition began in February with the promise that those in first place five months later would soon be enjoying sunnier climates south of the Isle of Wight. Little did they know that the destination was Montreux, a town located on Lake Geneva's shoreline, a part of Switzerland's Riviera.

Montreux is a French-style Edwardian town, considered by Fodor's world-famous tour books as one of Europe's most beautifully situated resorts. The town caters largely to foreigners and is known for its mild climate, lush vegetation and well-tended gardens. It is so well kept.

The United Kingdom's "Superleague" winners traveled to Montreux. 1. A tourist ship cruises past the famous Chateau de Chillon. 2. Dave Coates and Cheryl Gurr wave from the train on the way to Gstaad in the Swiss Alps. 3. Judy and Don Wineland. 4. Several U.K. execs and guests enjoy fondue. 5. RS John Rowley and his wife Beverley enjoy the sun while cruising Lake Geneva. 6. The group boards the cruise ship. 7. AVP/Marketing Alan Williams and Hans, the chef. 8. The men's choir, from left, Derek Scott, Terry Hunter, Mike Wortley, Paul Soper, John Rowley and Dave Burrows. 9. Arriving at the top of Les Diablerets Glacier. 10. Winners enjoy the Swiss Alps scenery.

FIG. 9–22 (below)

The Public Relations Journal. A double-page spread on Texaco in this newly redesigned professional publication shows how effective type alone can be in a layout. The primary graphic, the huge gothic signature (originally printed in red), immediately draws attention to the subtitle below. The reader is drawn back to the beginning of the story by an attractive initial letter and the ragged right lead-in paragraph. Column rules lend structure to the overall spread.

Note that although the wide titles and the top alignment of body type bridge the gutter, the capital X is placed directly on the fold. This makes the X a powerful accent graphic, letting it visually cancel the fold line and the vertical rules between pages.

Reprinted with permission from the April 1983 issue of the *Public Relations Journal,* copyright 1983.

FIG. 9–23 (top right)

This spread deals with the gutter just as effectively as the Texaco layout, but in a more traditional way. The gutter is simply overwhelmed by the black background, the white map, and a large line-conversion image of the subject of the story. All elements—including the direction the man is facing—lead us to the beginning of the story. The gutter is entirely forgotten.

Source: McLintock Main Lafrentz International, *Viewpoint.*

FIG 9–24 (bottom right)

Fluor Magazine. This eye-catching spread uses yet another technique to keep the gutter under control. The picture cluster on the right page is allowed to spill over to the left page along a horizontal axis even with the top of body type columns. This ties the layout together and provides pleasing places to insert captions. Note that pictures come all the way to the fold, cutting the gutter in half where captions appear. The flush-left axis for the main title, lead-in graf, and story help balance the full-bleed photo block on the right.

When an audit of Texaco's image indicated that the time was ripe to work with third-party groups on issues of mutual concern, the company launched a new approach to constituency relations

TEXACO
WORKING WITH PUBLIC INTEREST GROUPS

THE likelihood that one of the nation's largest oil companies and one of the oldest civil rights groups would march in tandem to Capitol Hill to lobby for the same legislation would have defied the imagination several years ago.

Yet, that is precisely what happened in the fall of 1982, when Texaco Inc. and the National Association for the Advancement of Colored People (NAACP) took the same side on an issue involving credit practices in the petroleum industry. Moreover, the NAACP was joined by several other large national constituency groups in supporting Texaco and the industry on the issue, a bill they perceived to be "anti-consumer."

The story of how the *Fortune* 500's fifth-largest company and those or-

This article is based on a presentation last month by Mary Ann Pires, APR, planning and constituency relations manager, Texaco Inc., White Plains, New York, before the Breakfast Roundtable of the Public Relations Society of America's New York Chapter.

ganizations, representing some 16 million people, became public-issue allies is the story of a major public affairs program begun by Texaco in 1980 to systematically reach out to influential third-party groups. It is a program that has the potential to significantly affect the corporate bottom line.

The program grew out of a 1979 audit of Texaco's image among a variety of constituencies, conducted for the company by Fraser Associates, formerly a Washington, DC, consulting firm. A key finding of the audit was that opportunities did exist to work with third-party groups on issues of mutual concern, and that the time was ripe to do so.

With that in mind, Texaco, early in 1980, established the position of consumer affairs (since renamed "constituency relations") manager in the Public Relations and Advertising Department. Mary Ann Pires, APR, was appointed to this position with the mandate to systematically open lines of communication with influential third-party groups; in short, to develop the interaction necessary for proactive public relations/public affairs.

The consumer affairs function was situated deliberately in the Public Relations and Advertising Department, rather than in Government Relations, since its focus was to be the broad one of developing sound two-way communication with major national constituency groups, as opposed to only searching out legislative allies. From that communication, it was believed, selected public-issue support would follow in the long term. This premise has since been validated—and well beyond Texaco's most optimistic projections.

Two-way communication

The key to why it happened lies in the words "two-way communication." For one of the underlying tenets of Texaco's program was that it would be keyed to *listening,* to learning about the various organizations, and not simply to talking *at* them. In this way, any information exchanges, program development and the like would arise from the interests and needs of the organizations, and not be imposed by the company. In essence, it was a way of demonstrating at the outset respect

for the organizations with which Texaco hoped to work.

Two-way communication was only one aspect of the philosophy underlying the program. The key points included Texaco's intention, from the very beginning, to develop *long-term relationships* with the groups contacted. Running contrary to the prevailing preoccupation with quarterly results, the company recognized that relationships between institutions, just like personal relationships, are not born overnight. While this may seem obvious, it is interesting to note that as Ms. Pires met with consumer and public interest group representatives, she heard all too frequently about someone from XYZ Corporation who had been by last year, but never returned.

It was Texaco's intention to return, to strive for ongoing contact, as much as resources would permit. The company was convinced that it could not expect people to have any interest in it, in issues of concern to it, if the only time it approached third parties was when it wanted something.

Texaco also decided that it *wouldn't*

promise anything it couldn't deliver, which often led to situations where the company was unable to respond to a given request on the spot. In the long run, however, the company's credibility was enhanced by the fact that when it said it would do something, it did.

Finally, the company resolved *not to develop* "*checkbook relationships*" with the various groups. By that, Texaco meant no contributions in situations where it could not work with the organization on a mutual objective. The company believed that only within that context would the interaction take place. In the company's view, the same learning process, the same breaking down of stereotypes, does not occur when a corporation simply mails off its annual check to a group.

Armed with a good feel for the program it wanted to develop, Texaco, again using outside public affairs counsel, examined broad population segments, *e.g.,* elderly, disabled and minorities, to determine which organizations were the foremost representatives of the constituencies the company wanted to reach. This research

covered current literature as well as conversations with consumer activists, industry representatives and government officials.

Ms. Pires and her department head, Paul B. Hicks, Jr., vice president of public relations and advertising, next determined priority contacts by asking questions concerning the membership base of each organization, whether it had any interest in energy issues, customarily worked with corporations, and had an influence on public policy. If the answer to a majority of these and similar questions was "Yes," they developed a detailed profile of the organization. Ms. Pires then contacted the group, not necessarily at the top staff level, but through the individual most closely associated with its energy activities.

Personal visits with 20 constituency groups nationwide were carried out throughout 1980 and the early part of 1981. The objectives were to become familiar with each organization, its goals and needs, and to enable the group to put a "face" to Texaco—to actually meet and talk with someone from a "Big Oil Company." Later,

A Successful Businessman
Takes On The Job Of
Improving Massachusetts
Economic Climate

from business to government

In mid-December, Massachusetts' newly-elected governor, Edward J. King, appointed George S. Kariotis to the sensitive and important position of secretary of economic affairs. Mr. Kariotis had been serving as chairman and chief executive officer of Alpha Industries, Inc., a Main Lafrentz client headquartered in Woburn, Mass.

The Office of Economic Affairs bears the responsibility for improving the state's business climate, thus encouraging the growth of employment in Massachusetts, which like most of the north eastern states has suffered a substantial loss of jobs to the sunbelt states over the past decade.

The appointment of Mr. Kariotis, a successful businessman who has never held a public office before, is viewed

as an embodiment of Governor King's campaign promise to make improvement of the state's business climate a priority of his administration.

Alpha Industries, which Mr. Kariotis helped found in 1962 (with four employees), has annual gross revenues of $15 million and employs more than 500 persons. This high technology company manufactures semiconductor devices (diodes), solid state control components, hybrid thick film RF/IF amplifiers

George S. Kariotis, secretary of economic affairs for the state of Massachusetts.

17

Orchestrating 200 Projects as One

Fluor's project-management skills come to the fore when coordinating activities on a massive modularized petrochemical project in Saudi Arabia. Huge modular units weighing up to 2,000 tons each are fabricated in Japan and shipped 6,500 miles to the jobsite.

The plan is ambitious—a first: package a major petrochemical complex into more than 200 transportable modules for shipment to Saudi Arabia.

The project: a massive multibillion-dollar grass-roots, ethylene-based facility for the Saudi Petrochemical Company (Sadaf).

Fluor, as managing contractor, is coordinating the fabrication and assembly of the modules which are being constructed at two shipyards near Nagoya, Japan. Approximately 625 workers at the Tsu yard of Nippon Kokan K.K (NKK) and 1,100 at the Aichi yard of Ishikawajima-Harima Heavy Industries (IHI) will be assigned to the project at peak construction next year. One of the reasons Sadaf chose modularization over conventional construction was that a great number of the total manhours required for the project would be expended in the fabrication yards, resulting in fewer workers at the jobsite.

As managing contractor, Fluor coordinates the activities of numerous material-supply companies and three engineering firms, who provide process design for the project. Fluor is also a process design contractor for the utility plant and offsite areas.

Initiating the flow of information, Fluor's design criteria and shipping sequence are given to the other process contractors (Badger, Braun and Dravo) who, along with Fluor, design and engineer each module.

For fabrication work at the NKK

yard, Fluor and Dravo design the off-site, utility, chlorine, and ethylene dichloride modules. Badger and Braun generate design drawings for IHI's work on the ethylene, ethylbenzene, styrene and crude industrial ethanol units. In coordination with Fluor, as managing contractor, each process contractor procures the major equipment to be installed in the modules they design.

Well in advance of the issue of design drawings for module construction, Fluor provides information to the module fabricators that permits them to purchase structural steel and pipe for a large group of modules. This advance procurement precludes piecemeal buying by the contractors and assures a ready start of fabrication once approved drawings are issued.

Upon receipt of approved drawings, the fabrication contractors prepare shop drawings and purchase any remaining bulk materials required for module construction.

In a sense, the Sadaf job is more than 200 small projects in one. "This is one of the most complex project management operations the industry has ever seen," said Jack Kirven, Fluor's executive project director for Sadaf's petrochemical complex.

"This experience has enhanced Fluor's expertise. We have the ability to evaluate the most appropriate construction techniques for projects anywhere in the world."

Distance and communications are critical elements of the project. ►►

A) *Worker at IHI's Aichi steel fabrication shop marks structural steel before torch cutting begins. This is one of the first steps in module fabrication.* B) *Finished modules will be transported from IHI's module assembly area to dockside by massive gantry cranes, each capable of lifting 400 tons.* C) *Fluor personnel examine isometric drawings.*

D) *A worker at NKK's pipe fabrication shop moves an overhead crane into position to relocate a length of pipe.* E) *A worker at the NKK pipe fabrication shop prepares pipe for welding.* F) *NKK assembly-yard workers steady pipe as it is loaded on a pipeway module.* G) *At NKK's assembly yard, a worker applies a final coat of paint to hard-to-reach sections of a structural steel module frame. The sections painted white are reinforcing struts that will be cut away once the long sea voyage is over and the modules are in place.* H) *Pipeway modules in the final stages of assembly at NKK's Tsu module fabrication yard.*

PHOTOGRAPHY: PAT ROLFE

24

FIG 9–25 (below)
CH2M Hill Reports. *A similar horizontal axis is used here, where the photo module is built around a dominant photograph. The irregular photo module, made up of fitted blocks, is linked together with alleys of equal size. Captions are fitted easily into the picture group. Open white space around the pictures balances the space found on the opposite page. In this layout the gutter is acknowledged, but the overall sense of open space makes it a minor problem.*

FIG. 9–26 (top right)
Dramatic display need not be so complicated that it interferes with reading. Here we find visual gutter bridges that don't require elements to be aligned. The elegant shape of the primary graphic in the center of the left page establishes a theme that is repeated in the rounded corners of the photographs on the facing page. All typography, including a beautiful swashed title, is understated, reserved. Note how easily the story can be read, and how little it detracts from the pictures.
From *Marathon World*, No. 3, 1980. Reprinted with permission of Marathon Oil Company

FIG. 9–27 (bottom right)
This spread from an article on the famous war correspondent, Ernie Pyle, reminds us that when content is serious and the audience interested, few gimmicks are necessary to enhance reading. A simple cluster of photos, all meeting in the gutter midway down, form a crosslike axis upon which all content can be positioned. Photos bleeding on three sides of the layout lend balance and have the effect of enlarging the images. This simple layout, effective but quietly distinguished, reminds readers of the man who inspired the story.
Courtesy of Panhandle Magazine, a Panhandle Eastern Corporation publication

NOAA

AQUACULTURE LAB
DESIGNED FOR TOTAL FLEXIBILITY

STORY BY JANE OSMER

Photographs by Gene Bonham

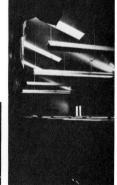

Below: Research at Newport emphasizes rehabilitation and enhancement of Pacific Northwest fish stocks, especially salmon.

Top: Tank room in the experimental wing. Working with the architect, CH2M HILL provided the HVAC systems for the lab.

Above: Laboratory research requires strictly controlled water quality and temperatures.

Above: Located in the test area, the greenhouse provides a controlled environment for experiments.

Left: A bottom fish gazes at visitors to the Newport aquarium.

Last spring, Jim Winton was performing a routine examination of chum salmon tissue at the newly completed Newport Aquaculture Laboratory of the National Marine Fisheries Service (NMFS) in Newport, Oregon. The samples were taken from 300 adult fish and airmailed from Japan as part of an importation certification procedure.

Winton, a doctoral candidate, noticed something unusual in the cell lines inoculated with material from the samples. Further testing revealed that the abnormality was in fact due to a virus, previously undescribed. Shipment of the eggs from these chum salmon was held up while the virus underwent additional testing to determine its effect on salmon.

This certification service, the only one on the West Coast, is just one of the many activities taking place at the Newport Aquaculture Lab. Owned by the National Oceanographic and Atmospheric Administration (NOAA) and operated by Oregon State University (OSU), the $2 million facility is used principally for a cooperative research program involving OSU and NMFS and for intensive teaching and research involving 11 of the university's academic departments.

Located on roughly two acres of land on Yaquina Bay, the new lab increased by a third the size of the existing Marine Science Center complex, which consists of an aquarium, offices, classrooms, laboratories and a library on a 50-acre site.

The architectural firm of Hewlett, Jamison, Atkinson and Luey, A.I.A. (HJAL) of Portland, Oregon, has been responsible for the design of all the buildings on the site. Each building complements the others, and the complex as a whole blends into Newport's seashore landscape.

The result . . .
one of the most advanced aquaculture laboratories in the country.

From the beginning, the architects worked very closely with the university. "OSU gave us a tremendous amount of valuable information which helped us outline the specific technical requirements for the facility," said Arthur L. Wilson, the HJAL project architect.

The 30,000-square-foot U-shaped building is divided into three main areas: the aquaculture development lab, the fish pathology lab and offices in a connecting wing.

The aquaculture development wing has several analytical labs and temperature-controlled rooms for specialized temperature-sensitive experimentation, as well as a large open area for fish tanks used in salmonid, bottom-fish, clam and oyster research. There are plans to include a food science and technology lab for the study of sea life nutrition. The entire second floor is devoted to study carrels and computers linked to the OSU campus in Corvallis.

The fish disease and pathology laboratory wing has separate entrances and can be converted to a complete isolation facility at a moment's notice. "We're the only fish research facility in the country to have this capability," noted Lavern Weber, director of the Marine Science Center.

Additional temperature-controlled rooms in the pathology wing store cell lines and other experimental components at 18°C. In one lab, there is a special transfer room sterilized by

the Grace of Glass

Spanning the ages, the superb collection at the Toledo Museum of Art demonstrates the power of glass to dazzle and charm

Nearly 3,500 years of glassmaking are embraced within the Toledo Museum of Art's 5,000-odd works of art in glass. Upper left: American blown-glass goblet by German expatriate John Frederick Amelung, between 1780-1790, noted for wheel-engraved inscriptions. Lower left: Gold-glass plaque, a fragment of a bowl from Italy, 4th century A.D. Latin inscription reads, "The Lord gives the law"; recipients are SS. Peter and Paul. Center: Blown iridescent goblet by Louis Comfort Tiffany, 1900-1905, the leading American exponent of Art Nouveau. Upper right: Baroque drinking glasses, Holland, 17th century, showing exaggerated artistic forms and calligraphic flourishes. Lower right: Mold-pressed necklace and headband remnants, Mycenaen Greece, 13th century B.C.

In many ways the glass collection at the Toledo Museum of Art is a tribute to mankind's achievement in this wondrous substance, so much akin to crystal—the pure creation of nature.

Of course, the glass industry of the city of Toledo, Ohio—headquarters of Owens-Illinois, Inc., Owens-Corning Fiberglas Corporation and Libbey-Owens-Ford Company—is in large part responsible for making glass a product of utility and abundance as well as of beauty.

"When one thinks of Toledo, one thinks of glass," says Otto Wittmann, director emeritus of the museum and a man who presided over the design and installation of the superb glass gallery there. "There is no better home for such a collection than the Glass City itself."

Certainly it was the capable and enthusiastic assistance of glass industry leaders like Edward Drummond Libbey and Harold Boeschenstein that first created, then nourished the museum's collection. In addition, there has been broad-based support for the museum among other businesses and individuals. Marathon Oil Company has been a consistent backer, by way of contributions made through the Marathon Oil Foundation.

The collection is virtually without peer, a source of pride and distinction for the city. "The permanent glass collection in Toledo is generally recognized as one of the finest in the United States," writes Connoisseur magazine. The museum's gifted young director, Roger Mandle, says unhesitatingly, "Our glass, so rich and varied, is an important international resource." Certainly the collection figured in a recent commendation of the museum by ARTnews, for consistently discriminating connoisseurship.

But as this magazine recognized, the Toledo Museum of Art is much more than a gallery of glass. It is an imposing example of neo-classic architecture, a unique teaching facility, and the repository of a wide range of historical art styles and media.

Having said this, it's a temptation to put the museum on a pedestal. And yet there is nothing exclusive or aloof about Mandle's stewardship of the facility. There is no admission charge and in all things Mandle has championed a grassroots approach to art.

"This is what first kindled my interest in the museum," says Elmer A. Graham, Marathon's senior vice president, Finance and Administration. "It is a rare thing when one of the top 10 museums in the country deviates from tradi-

Left: Mold-blown head vases, Rome, 1st to 3rd centuries A.D. Note the vase in the center, supported on a pedestal with a mirror underneath to reveal a face wheel-cut on the bottom. Right: Unguent jar, Egypt, 1570-1340 B.C. Representative of Egyptian sand-core vessels, among the earliest reliably dated.

tional emphasis on display and undertakes a program of outreach to area communities."

Graham, who now serves on the museum's board of trustees, Mandle and others helped shape a program that developed a satellite constituency of art patrons in communities as far as 50 miles from Toledo. These patrons are at once informed and entertained by lectures, tours and showings designed with their particular interests in mind. Unfailingly, there is an interest in the museum's glass.

The glass collection has been lovingly compiled from bequests and purchases that range from a single vase conferred by Colleen Moore, an actress of the silent screen, to a rare Islamic vessel used for years as a family cookie jar, to the

acquisition in 1959 of 55 choice pieces from the George McKearin Collection, widely held to be the most famous group of American glass.

The sensitively designed exhibit area itself befits the superlative nature of the museum glass. While the museum was founded in 1901, it was not until 1970 that Toledo had a gallery worthy of its rich endowment in glass.

The glass gallery, says Wittmann, is designed to invite and instruct. "It has a welcoming aura; it's beautiful and fluid like glass itself."

Appropriately enough, a visitor's first glimpse of glass here is heralded by a unique, monumental polychrome mural, executed especially for the museum by Dominick Labino, honorary curator of glass. To observe the symphonic succes-

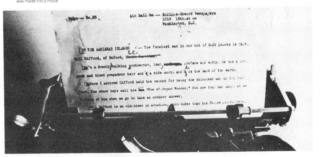

Left: Enameled mosque jar, Egypt, 14th century A.D. Ancient glass was owned only by the wealthy. Right: Mold-blown, cut-glass punch bowl set, United States, 1903-1904, by John Rufus Denham. The largest punch bowl of its kind, the piece is from the "Brilliant" period, 1880-1915, which favored spoked, heavy-cut wares.

21

Pyle's boyhood home (at left) has been established as a memorial to Ernie. The ground floor is filled with family and period furnishings. There is a basement museum built around displays of memorabilia from Ernie Pyle's wartime activities and boyhood.

An original manuscript and the typewriter it was written on (below) evoke the memory of war correspondent Ernie Pyle.

Ernie (bottom) at work at his "office" in Normandy, France. Pyle's writings won him a Pulitzer Prize. He was also the author of several books, one of which, "The Story of G. I. Joe," was made into a movie.

Ernie's travels took him into every state of the union at least three times, and to Canada, Mexico, Alaska, the Hawaiian Islands and Central and South America, usually with Jerry at his side.

In 1940 Scripps-Howard sent Ernie to England to report on the war, and his popularity began with his columns on the fire bombings of London. Following a visit back to the states to visit his ailing wife, Pyle returned overseas, joining our troops who were training in Northern Ireland and England. Then came the invasion of North Africa. Ernie arrived on the scene shortly afterward.

It was in Tunisia that Pyle's love affair with the infantry bloomed. Here he saw for the first time the incredible conditions they endured; the loneliness, the fear, the cold, the boredom — and the courage with which they faced up to all of these, and a well-trained German army as well.

Like the boys he wrote about, Ernie experienced the ambivalence of war. He wrote: "I can't deny that war is exhilarating. The whole tempo of life steps up, both at home and on the front. There is an intoxication about battle, and ordinary men can sometimes soar clear out of themselves on the wine of danger-emotion." And later he would confide to a friend: "I hate to go back to the front . . . I dread it and I'm afraid of it."

But back he went. All in all he spent 2½ years in Europe, from North Africa to Sicily, to Italy, to the beaches of Normandy on D-day plus one, and on to France where he saw the liberation of Paris before returning to his home in Albuquerque for a well-earned rest.

Pyle was constantly trying to convince his friends and other

correspondents that he was "through" — drained of his creative juices, ready to call it quits because he just couldn't stand the sound of another artillery shell or the sight of another bleeding guy. And then a short time later he would turn up with another assault force — all the time protesting that he just couldn't take any more of it.

In 1945 Ernie reluctantly went back to the war again, this time to report on the Pacific Theater. As usual, he complained, but rationalized his action with the logic he had used throughout the war: "What can a guy do? I know millions of others who are reluctant, too, and they can't even get home." He island-hopped for awhile and went in with the seventh wave of Marines in the Okinawa landing. On April 16 the 77th Infantry Division assaulted an outlying island, Ie Shima, a 10-mile-square piece of geography that merits a

footnote in history only because of Ernie Pyle.

Fighting on the island had been furious. Two days after the landing Ernie and other correspondents took a boat to the island, and the following day Pyle, two officers and a pair of enlisted men took off in a jeep for a forward post. At a junction on the narrow road a Japanese sniper opened fire with a machine gun. The men quickly abandoned their vehicle and dove into shallow ditches along the side of the road. When Pyle raised his head to locate his companions the sniper's bullet found its mark in Ernie's left temple.

Pyle was buried on the island. After the war the body was moved to an Army cemetery on Okinawa, then to the National Memorial Cemetery of the Pacific in Punchbowl Crater, near Honolulu.

The dogfaces didn't forget their writer-friend. At the site of his

death the GI's put up a crude marker bearing the words: "At This Spot The 77th Infantry Division Lost a Buddy, ERNIE PYLE, 18 April 1945." Later the marker was replaced with a monument, similarly inscribed.

Dana, situated in far western Indiana barely a half mile north of Panhandle Eastern's mainline transmission system, also remembers its favorite son. In 1973 when Pyle's birthplace — abandoned and badly vandalized — was in danger of being razed, the Indiana American Legion launched a money-raising campaign to preserve the two-story structure. Two years later the house was moved to the main street of Dana near the edge of the business district and restored to near its original condition. It was formally dedicated as a state memorial in July 1976, and is open to the public year round.

23

24

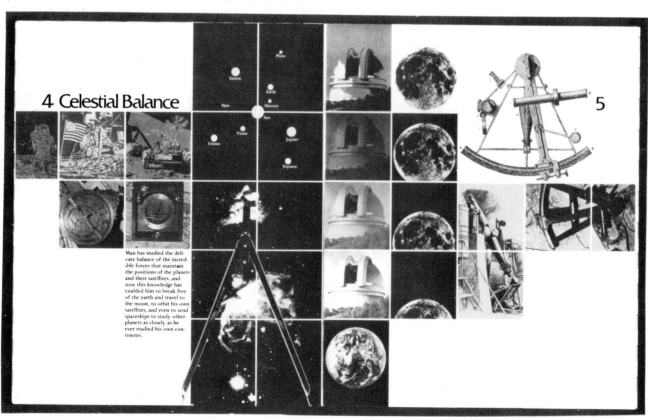

FIG. 9–28
"Celestial Balance" brochure grid.

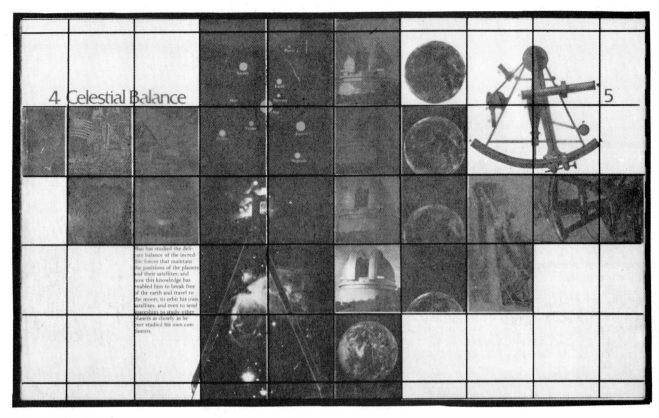

FIG. 9–28 (continued)

Extending the Principles to Other Publication Formats

Nonmagazine formats, such as those used for brochures and booklets, require many of the same design considerations as magazines. They must present material clearly in spreads, with accent on graphics and typography. Usually, each spread has its own titles and subtitles, one or more copy blocks, illustrations or photographs, and captions.

The grid patterns in Figures 9–28 and 9–29 show two completely different approaches to organizing brochure content by the grid method. The first, a booklet by a major paper manufacturer, employs a 5- by 7-unit grid per page. The design allows 20 full units per page, plus partial units on all sides for margins. The main objective of the brochure is to highlight how a large number of different graphic effects will reproduce on a certain text paper. This requires a very flexible grid that places almost no restrictions on shapes and sizes.

The second booklet, printed with both gloss-coated book and text paper, requires a grid with split

personalities. Because of wide variation in the sizes of product photographs, a 3-column, 12-unit grid was selected for pages printed on gloss paper. The grid actually is broken down farther—into a 6 by 12, 72-unit structure. This allows the use of both a 2-column square layout and one with three vertical units instead of four.

The text paper pages opposite the photographs on the gloss paper contain the body copy in an easy-reading, 2-column format with wide outer margins and ragged right type. The contrast between the crisp photographs and the soft type pages is reinforced by both type and paper. Since the text and gloss paper switch sides on half of the spreads, the design grid is simply flopped over to match.

Grids Are the Beginning, Not the End

Content for brochures is best selected for the way in which it meets certain advertising and marketing objectives, not for how it fits into some predetermined design pattern. One of the most important things to remember about brochures is that they usually work best when they are tied together with some kind of

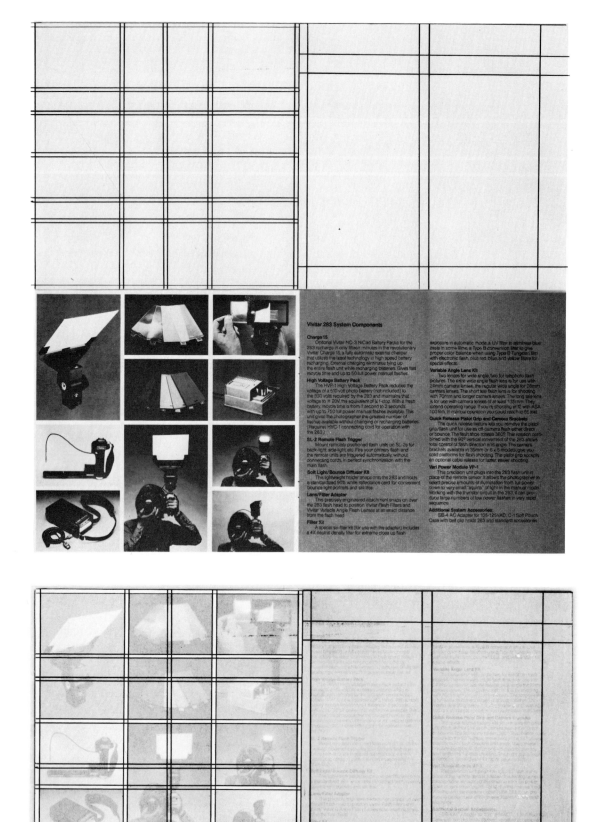

FIG 9–29
''Flash'' brochure grid.

design thread—a continuous visual link between all spreads and pages. A grid may help do this, but other things may do the job as well. Consistent styles and sizes of type, repeated color bars or tints, common backgrounds, and similar rules running at the top or bottom of pages or between columns provide the continuity a brochure requires to do its job.

Figures 9–30 to 9–37 show several other ways to achieve design continuity in brochures.

Newsletters—Personalizing the Design

Of all the design formats we have mentioned, the one most likely to be attempted by the novice designer is the newsletter. This special category of publication requires a few reminders about content. Most newsletters:

• are just that . . . a personal medium of news communication designed to reach an especially interested audience;

• attempt to convey their content in brief, sometimes friendly, always informative units;

• provide "inside" information that readers can't get elsewhere;

• are neither newspapers nor magazines, and don't have to look like either.

Some Design Tips for Newsletters

Since nearly all newsletters use the standard 8½ × 11 inch paper page format, it is easy to adapt magazine grid designs to fit their layouts. This may not be wise, however.

With newsletters it is especially important to consider the reading audience and how it might best be reached. Slick is not always better. Pictures and fabulous graphics will do little to inform readers, if it takes weeks to produce them. Never forget—this is a NEWSletter. An up-to-the minute, well-written, highly informative 2-pager is infinitely better than a 16-page design prizewinner built around last month's "news."

FIG. 9–30

The repetitive graphic. *In a delightful rerun of the Orient Express, Weyerhaeuser Company never lets us forget the theme. Spread after spread, the cars of the famous train stream by the bottom of the pages alongside a map of the route. A single-column format presents both ornate typography and multicolor illustrations to good advantage.*

FIG. 9–31
The promotional brochure for a daily newspaper uses a similar technique, the repetition of two horizontal rules against which design elements can be positioned. Rules bleed on both sides.
Courtesy of The Register, Santa Ana, CA

FIG. 9–32 (bottom)
The Clothesline. *An effective technique that organizes but doesn't restrict is the use of a clothesline, a horizontal axis onto which columns of content can be "hung." In the ballooning brochure, all elements in the spread align across the top (except for one figure in the background). The bottoms of columns are ragged, but this doesn't matter. The horizontal alignment completely organizes the spread and allows the white space so necessary for this subject.*
Raven Industries, Inc., Sioux Falls, SD

FIG. 9–33 (opposite, top)
This spread depicting applications for an electronic records system for restaurants employs the clothesline technique most attractively. Columns hang neatly over a display of succulent dishes, barely intruding on the dining atmosphere.
ATV Systems, Inc., Santa Ana, CA

The Register reaches more spenders.

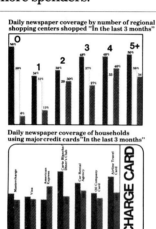

The Register reaches 44-percent of Orange County households who use credit cards. The Times reaches only 36 percent.

And whether you're reaching for active shoppers—ones who shop one, two or five shopping centers—The Register reaches more of them than The Times.

The Register also reaches more of the people who say they don't shop any shopping center.

No matter what measure of spending propensity you use, you'll find The Register reaches more Orange County spenders than The Times.

Key: The Register (total) ■
The Register (exclusive) □
Los Angeles Times ■
The Register (Exclusive) denotes that portion of
The Register's readership which is exclusive of
the coverage of any other newspaper.

Source: The Register's 1980 Consumer Attitude Survey
conducted by Facts Consolidated.

Because many of the people 45 and older have minimal mortgage payments and have long since acquired the "essentials" of suburban life, much of their spending is elective.

The wealthier ones, those with an annual income over $45,000 have a very high propensity for high ticket spending (expensive jewelry, cameras, travel). And to a greater extent than any other age group, they tend to patronize the finer fashion department stores.

The Register reaches 44 percent of the households with over $35,000 annual income. And more than 43 percent of the households with $50,000 or more annual income.

Raven puts your company above the rest of the world.

The mystical lure of a balloon floating lazily over a public gathering COMPELS attention. When it carries your company's message, a Raven balloon becomes your powerful billboard in the sky. You'll reach thousands of prospective customers with every flight. Our specialists can reproduce almost any logo, slogan, or design in full color. In terchangeable messages can be attached with Velcro fasteners. With our exclusive Superpressure System (optional), you'll tether in winds 15 mph and higher. We've put more companies above the rest of the world than anyone else. Why not yours?

These (and many other) national and regional companies have chosen a Raven Flight System:

Air Canada • Anheuser-Busch, Natural Light • Arby's • Budweiser • Canada Dry • Century 21 Realtors • Coca-Cola • Columbia Savings • Coors • Country Kitchen • Design Master Homes • Diamond Shamrock • Disney World • Drummond Brothers Beer • Export A Cigarettes • Fafca • Falls City Beer • Forbes Magazine • Freedom Federal • General Electric • Git-N-Go • Kawasaki • Kool Cigarettes • Labatts Beer • Lark Cigarettes • Male Slacks & Jeans • Marrocaine Swiss Cigarettes • Massey Ferguson • Monroe Shock Absorbers • Montana Mining Co. Restaurant • Mr. Pibb • National Industries • Paine-Webber • Pentax • Porter Paints • Purity Bread • Schlitz Light • Sprite • Toyota of Canada • United Jersey Banks • United States Navy • U-Totem • Wendy's • Yellow Cab • Yellow Pages

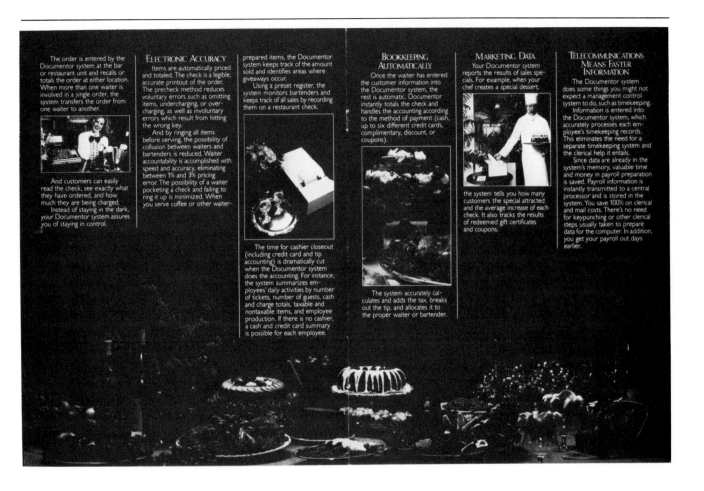

The order is entered by the Documentor system at the bar or restaurant unit and recalls or totals the order at either location. When more than one waiter is involved in a single order, the system transfers the order from one waiter to another.

And customers can easily read the check, see exactly what they have ordered, and how much they are being charged.

Instead of staying in the dark, your Documentor system assures you of staying in control.

ELECTRONIC ACCURACY

Items are automatically priced and totaled. The check is a legible, accurate printout of the order. The precheck method reduces voluntary errors such as omitting items, undercharging, or overcharging, as well as involuntary errors which result from hitting the wrong key.

And by ringing all items before serving, the possibility of collusion between waiters and bartenders is reduced. Waiter accountability is accomplished with speed and accuracy, eliminating between 1% and 3% pricing error. The possibility of a waiter pocketing a check and failing to ring it up is minimized. When you serve coffee or other waiter-

prepared items, the Documentor system keeps track of the amount sold and identifies areas where giveaways occur.

Using a preset register, the system monitors bartenders and keeps track of all sales by recording them on a restaurant check.

The time for cashier closeout (including credit card and tip accounting) is dramatically cut when the Documentor system does the accounting. For instance, the system summarizes employees' daily activities by number of tickets, number of guests, cash and charge totals, taxable and nontaxable items, and employee production. If there is no cashier, a cash and credit card summary is possible for each employee.

BOOKKEEPING AUTOMATICALLY

Once the waiter has entered the customer information into the Documentor system, the rest is automatic. Documentor instantly totals the check and handles the accounting according to the method of payment (cash, up to six different credit cards, complimentary, discount, or coupons).

The system accurately calculates and adds the tax, breaks out the tip, and allocates it to the proper waiter or bartender.

MARKETING DATA

Your Documentor system reports the results of sales specials. For example, when your chef creates a special dessert,

the system tells you how many customers the special attracted and the average increase of each check. It also tracks the results of redeemed gift certificates and coupons.

TELECOMMUNICATIONS MEANS FASTER INFORMATION

The Documentor system does some things you might not expect a management control system to do, such as timekeeping.

Information is entered into the Documentor system, which accurately processes each employee's timekeeping records. This eliminates the need for a separate timekeeping system and the clerical help it entails.

Since data are already in the system's memory, valuable time and money in payroll preparation is saved. Payroll information is instantly transmitted to a central processor and is stored in the system. You save 100% on clerical and mail costs. There's no need for keypunching or other clerical steps usually taken to prepare data for the computer. In addition, you get your payroll out days earlier.

FIG. 9–34

The inset. *Visual impact can be enhanced where illustrations are superimposed, especially where the background image is symbolic or creates an interesting pattern. The insets in this business computer brochure enliven the design.*
Microdata Corporation

data. The manipulation, reporting and analysis of data held locally can be accomplished quickly and at a low cost with Sovereign. Only the results of processing may need to be consolidated with that of other business units via data communications or other exchange media.

The Processing Terminals in a Sovereign system can work independently or they can work together to perform a wide range of processing assignments. The File Processor in the Sovereign system has its own operating software for main-

taining, sorting, updating, retrieving and outputting records and files. Each terminal can be devoted strictly to assigned applications, working with the File Processor and a common data base of information.

User-programming is accomplished with either Sovereign's enhanced version of the BASIC programming language or an advanced version of ANSI 74 COBOL, specifically designed for small computers. These two internationally recognized languages open the way to local programs, from the simple summary to sophisticated business application development.

LOCAL DATA BASE INQUIRY.

Sovereign offers another unique feature, an inquiry command language that allows the non-technical user to access a common data base of business information. This terminal-oriented language consists of relatively free-form English sentences containing verbs, file-names, data selection criteria and control modifiers. This inquiry capability provides for selective retrieval of information. It also provides for automatic report generation whereby output may be sorted in any defined sequence.

COMMUNICATIONS TO MAINFRAME COMPUTERS.

The Sovereign system is designed to give the best of both worlds. On the one hand, it is fully capable of independent data entry, processing and inquiry, while on the other, the system provides for batch and/or interactive communications with mainframe computers. Popular IBM disciplines are available and those of other CPU systems have been implemented as well. Further, Sovereign can communicate concurrently to more than one mainframe computer or other Sovereign and Microdata systems.

Each Processing Terminal in a Sovereign system can act as a communications processor. This allows data to be transmitted to and received from other remote computer systems.

Verified and corrected data, entered and stored in the Sovereign system, can be transmitted to remote mainframes. Similarly, print files from the mainframe can be transmitted back to the Sovereign Processing Terminal. Depending on the protocol, Sovereign also allows data within the central computer master files to be accessed in batch or interactive transaction mode from any of its Communications Processing Terminals.

The result is a virtually instantaneous exchange of business data between Sovereign and the central processing facility.

FIG. 9–35

TEMPO ACTION. *This political action group selected a newspaper-like format for its publication. The use of column rules, a bold nameplate, flush-left heads, and a political cartoon makes this a no-nonsense medium.*
Tiger International

FIG. 9–36

WIIS NEWS. *A nice variation on the standard newspaper format is this insurance newsletter. Note the flush-right side headings that help to index the content. An attractive floating nameplate is placed beneath the index at the top.*
Western Insurance Information Service

FIG. 9–37

DAYTIME: PM. *This insurance newsletter uses another variation on newspaper design—the horizontal module. This unusual example with its vertical nameplate (and vertical section headings inside) is a delight to read.*
Pacific Mutual Life Insurance Company

NOTES

[1]Roger F. Fidler, "A Checklist for Functionally Integrated Design," *Newspaper Design Notebook*, 2, no. 1 (January–February, 1980), p. 9.

[2]The term "optimum" format, introduced by the well-known newspaper designer Edmund C. Arnold, in his book *Modern Newspaper Design* (New York: Harper & Row, 1969), refers to column widths which provide the best possible readability for newspaper body type.

[3]Mario R. Garcia, *Contemporary Newspaper Design: A Structural Approach.* (Englewood Cliffs, N.J.: Prentice-Hall, Inc., 1981), pp. 4–5.

[4]Edmund C. Arnold, *Functional Newspaper Design* (New York: Harper & Row, 1956), p. 166.

[5]Garcia, *Contemporary Newspaper Design*, p. 40.

[6]Garcia, *Contemporary Newspaper Design*, p. 45.

[7]Garcia, *Contemporary Newspaper Design*, pp. 45–48.

[8]Allen Hurlburt, *The Grid: A Modular System for the Design and Production of Newspapers, Magazines, and Books* (New York: Van Nostrand Reinhold Co., 1978), p. 36.

[9]Ruth Clark, *Changing Needs of Changing Readers: A Qualitative Study of the New Social Contract Between Newspaper Editors and Readers.* Yankelovich, Skelly & White, Inc., May, 1979, pp. 29–33.

[10]J. W. Click and Russell N. Baird, *Magazine Editing and Production*, 3rd ed. (Dubuque, Iowa: Wm. C. Brown Company Publishers, 1983), pp. 5–6.

[11]Hurlburt, *The Grid*, pp. 58–60.

[12]Donald E. Hill, *Techniques of Magazine Layout and Design*, 2nd ed. (Huntsville, Ala.: Graphic Arts & Journalism Publishing Company, 1972), pp. 11–13.

10 Pasteup
Assembling Text and Graphic Images for Reproduction

Rapid changes in the world of graphics production have reduced the need for pasteup work by hand. Still, there is no substitute for the sense of control and the satisfaction you derive from physically handling the task. Every headline, every illustration, every column of type yields to the slightest adjustment. Finally, the finished assembly appears, looking very much like the printed piece that will follow. In this chapter we provide a few tips that should help make pasteup easier.

Introduction

Once the design for a printed piece is established, type is set, and illustrations are produced, the pages are assembled and every element is affixed in its proper position.

A page, assembled on paper or board, may be called by various names—a pasteup, a flat, a keyline, or camera-ready art. Whatever its name, it normally consists of a mechanical assemblage of page elements into a form suitable for printing.

Despite the recent strides in electronic page assembly discussed in Chapters 6 and 7, traditional methods will continue to be used for some time, especially for inexpensive, low-volume work. Fully assembled pages contain some or all of the following:

Galleys—paper strips of body type, usually typeset in columns;

Display type—headlines or titles, usually typeset but sometimes lettered by hand or with transfer type, and photographed for pasting;

Photograph areas—windows, formed by red or black patches or outlines, where photogrpahs or other continuous tone images are to appear. In lower quality publications, diffusion-transfer or velox prints containing the screen-dot image are pasted directly to the page;

Line images—nonhalftoned art, borders, rules, and other images in addition to type and halftones. These are usually solid, but sometimes consist of areas containing screen patterns and other mechanical tinting film or tapes;

Film overlays—usually cut from peel-apart film or acetate, and used to separate added ink colors and tints from the predominant color (usually black);

Tissue overlays—vellum or tracing tissue sheets, placed over finished pages. Printing instructions are marked on these overlays;

Cover sheets—outer overlay of heavy kraft or cover paper, for neatness and protection.

Determining the Appropriate Pasteup Standard

The quality of a printed image depends greatly upon the care with which the original was prepared and processed. Poorly pasted pages yield unprofessional results. As a rule, trim and affix every element to the page neatly and carefully. Keep the working surface clean and place protective overlays over finished work.

Not all jobs require the same standards of preparation. For comparison, it might be helpful to imagine three levels of quality, based on the intended end use of the product and time allotted for production:

Level 1

Typical products: Small newsletters, resumes, reports, flyers, posters.

Common materials and procedures: Rubber cement or spray adhesive on heavy board or blue-lined graph paper. Text is typewritten with film ribbon, headings are hand-lettered (in dense black) or transfer type. Line art is inked, borders and rules are applied from sheets or rolls of tape. Tissue overlay is optional. Usually done quickly at low cost for relatively small audiences. Often photocopied for economy rather than printed.

Disadvantages: Because less expensive tools and materials are commonly used, several problems may appear in the final product. These include uneven density of reproduction, crooked pasteup, shadow lines resulting from pasting materials of differing thickness, or chipped transfer type.

Advantages: Fast, easy, economical, and yields good quality with careful work.

Level 2

Typical products: Newspapers, commercial flyers, single-color brochures, inexpensive catalogs.

Common materials and procedures: Wax or liquid adhesive on sheets or boards often preprinted with blue grids or column guidelines. Body type and headlines often are set on the least expensive photographic typesetting paper. Halftones appear as prescreened positive prints on photographic paper. Line art is photographed and pasted in positive printed form. Borders or rules are applied as tape or from preprinted border sheets. Tissue overlays are rare for newspapers due to speed required for handling. Products are intended for mass audiences, relatively brief useful life, or where an economical look is acceptable.

Disadvantages: Frequently crooked pasteup and uneven image density (especially correction lines), crooked rules and borders, unequal spacing between headlines, body type, borders, and rules. Newspaper pasteup tends to be of lower quality because of deadline pressures.

Reproduction detail of photos may suffer due to coarse (65- to 100-lines per inch) screens required when prescreened prints are rephotographed from pasted pages and printed on newsprint.

Advantages: Speed, adequate quality, relative ease of production training.

Level 3

Typical products: Magazines, color brochures, annual reports, promotion pieces, and other "slick" publications.

Common materials and procedures: This level of page assembly is often accomplished today by area composition or pagination. Full pages are typeset completely as single pieces after having been composed on the VDT screen. Pages may be assembled as film, not paper, on film grids.

Quality varies from very good to excellent, depending upon the nature of the publication. Advertising and public relations agencies and design houses usually furnish high-quality duplicates of complete ads ready for pasteup, or they may produce a copy on film ready to expose to a printing plate.

Type usually is set on high-quality paper or film. Due to longer production schedules, most type corrections are accomplished electronically before the final page is output, eliminating most paste-overs. Windows or blanks are provided in picture areas, allowing finely detailed film halftones to be exposed separately to the printing plate in those areas. All pasted matter is absolutely straight and is spaced according to well-planned formats or grid designs.

Disadvantages: Higher costs, materials furnished to printers by customer—especially prescreened photographs and art—sometimes do not meet reproduction standards of publications or requirements for the printing process used.

Advantages: Highest quality, good preservation of assembled materials due to care used in their production.

Tools and Materials

A wide variety of tools are used by pasteup professionals. When deciding what pasteup tools are needed, we are like the do-it-yourself mechanic. Simple or infrequent jobs may require only the average assortment of items found in a home toolbox. The more work we do, however, the more we realize how much easier the jobs become with the tools especially designed for each task.

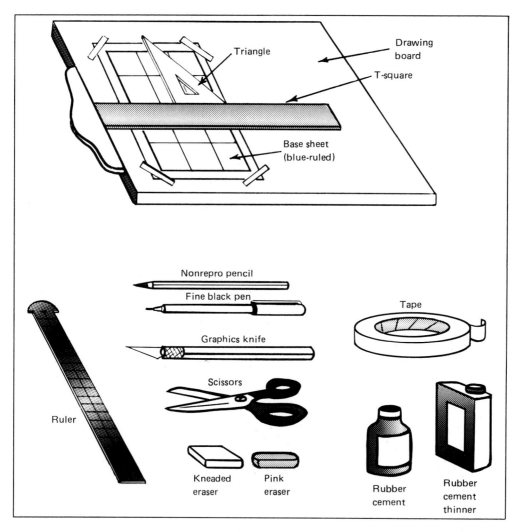

FIG. 10–1
"Home toolbox"
pasteup items.

Basic Materials

The beginner can paste up simple jobs with a few "home toolbox" items shown in Figure 10–1. They include the following basics:

1. Graphics knife. The X-acto No. 1, with No. 11 blade, is popular, but other brands and styles exist. Some people resharpen blades on a whetstone, but most users keep plenty of fresh blades handy.

2. Adhesive. Either rubber cement plus thinner, or spray adhesive is most common. Avoid white glue or any waterbased adhesives which may cause paper to curl. Cement can be thinned one part thinner to two or three parts cement for two-surface (dry) pasting. Slight thinning may be required for single-surface (wet) pasting. When dry pasting is done, the thinned cement is applied to both surfaces and allowed to dry. When the surfaces are then carefully brought into contact, a flat, even bond results.

3. Scissors (shears). A new pair with sharp, 8-inch blades is a good investment.

4. Metal ruler. This is helpful for measuring and cutting with a graphics knife. A 12-inch size marked with inches and picas is fine, but an 18-inch size will come in handy occasionally.

5. Erasers. Both hard (pink) and kneaded (gray, moldable) varieties are needed.

6. Masking or drafting tape. The 1-inch or ¾-inch width is fine. White art tape gives your work a more professional look.

7. Blue pencil. This must be light, nonreproducible blue. Some so-called "nonrepro" pencils or liquid pens produce dark blue lines that may show up when a page is photographed. If in doubt, buy the lighter varieties.

8. Baseboard. Most smooth, white or blue-lined board and papers will work; however, for best handling and permanence, a substrate of at least postcard thickness is best. The new ruling and mechanical boards now available are excellent, but hot press illustration board is also popular.

9. T-square. Inexpensive 12-inch or 18-inch wood or plastic types are readily available. The cheapest plastic varieties can be pretty inaccurate, so spend a little extra and get a solidly built style. Many people like the transparency of some plastic squares. Metal T-squares are the best and also the most expensive. Check for squareness of any T-square on the "true" edge of a table or board designed for pasteup.

10. Triangle. A 12-inch, 30-, 60-, 90-degree plastic triangle is recommended. Some are tinted to show up well on a table, others have measuring scales etched into the sides, and some even have handles. Again, metal versions which provide a cutting edge are top-of-the-line—and expensive.

11. Square-edged board or table. Most people required to do pasteup have access to some form of drafting table or light table with a square edge. If you do not, a small drawing board is needed. Portable wooden or plastic versions are very handy. Several styles now include a built-in, sliding T-square or parallel bar.

12. Fine-tipped black pen. Expensive technical pens capable of inking very fine lines in india-type ink are standard tools of the professional. However, many high-quality ball-tipped "designer" pens or liquid markers are now made which produce very fine, reproducible black lines for a fraction of the cost of technical pens.

These twelve items above represent a reasonable minimum of materials for basic pasteup. Adequate pasteup of uncomplicated material is done every day on blue-lined graph paper with only scissors, rubber cement, clear tape, and a plastic ruler. The items suggested in the basic list make the task easier and the results more suitable for printed reproduction.

Beyond the Basic Tools

For those who do considerable pasteup of camera-ready work, most of the following items, depicted in Figure 10–2, are necessary to save time and produce more professional results.

FIG. 10–2
Some professional tools.

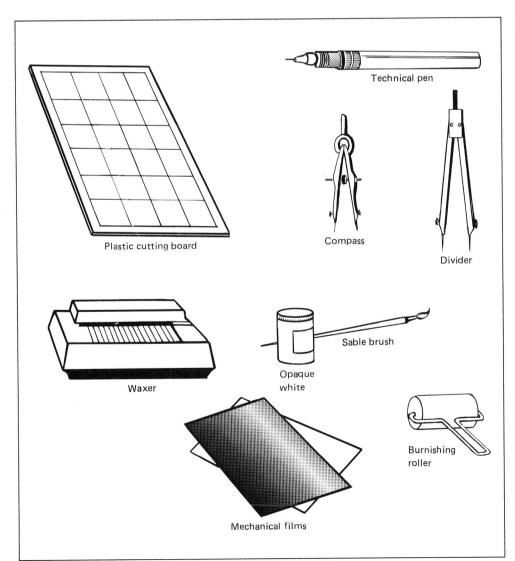

Technical pen

Compass

Divider

Plastic cutting board

Waxer

Opaque white

Sable brush

Burnishing roller

Mechanical films

1. Cutting board. One of the new self-sealing molecular plastic cutting boards will help keep knife blades sharp and eliminate problems of replacing other cutting surfaces which tend to retain slice marks.

2. Technical pens. While often used for general drawing, these pens are also most useful in pasteup where clean, consistent black rules, solid areas, and guidelines must be inked. They are filled with drawing ink and have interchangeable tips which allow many thicknesses of rules to be drawn. Tips which produce a range of thicknesses from very fine (00) to medium (2) will suffice for most purposes.

3. Mechanical drawing instruments. A couple of these are good to have: compass—for drawing circles in pencil or ink, and a divider—for measuring and transferring precise distances from layout to pasteup.

4. Waxer. This is a device that applies a thin coating of melted wax to the back of type galleys and other pieces to be pasted to the page. The wax holds down the strips securely and allows repositioning more readily than rubber cement.

5. Opaque white paint or correction fluid. Used for covering up unneeded lines and blemishes, and for "painting" around the edges of pasted pieces to prevent shadow lines in photocopying.

6. Mechanical films. These special films are used where large solid areas, windows, or mechanically produced images are needed and otherwise would have to be drawn in ink. Two types are especially useful:

> **a.** Red masking film (Ulano's Rubylith is one widely used brand). It is preferred by some printers over the orange (Amberlith) type because they believe it photographs more easily. The two-piece film is made up of a soft, translucent red layer on top of an acetate base. The user cuts the top (red) layer to the desired shape using a straight or swivel-tip graphics knife, then peels away the excess red material.

> Because the red layer is translucent, it allows us to follow the outlines of any artwork below in cutting out the proper shapes of images. The result is an acetate overlay containing one or more red images that now can be photographed in the same manner as original art. Since red photographs as easily as black with high contrast orthochromatic graphics films, a clear, sharp negative image is produced.

> Once a plate is exposed from the negative, any color of ink can be used to print the images.

Chapter 12 discusses the use of these films for mechanical color separation.

> **b.** Blockout film—This self-adhesive, opaque red or black film is cut and attached directly to a pasted page to create clear negative "windows" for halftones. This is done when photographs are to be made into film halftones rather than into screened positive prints. When the pasted page is photographed, the blockout film creates clear areas behind which film halftone negatives are taped. Blockout films can be used to create any large solid area as well, making tedious inking of these areas unnecessary.

7. Pasteup table. Many types are sold. Light tables with fluorescent bulbs to illuminate the pasteup sheet from beneath are especially good where pasteup is being done on gridded paper or film, since the grid pattern shows through for easy alignment. Pasteup tables often contain backboards for attaching layout copy and drawers for filing. Figure 10–3 shows several common types of tables used for pasteup. Some highly adjustable pedestal models offer stylish comfort at a premium price.

Look for features which will make production easier and more comfortable after a few hours at the board. Some tables have built-in T-squares, others allow easy attachment of a drafting machine. Perhaps a parallel board—a drawing board with built-in, sliding horizontal bar—is best for you. They are cheaper and can be placed on any table. A good adjustable-arm lamp will make any table a more efficient tool.

Step-By-Step Pasteup Procedure

Figure 10–4 shows the basic steps required to paste up a typical newsletter page. This is only one way to do the job. You may find ways that work better for you. Many of these steps apply to other kinds of pasteup as well.

Step 1. *Establish a plan.* Draw a full-size layout or dummy. A well-drawn layout is very helpful in guiding both typesetting and pasteup. It can be drawn on layout bond or tissue, with pencil or markers. It should be detailed enough to guide the accurate spacing of type as it is set and the positioning of elements when they are pasted. It should NOT be a work of art. It is like a blueprint for the job. Mark sizing, spacing, and positioning of each element clearly enough that it can be followed without additional instructions. Review the layout procedures in Chapter 8.

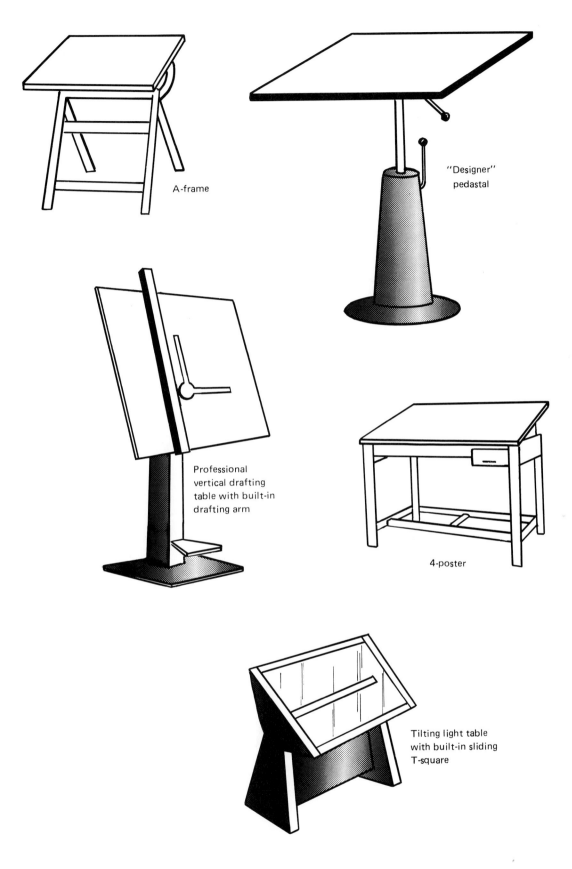

A-frame

"Designer" pedastal

Professional
vertical drafting
table with built-in
drafting arm

4-poster

Tilting light table
with built-in sliding
T-square

FIG. 10–3
*Five popular styles of tables for drafting, layout, and
pasteup.*

1. Establish a plan.

2. Organize the work.

3. Square the board.

4. Mark the page area.

5. Trim and wax.

6. Position elements.

Blue lines drawn on type help position it on the grid.

NEWSLET

7. Recheck.

Temporem autem quin saepe eveniet ut er rep earud rerum hic tenetur asperiore repellat. Han ad eam non possing acc memorite it tum etia er fier ad augendas cum civiuda. Et tamen in bu ned libiding gen epular umdnat. Improb pary dodecendense videant

omning null siy caus p facile explent sine julla sit amet, consectetur a incidunt ut labore et do minimim veniami quis nisi ut aliquip ex ea co in is prehenderit in vol dolore eu fugiat nulla dignissim qui blandit

Plant Dedic

8. Apply finishing touches/burnish.

NEWSLETTER

PRO
GOA

Cover sheet

FIG. 10–4
Steps in pasting up a page.

Step 2. *Organize the work.* Arrange all tools and materials for pasteup. Try to have all elements typeset or photographed prior to beginning the work, but don't be surprised if a few modifications in type or art are needed along the way.

Step 3. *Square the board.* Place your base sheet, board, or grid on your table and straighten it, using a T-square or other parallel slide. If you are working with a blank sheet, simply straighten the top edge. If you are working with prelined sheets, position it so that horizontal lines are parallel to the T-square. Attach the corners of the sheet to the table with masking tape.

Step 4. *Mark the page area.* Use a T- or slide-square and triangle to draw a blue rectangle defining the full working area of the page. This is sometimes referred to as the paper page or the live area. If the sheet is preprinted with a blue grid, it's still a good idea to draw a blue rectangle over the grid to define your exact working area.

Draw a set of thin black marks at each corner. These marks are extensions of each blue-lined side, but are placed about ⅛ inch away from the corners of the rectangle. Later, when the completed page is photographed, the blue rectangle will disappear, but images from the black marks will show up on the negative. They define each corner and help the printer position the negative on a masking sheet for plate burning.

Some printers prefer that a red rectangle, not blue, be drawn to define the page. They reason that negatives can be positioned on a masking sheet more easily when the full rectangular image, left by the red lines, is present. These lines on the negative can be covered easily after positioning so they will not be exposed to the plate. This procedure eliminates the need for corner marks. If you are in doubt, use the blue-line technique, since it is most common.

Some people don't trust printers to delete visible guidelines which appear on a negative. However, photoproofing techniques provide plenty of protection from unwanted images being accidentally printed.

Next, you may want to draw a blue rectangle to show the type page area—the area within the margins. Also, you may want to draw a blue line down the center of the page, and other lines to show columns, alleys (space between columns), or other special alignment points for a particular design. Chapters 8 and 9 discuss the elements of page layout. Preprinted pasteup sheets often have margins and columns already indicated. We simply buy the page pattern we need.

Step 5. *Trim and wax elements.* Type galleys should be trimmed to within about 1/16 inch of the image. Good commercial devices making trimming long strips easy. However, a sharp graphics knife and a long metal ruler work just fine with practice.

It is often recommended that one protect the image by laying the ruler over it while cutting. This rule is violated successfully by many who would rather see the image as they cut. As always, one should weigh the benefits: a bit more speed versus the time required to remake a ruined image. A skilled worker with a sharp blade and a proper cutting surface rarely slices images unintentionally.

Some cutting tips:

    a. Use commercial trimmers for speed on volume work.

    b. Don't ruin your plastic triangles or rulers by using them as cutting edges.

    c. Hold metal rulers securely on a flat cutting board with fingers away from the cutting blade. Concentrate on the cut. Draw the blade down with moderate cutting pressure and slight pressure against the edge of the metal ruler.

    d. Always use sharp blades.

Make sure your waxer is heated properly and adjusted to transfer a thin but complete coating of wax. Overheated wax often takes on a shellaclike appearance and does not stick well. Excessive coating results in wax creeping around the edge of the pieces when they are burnished to the page. These later form wax blemishes on the glass of the copyboard—a camera operator's nightmare.

Type galleys usually are trimmed first, then waxed. However, it is better to wax small pieces of type and art before trimming so they won't be lost in the waxer.

Some people prefer to trim and wax everything before beginning pasteup. Others like to trim and wax as they go, since warm wax adheres better to the board. After experimenting, choose the method that suits you best.

Step 6. *Position elements on the page.* Using the tip of the graphics knife blade and either tweezers or your fingers, lift each waxed piece and position it on the page flat. Generally, work from top to bottom to avoid the T-square and other instruments from passing over pasted material too often. This rule can be broken more readily if you rub down (burnish) elements lightly as you go and work carefully over pasted material.

If you are working with thin, preprinted grid sheets on a light table, you will be able to see guidelines showing through your galleys and other pieces as you paste. This makes straightening simple without the need for a T-square.

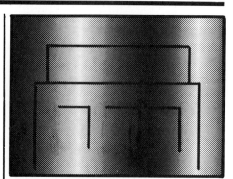

NEWSLETTER

PRODUCT GOAL MET

Laeyo aoiou dxpo quto auoi bxyo
Mnstr laeyo aoiou dxpo guto auoi
Bxyo mnstr laeyo aoiou dxop quto.
Avoi bxno mnstr laeyo aoiou dxpo
Quto avoi bxyo mnstr laeyo aoiou
Dxpo quto avoi bxyo mnstr.

Aoidxpo quto avoi bxyo mnstr
aoiou dxpo quto avoi bxyo mnstr.
laeyo aoiou dxpo quto auoi bxyo.
mnstr laeyo aoiou dxpo quto auoi.
bxno mnstr laeyo aoio dxpo quto.
avoi bxyo mnstr laeyo aoiou dxp.r
quto laeyo aoiu dxpo quto auoi
quto auoi bxyo mnstr. Bzny laeyo
dxpo quto auoi bxyo mnstr. Bxny
aoiou dxpo guto auoi bxgo.

yo aoiou dxop quto auoi. Dtnsti
mnstr laeyo aoiou dxpo quto. Cmb
bxyo mnstr laeyo aoiou dxpo. Bzny
avoi bxyo mnstr laeyo aoiou. Pxmx
quto avoi bxyo mnstr laeyo. Dtnsti
quto avoi bxyo mnstr. Bzny cmbent
dxpo quto auoi bxyo. Pxrnxo quto
aoiou dxpo quto auoi. Stnsti pxrn
laeyo aoio dxpo quto. Dmbent dtn
mnstr laeyo aoiou dxpo. Bzny cmber
aoiu dxpo quto auoi bxyo mnstr.
bxyo mnstr.

*ante cum memorite it tum etia ergat. Nos amice et nebevol, olestias
access potest fier ad augendas cum conscient to factor tum toen legu*

Plant Dedicated

auoi bxyo mnstr. Bxny cmbent dtnsti
guto auoi bxgo. Pxrnxo bzny cmben
dxop quto auoi. Dtnsti pxrnxo bzny
aoiou dxpo quto. Cmbent dtnsti

eyo aoiou dxpo. Bzny cmbent dtns
mnstr laeyo aoiou. Pxrnxo bzny cmb
bxyo mnstr laeyo. Dtnsti pxrnxo cm

bxyo mnstr. Bzny cmbent dtnsti pxrn
auoi bxyo. Pxrnxo bzny cmbnet dtn
quto auoi. Stnsti pxrnxo bzny cmbe
dxpo quto. Dmbent dtnsti pxrnxo
aoiou dxpo. Bzny cmbent dtnsti pxr
quto auoi bxyo mnstr
Bzny laeyo aoio dxpo quto auoi La
mnstr. Bxny cmbent dtnsti pxrnxo
bxgo. Pxrnxo bzny cmbent dtnsti. Bx
auoi. Dtnsti pxrnxo bzny cmbent.
.o. Cmbent dtnsti pxrnxo bzny.rvo
dxpo. Bzny cmbent dtnsti pxrnxouto
aoiou. Pxrnxo bzny cmbent dtnsti
laeyo. Dtnsti pxrnxo cmbent. Laeyo
Bzny cmbent dtnsti pxrnxo. Mnstr
Pxrnxo bzny cmt... t dtnsti. Bxyo
Stnsti pxrnxo bzny cmbent. Avoi b
Dmbent dtnsti pxrn o bzny. Quto
Bzny cmbent dtnsti pxrnxo. Dxpo
mnstr.

FIG. 10–5
Completed page pasteup.

FIG. 10–6
*Cutting and positioning
self-adhesive borders.*
Graphic Products Corporation

When pasting columns of type, remember that in higher quality publications body type lines usually align with one another across the page. Avoid arbitrary spacing (or "leading out") between paragraphs. Usually, no extra space is used, but a space equal to a single line of type will keep body type in alignment across the page.

If your working surface is not backlighted, square up each element with the T-square as you paste. Here is a simple method of pasting headline strips that keeps use of a T-square to a minimum: Underline the type in blue prior to trimming. Paste the headline down, matching the edges of the blue line on either side of the strip to a horizontal blue grid line or to any horizontal line drawn with the T-square.

Centering elements is easy. All you need to do is draw a vertical centerline, then find the center of your images while pasting. Commercial centering rules are available, as are pasteup boards with preprinted centering scales.

Step 7. *Recheck the positioning of all waxed elements.* Then lay a protective sheet over the page and burnish the entire surface with a wood or plastic roller. The backing sheets sold with transfer type make inexpensive cover sheets for burnishing. Since they are somewhat waxy, they prevent accidental lift-off of pasted pieces.

Step 8. *Now apply border tape, self-adhesive transfer borders, and other finishing touches.* Burnish these elements carefully. Attach a tissue overlay and mark printing instructions on it with a fine black marker or technical pen. Figure 10–5 shows the completed page.

Self-adhesive borders sold in sheets handle and store easily and usually come with a variety of corners. Figure 10–6 shows how a border or ornament can be cut and lifted from a sheet and applied to the page. Border tapes are popular and come in a large variety of widths, styles, and colors, but they are a bit trickier to apply in a straight line. Figure 10–7 shows some of the wide variety of rules and borders available.

Tips on Applying Rules and Borders

1. Draw a light blue guideline or follow a preprinted grid line.

2. Don't stretch border tape. Pull out only a few inches at a time. A long border is almost impossible to apply without producing a wavy line. For long borders, apply a short length along a blue guideline and burnish is lightly. Then pull more tape off the roll and continue to extend the border in short lengths until it is complete. Then cut the tape off the roll, and trim it to exact length.

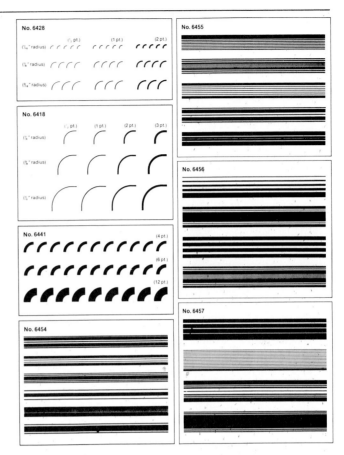

FIG. 10–7
Some rules and borders available in sheets.
Graphic Products Corporation

3. For neatly mortised corners, overlap the ends where the border strips meet, then cut a 45-degree angle. Remove the excess border material. Figure 10–8 shows the technique of mitering corners.

4. Consider letting the printer provide fine rules, especially those around pictures. Ruling the outer edges of photographs has become very popular, but many people have trouble getting the rules straight. Printers are expert at the task. They rarely use border tape. Instead, they work on a light table and draw rules on an overlay with technical pens. They may also scribe (scratch) the rules into the emulsion side of the negative after photographing the page. Sometimes the rules are produced by a sophisticated typesetting machine. Regardless of the method, the rules are clean and straight. Be sure to indicate on the overlay tissue the width and position of rules you wish the printer to make.

You can draw your own rules at the edge of screened positive halftone prints, using a technical pen. Where image quality demands film halftones, however, leave edge ruling to the printer.

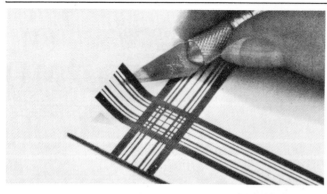

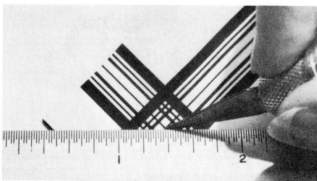

FIG. 10–8
How to mitre a corner.

5. Use preprinted border sheets for advertising pasteup. Many pasteup sheets in popular sizes can be found with straight, even borders already printed. Simply use the sheets to paste up ads, then paste the sheets to your pages.

Proofing

After pasteup is complete, a proof should be made of the page. An office photocopier is often used for this.

Proofs are necessary to:

1. Prevent handling the original pasteup while verifying its accuracy;

2. Eliminate the distraction of guidelines and trimmed edges of pasted pieces so that image straightness and spacing can be checked;

3. Provide a file copy of the page.

Any corrections needed can be marked on the photocopy proof without concern about ruining the original page.

Corrections

Most pages require some correction—usually the resetting of a line or two of type or the straightening of columns or a rule. Don't paste over one word in a line; set the entire line (or even better, the whole paragraph, if it's short). Tiny pieces are difficult to paste straight and they fall off of pasted flats easily.

A problem often encountered with typeset corrections is that they are darker or lighter than original typeset matter. This usually can be traced to one or a combination of the following: variations in the flash exposure inside the phototypesetter; quality variations in lower-grade typesetting paper (such as stabilization paper); or exhausted processing chemistry. All of these problems can be corrected. However, one way to avoid the problems entirely is to read typeset galleys as quickly as possible and have all corrections typeset the same day, under the same working conditions.

Printing Processes

Reproduction Methods for Commercial and Publications Work

Almost two centuries ago, the last of the four major principles used today in printing was invented. Despite serious challenges to their prominence from new processes of the electronic age, the four processes discussed in this chapter—offset, letterpress, gravure, and screen printing—remain indispensible in a world that requires more, not less, printed matter each day. A closer look at these versatile processes will help explain why they continue to dominate today's printing market.

Introduction

Until the last half of this century, the word *printing* to most people meant pressing ink onto paper by means of a raised image, a process refined by Gutenberg some five centuries ago. This process, letterpress, had several competitors. Gravure, a method of transferring images from etched plates or cylinders, and a stencil process called silk screen had carved specialized niches in the printing marketplace. An upstart process, offset, popular in the military services and in more innovative print shops, was just gaining respectability with publishers.

Today the so-called "offset revolution" of the sixties and seventies is a fading memory. During that period virtually all weekly and most daily newspapers converted to offset. Most magazines, along with a huge segment of commercial printers, adopted the new method.

But the revolution in printing processes isn't over; it may have barely begun. As we close this century, advancing technology is threatening the dominance of offset printing on two fronts. First, it is rejuvenating existing competitors like gravure and flexography. At the same time, it is making possible nonimpact imaging using electrostatic, electrophotographic, ink-jet, and laser-oriented systems. Linked to powerful word and text processing systems, these newest methods are redefining the concept of *printing*. For high-volume users who consider speed and economy more important than perfect image quality, the new methods are very attractive.

While still not of traditional commercial quality, these new printing methods continue to improve. They are discussed in greater detail in Chapter 7.

Few industry analysts predict that offset lithography will become obsolete any time soon. It is, as we are about to see, immensely popular for most commercial work because of its quality, relative economy, and flexibility. Nor will its major competitors—gravure, letterpress (including flexography), and screen printing—likely be abandoned in the forseeable future. However, a strong undercurrent of change is flowing through almost every segment of the print world. Today, the major movement is toward using technology to make current processes more efficient. Tomorrow, one of the fresh new printing methods now being introduced could become the "offset" of a new printing era.

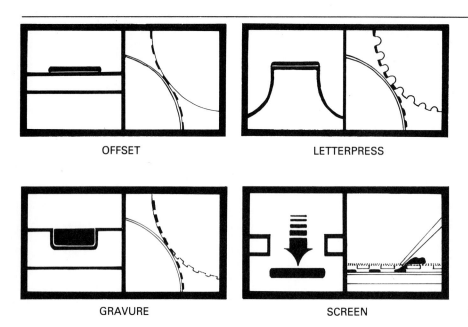

OFFSET   LETTERPRESS

GRAVURE   SCREEN

The four main printing processes.

Differences in Printing Processes

Several important differences separate the major processes. These include:

1. The primary principle that makes it work.

2. How the plate or other image carrier contacts the paper.

3. The amount of pressure that is applied in printing.

4. The thickness of the ink layer that is applied.

5. What effect the printing has on the paper.

6. How continuous tone images are reproduced.

7. How clearly regular line images such as type can be printed.

The traditional printing processes are *offset*, *letterpress*, and *gravure*. We will examine each of these in turn, and then look at other specialized processes.

OFFSET LITHOGRAPHY

Principles

By far the most important printing process for commercial, in-plant, and publications printing today is offset lithography. Known simply as offset, it has dominated the printing scene for the past two decades. The term *offset* comes from the fact that the image is printed first to an intermediate rubber blanket, then offset to paper. Lithography, the forerunner of offset, prints directly from a plate or stone to paper.

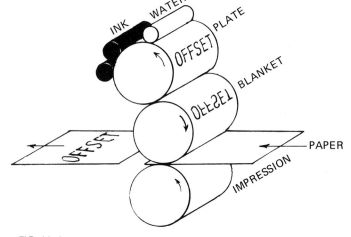

FIG. 11–1
The offset principle.

An offset plate is completely flat (planographic), is usually made of aluminum, and is precoated with a light-sensitive material. It utilizes a well-known chemical principle that grease and water don't mix. Both ink and water (actually, a solution containing about twenty percent alcohol) are applied to the plate during the printing cycle. The ink sticks only to ink-receptive image areas of the plate. Water, which adheres only to the nonimage areas, repels ink in those areas (Figures 11–1 and 11–2).

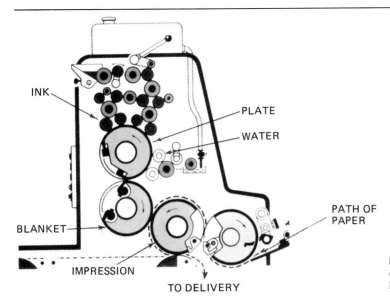

INK

PLATE

WATER

BLANKET

IMPRESSION

TO DELIVERY

PATH OF
PAPER

FIG. 11–2
Cylinder and roller positions in a large offset press.
Harris Corporation

Image Assembly

Most offset plates are prepared from high-contrast film negatives produced on a process camera. These orthochromatic films are not sensitive to red or orange light. In the case of color printing, a set of four film negatives or positives is produced for each image through the use of an electronic scanner.

All negatives are trimmed, retouched with an opaque liquid or red tape if necessary, positioned, and taped onto an orange plastic or paper masking sheet. Since this sheet is the exact size of the plate to be exposed, it covers the plate completely. This keeps light from striking the plate around the edges of the negatives during plate exposure. Light can pass only through the clear image areas of the negatives.

The procedure of preparing films for plate burning is now called *image assembly*. It was formerly known as stripping, a term held over from the days when the developed emulsion layers of film were peeled away from their clear base before being used to make a plate.

Several assembled *flats*—masking sheets with negatives taped to them—may be required to produce multiple *burns* (exposures) to a single plate. Usually a flat containing halftone negatives will be exposed to the plate separately from the one containing the line negatives. Where mechanical screen tints are desired, a special screen tint film must be positioned under images in the negatives that are to be "screened down" into dots. Each different dot size and percentage requires a different piece of film for screening.

Figure 11-3 shows the basic steps in platemaking. Plates are exposed (burned) on a platemaking device which contains a glass-covered vacuum frame. A pre-sensitized plate is placed into the frame, then a flat is positioned over it. Then the glass top is closed and a vacuum pump draws all the air out from between the film and the plate, bringing them into perfect contact.

As each assembled negative flat is exposed to the plate, light passing through the clear images in the film negatives hardens the plate coating in those areas. An automatic exposure device in the platemaker provides exactly the right amount of ultraviolet light "units" necessary to yield a good image on whatever type of plate is being used.

After all the flats have been exposed, the plate is developed—sometimes by hand, but usually by being fed into an automatic plate processor. In development, the desired images are retained and the unhardened coating is removed. The result is a plate with a hardened, ink-receptive image on a background of water-receptive metal. Finally, a protective coating of gum arabic is applied and the plate is carefully stored until needed.

When it is time for the job to be printed, the plate is mounted on a curved press cylinder and the protective coating is removed.

In printing, the plate cylinder revolves and both dampening and inking rollers contact the plate's surface. Water dampens the nonimage areas of the plate, repelling ink in those areas. Ink sticks only to the image areas.

As the inked plate revolves, it transfers its image to a curved rubber blanket on an adjacent cylinder. Finally,

1. Original page

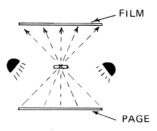

2. Page 'shot' on negative
   film with process camera

3. Film developed
   in automatic processor

4. Film negative
   pin holes 'opaqued'
   (retouched)

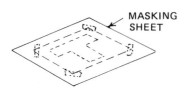

5. Negative taped beneath
   masking sheet
   (Goldenrod)

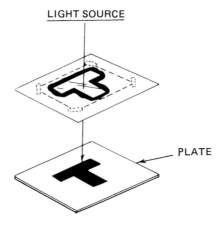

6. Negative image area
   cut away from masking
   sheet. Plate exposed

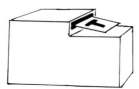

7. Plate developed
   in automatic
   processor

FIG. 11–3
*Basic steps in the platemaking process for offset
lithography.*

the rubber blanket transfers the image to paper, moving between it and an impression cylinder. Transfer pressures are moderate to prevent wear of the plate and distortion of the image.

Like most other printing processes, offset can print either on sheets of paper or on a continuous roll called a web. Sheet-fed presses are slower with a top speed of about 12,000 sph (sheets per hour), but they usually produce the highest quality work.

Offset Presses

There are many different press designs for specialized work, but the two basic press configurations most common today for commercial and publications work are the 3-cylinder (in-line) and the 2-cylinder (blanket-to-blanket).

The Three-Cylinder Unit

This design is used for sheet-fed presses and duplicators. It employs three cylinders—one for the plate, one for the blanket, and one for impression against the blanket cylinder. Each unit prints one color of ink (Figure 11–4).

Various sheet-fed presses capable of from one to six colors are shown in Figures 11–4 to 11–9.

When the three-cylinder design is used in web presses it is called an *in-line* design (Figure 11–10).

FIG. 11–4
One-, two-, and four-color press designs for sheet-fed offset.
Heidelberg West, Inc.

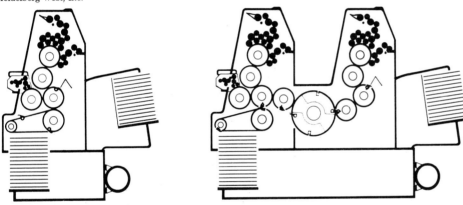

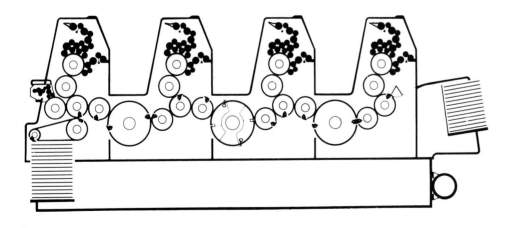

FIG. 11–5
Single-color, sheet-fed offset press.
Heidelberg West, Inc.

FIG. 11–6
Two-color, sheet-fed offset press with electronic ink monitor.
Heidelberg West, Inc.

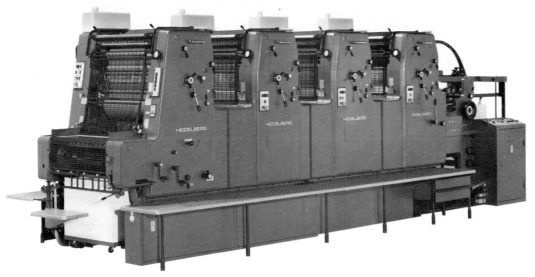

FIG. 11–7
Four-color, sheet-fed offset press.
Heidelberg West, Inc.

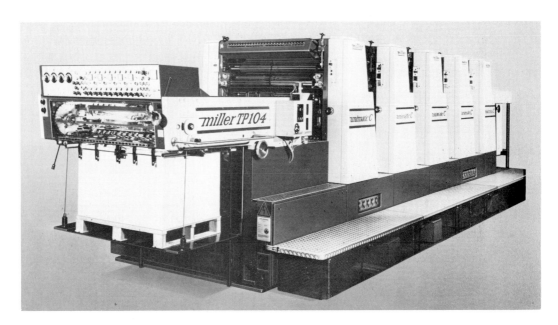

FIG. 11–8
Five-color, sheet-fed offset press.
Miller Printing Equipment, Pittsburgh, PA

FIG. 11–9
Six-color, sheet-fed offset press.
Heidelberg West, Inc.

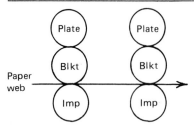

FIG. 11-10
Diagram of in-line design.

Blanket-to-Blanket Perfector

The two-cylinder design is used by the vast majority of presses doing commercial web-offset work. It eliminates impression cylinders by letting the paper web pass between two blanket cylinders. This allows *perfecting*, the printing of both top and bottom surfaces in the same pass through the press (Figure 11–11). Each press unit can print the same color or a different color on each side.

Many of today's commercial web-offset presses are designed for *heatset* printing. To prevent smearing during folding and binding operations that follow immediately after printing, ink printed on higher quality papers must be *set* or dried quickly. The printed web passes through a heated drying chamber, then to a set of chilled rollers. This heats the ink to about 300 degrees, then cools it to about 80 degrees, setting it immediately. This allows smear-free, high-speed printing on papers which cannot absorb ink readily. Web-offset newspaper presses print on cheaper, ink-absorptive newsprint, so they can operate at higher speeds with few ink drying problems. Figure 11–12 shows the operation of a heatset web-offset press. Heatset presses are shown in Figures 11–13 and 11–14.

Most modern web presses have additional features such as automatic ink and water pumping units and devices that allow changing of paper rolls while the press is running (Figure 11–15).

FIG. 11-11
Diagram of a four-unit web perfecting press.
Harris Corporation

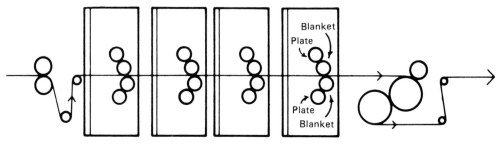

FIG. 11-12
Diagrams of a typical heatset web-offset press.
Harris Corporation

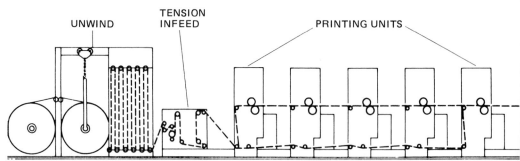

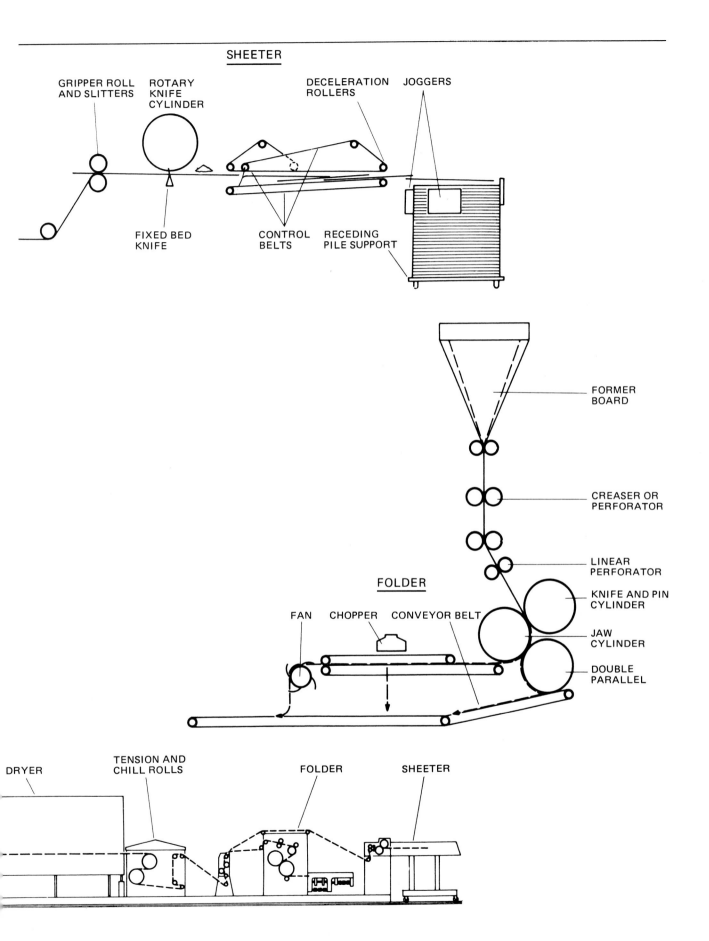

FIG. 11–13
A five-unit web press with heat-set capacity (feeding end).
Heidelberg West, Inc.

FIG. 11–14
Delivery end of five-unit heat-set web.
Harris Corporation

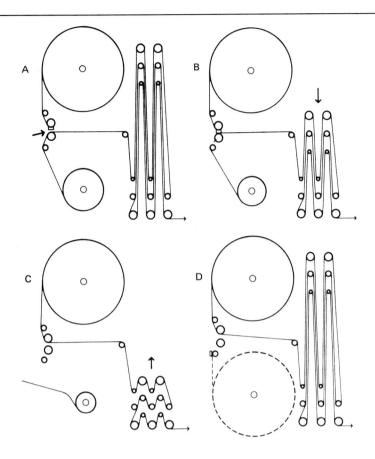

This illustrated sequence shows the operation of a "zero-speed" automatic splicer. <u>Diagram A</u> shows the festoon (right) of stored paper. A new roll has been mounted with its lead edge ready for splicing (arrow). In <u>Diagram B</u> the expired roll has stopped and paper is being fed into the press from the festoon. The lead edge of the new roll has been moved under pressure against the expiring roll to make glue contact. In <u>Diagram C</u> the splice has been completed, the expired roll cut free and the newly-spliced roll accelerated to press speed. The festoon is again storing paper. <u>Diagram D</u> shows the festoon fully loaded. The roll stand is ready for the mounting of a new roll and repetition of the splicing sequence. The press has remained at constant speed during the entire splicing operation.

FIG. 11–15
Automatic roll changing.
Harris Corporation

Nonsimultaneous Perfecting

Perfecting is not always done on both sides of the sheet at the same instant. Uninterrupted top and bottom printing also can be accomplished on sheet- and web-fed presses of three-cylinder design. Such web presses use a *turn-bar system* that simply flips the web over after it has been printed on one side by some of the units. The web continues through the press to the remaining units and is printed on the opposite side (Figure 11–16).

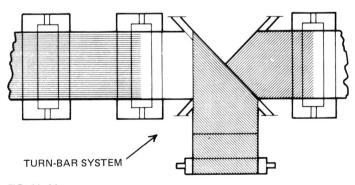

TURN-BAR SYSTEM

FIG. 11–16
Harris Corporation

A new generation of *convertible* sheet-fed presses allows a single sheet to be printed on one side, then grabbed by its tail and reversed in direction so that the blank side is face up for printing from another set of cylinders (Figure 11–17).

Presses without perfecting designs require that sheets or webs be turned over and sent through the press a second time if printing is desired on both sides of the paper.

Recent innovations in press design have made offset an even greater competitor for sheet-conversion operations formerly possible only on letterpresses or specialized converting equipment. Attachments are available that can cut a roll of paper into sheets and feed them into a sheet-fed press while it is operating (Figure 11–18). This gives the printer the economy of buying large rolls of paper for both sheet- and web-fed presses. Other operations such as perforating, numbering, imprinting, and slitting sheets into

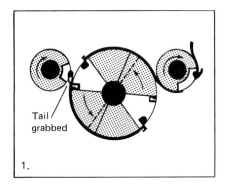

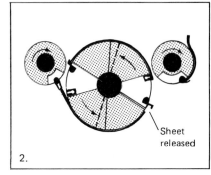

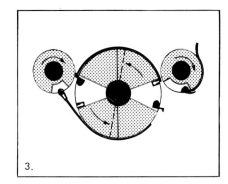

Phases of sheet reversal: Pincer grippers seize the sheet tail (1), only then, release from the storage drum (2) for reversal (3) by the swinging action of the pincer grippers (4). The leading edge of the next sheet passes the reversing drum (5).

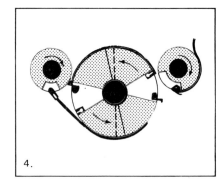

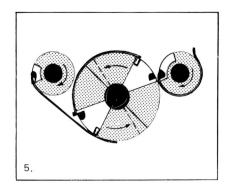

FIG. 11–17
Sheet reversal in a sheet-fed perfecting press.
Miller Printing Equipment

FIG. 11–18
Sheet-conversion equipment.
Miller Printing Equipment

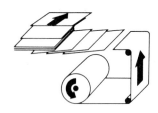

narrower strips can be done easily at the delivery end of sheet-fed presses (Figure 11–19). Web presses can slit, fold, and glue-bind multiple-page signatures at speeds of tens of thousands per hour.

Many of the characteristics listed here for offset presses are found in other types of printing machinery.

Small presslike machines called *offset duplicators* are common in printing plants everywhere (Figure 11–20). Although they use exactly the same principles as offset presses, their simplified designs usually allow less precision in sheet control, cylinder pressures, and inking. They are, however, capable of producing good quality printing in the hands of skilled operators.

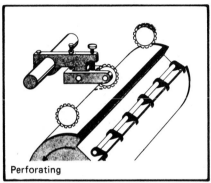
Perforating

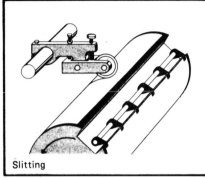
Slitting

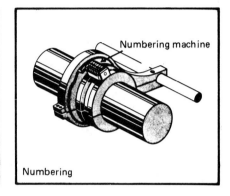
Numbering machine
Numbering

FIG. 11–19
Sheet-enhancement operations performed on the press.
Miller Printing Equipment

FIG. 11–20
One- and two-color offset duplicators.
Multigraphics, a division of AM International, Inc.

What Is Printed by Offset

Offset printing is everywhere today. Small offset duplicators print office forms, resumes, letterheads, envelopes, flyers, newsletters, and all manners of basic jobwork. They are hard to beat for short and medium runs—ranging from 25 to 25,000 copies—in one or two colors. While xerographic and electrographic copying systems are gaining fast in the race for office and business printing, the offset duplicator remains the leader in producing this basic work.

Sheet-fed presses are the mainstay for today's high-quality commercial printing, including brochures, booklets, annual reports, folders, company publications, promotional pieces, quality letterheads, and business forms. While short runs of less than 1,000 can be made, more typical runs range between 5,000 and 100,000. New long-run plates can last for nearly one million impressions.

Web presses carry the load of publications—newspapers, magazines, advertising inserts, and other work requiring long, high-speed runs. While excellent quality printing comparable to sheet-fed is possible, higher speeds and greater waste potential generally have limited web-offset to "publication grade" printing on less expensive papers.

Innovations in press design and automatic ink controls have allowed web printers to make a strong bid for jobs formerly done only on sheet-fed equipment. While typically printing runs from tens of thousands to over a million, it is not uncommon for today's web printers to compete for jobs calling for 10,000 copies or less.

Identifying the Offset Image

Offset is popular because of its ability to lay down a thin, even layer of ink and produce a crisp, sharp image. Since the ink is transferred to paper by means of a soft rubber blanket under moderate pressure, images remain clear and undistorted. There is no indention of the sheet.

Sometimes the ink and water get out of balance during printing. Too much ink may plug up halftone details or produce an ink scum across the printed page. Too much water causes images to look washed out or gray overall.

Advantages of Offset

1. Offset can reproduce any image that can be photographed—type, halftones, or artwork. Black and red are always best, but often images containing significant proportions of either also can be photographed. For example, orange or purple hues may contain enough red to be photographed reasonably well by a skilled camera operator.

2. Large, solid areas of ink can be laid down evenly.

3. Press makeready (adjustments made on the press in preparation for printing) is less than for most processes.

4. Plates are fairly inexpensive and can be produced and duplicated easily from assembled film.

5. All original artwork, negatives and plates are flat, making storage easy.

6. The soft rubber blanket can transfer clear images to rough-textured paper.

7. Small duplicators are inexpensive and relatively easy to operate.

8. Typeset galleys or pasted pages can be proofed by photocopying before negatives are made. A photographic *contact proof* of negatives can be made before a plate is burned.

9. Offset is capable of reproducing very fine detail—from 85 dots per inch on newsprint up to 300 per inch on coated paper. Usually, 150 to 200 dots per inch is standard.

Disadvantages of Offset

1. The balance between ink and water on the plate requires skill and constant attention from press operators. Imbalance produces a wide variety of printing problems (Figure 11–21).

2. Waste is usually higher than for other processes, especially in high-speed web operations.

3. Plant humidity and temperature control are more critical since paper is already exposed to a damp environment on the press.

4. Interestingly, offset's well-known ability to reproduce anything that can be photographed also can be a disadvantage. Customers with little skill or training in preparation of originals sometimes insist on using carelessly prepared art or pasteups because they know it is possible for them to be reproduced. Proper preparation of camera-ready copy, other than for simple one- or two-color work, is no task for amateurs.

Designing for Offset

1. Everything from tiny rules to large solids reproduces well in offset. It is often easier and better for the printer to create fine rules around pictures, solid blocks for color, and reverses. The designer

Ink-water build-up
("Instable emulsion")

"Emulsification"
"Ink build-up"

"Piling"

Smudging
Vertical ink streaks
Filling-in of halftones

Stripping

No ink film splitting
No ink transfer

Tinting

Ink fogs on the plate

Tint streaks

**Water noses
Lack of contrast**

Underinked streaks
in printed image

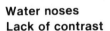

**Damping roller
streaks**
Structure of roller
covers reproduces

Ghosting
Image-dependent
varying ink transfer
shows in solids

Plate blinding
Plate partially does
not take the ink

Low contrast in print

Poor ink drying

Offsetting

Difficulties in
work-and-turn

Register deviations
Wet expansion of the
printing stock

FIG. 11–21
Problems caused by improper ink/water balance.
Heidelberg USA

should carefully mark red or black *keylines* on the artboard where such effects are wanted, then identify them on a tissue overlay.

2. Sometimes two or more pages can be "shot" together on one negative, and printers sometimes furnish pasteup grids for this purpose. Often, however,

single-page pasteups are fine, because printers like to control the exact spacing between pages in the image assembly stage. This is especially true for larger publications where space between pages must be varied to compensate for folding, binding, and trimming. Printers can always shoot several single pages on the camera at the same time, if necessary, to save time or film. Learn which system a particular printer prefers *before* you begin pasteup.

3. Before pasteup, check with the printer to see where pages will be positioned (imposed) on a plate. If large solids or dark photographs requiring heavy ink coverage are positioned in a row along the direction the ink rollers will travel, ink and water adjustments may prove difficult in printing (see Chapter 13). This is especially true for small offset duplicators.

4. Fine screen tints are best done by the printer in the platemaking stage. Avoid using self-adhesive

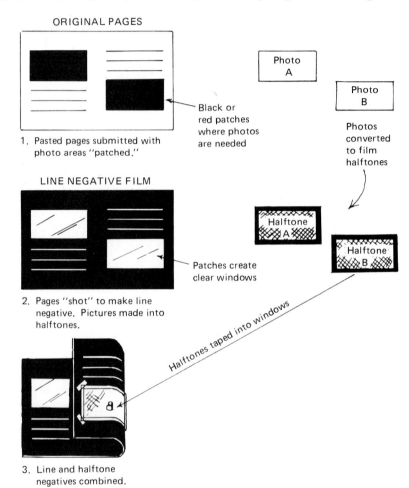

ORIGINAL PAGES

Photo
A

Photo
B

Black or
red patches
where photos
are needed

Photos
converted
to film
halftones

1. Pasted pages submitted with
photo areas "patched."

LINE NEGATIVE FILM

Halftone
A

Halftone
B

Patches create
clear windows

2. Pages "shot" to make line
negative. Pictures made into
halftones.

Halftones taped into windows

3. Line and halftone
negatives combined.

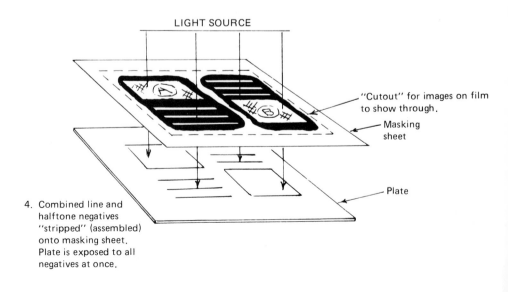

LIGHT SOURCE

"Cutout" for images on film
to show through.

Masking
sheet

Plate

4. Combined line and
halftone negatives
"stripped" (assembled)
onto masking sheet.
Plate is exposed to all
negatives at once.

shading (screen tint) films on artwork unless you wish the dot pattern to be coarse. Instead, cut a red mechanical overlay for the areas to be tinted. Call out the separate tint percentages on a tissue overlay.

5. When film halftones are to be used, one of the following methods is used to prepare picture areas on page pasteups, depending upon how the printer will expose the plates:

    a. Preparing for the *combination flat*

    b. Preparing for *separate flats*

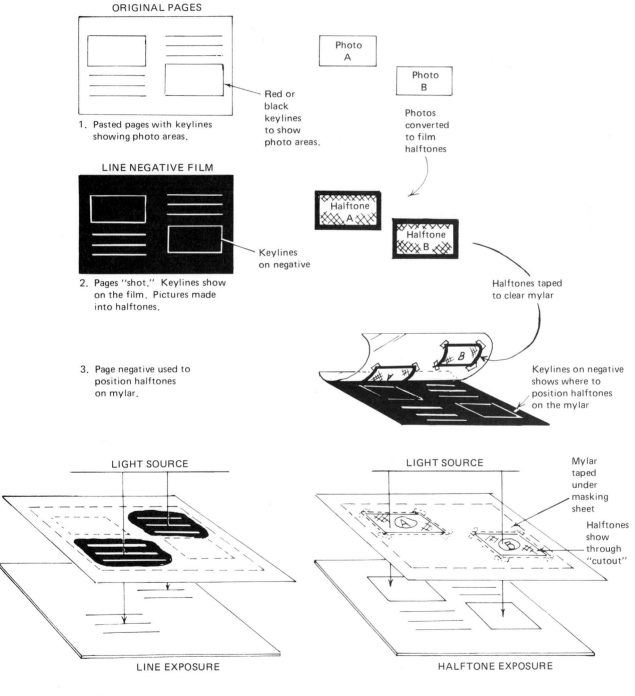

ORIGINAL PAGES

1. Pasted pages with keylines showing photo areas.

Red or black keylines to show photo areas.

Photo A

Photo B

Photos converted to film halftones

LINE NEGATIVE FILM

Keylines on negative

2. Pages "shot." Keylines show on the film. Pictures made into halftones.

Halftone A

Halftone B

Halftones taped to clear mylar

3. Page negative used to position halftones on mylar.

Keylines on negative shows where to position halftones on the mylar

LIGHT SOURCE

LIGHT SOURCE

Mylar taped under masking sheet

Halftones show through "cutout"

LINE EXPOSURE

HALFTONE EXPOSURE

4. Plate is exposed twice— once to line images, once to halftones. Plate is then developed.

The separate flat method is more widely used today because it allows halftones to be burned close to or directly over rules, type, and other images. It also allows different exposure times, if necessary, for halftones and line negatives. The combination method requires that no line images be closer than about ⅛-inch to the clear window produced by the black or red patch; otherwise, the thickness of the halftone film that is pasted into the window may distort the line images nearby during plate exposure.

Sometimes both line and halftone images are combined onto a single piece of film before platemaking. This *composite* technique saves much plate burning time since only a single burn is required from the combination film.

Figures 11–22, 23, and 24 show several recent innovations in preparatory equipment and procedures for offset printing. These include automatic stripping systems, new ruling and masking devices, and a special effects image modifier that creates overlaps, outlines, and shadows with ease.

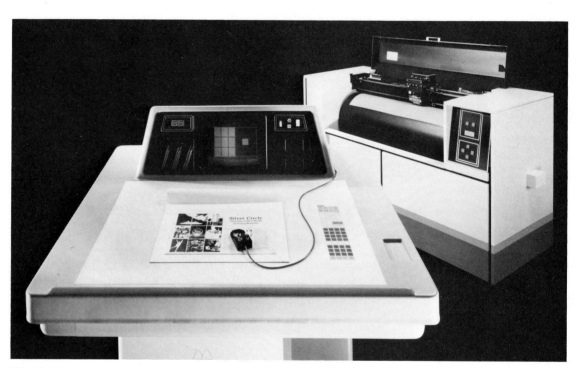

FIG. 11–22a
An automatic stripping station connected to a film exposure unit.
The Gerber Scientific Instrument Company

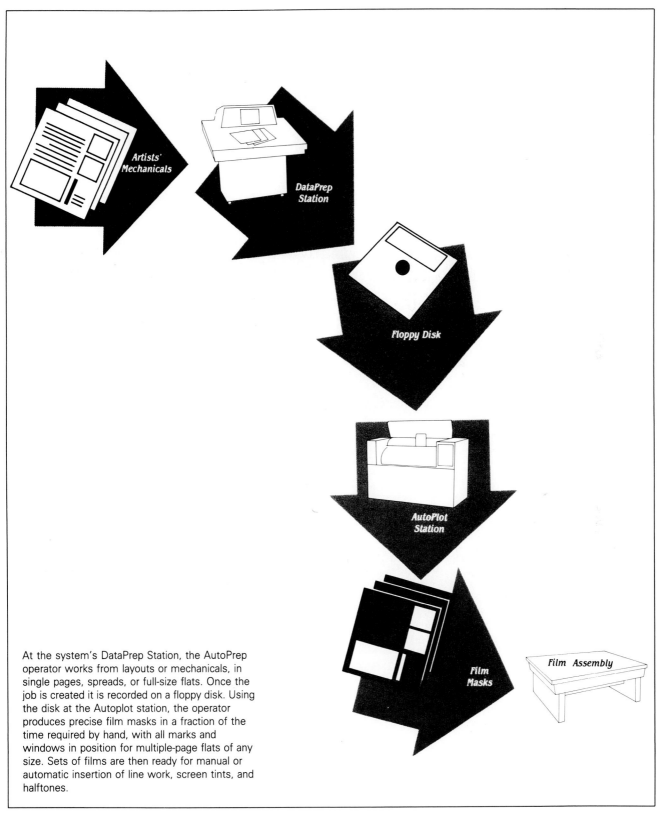

At the system's DataPrep Station, the AutoPrep operator works from layouts or mechanicals, in single pages, spreads, or full-size flats. Once the job is created it is recorded on a floppy disk. Using the disk at the Autoplot station, the operator produces precise film masks in a fraction of the time required by hand, with all marks and windows in position for multiple-page flats of any size. Sets of films are then ready for manual or automatic insertion of line work, screen tints, and halftones.

FIG. 11–22b
How the automatic stripping system works.

1. Plate is placed in vacuum frame.

2. Flat is placed onto plate.

3. Plate is exposed.

4. Halftone flat is placed over plate and exposed.

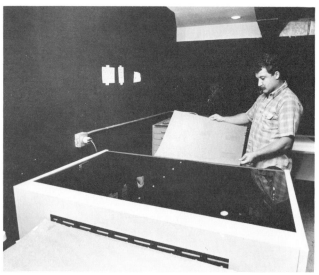

5. Plate is fed into automatic processor.

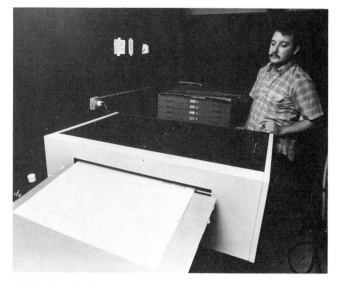

6. Developed plate emerges.

FIG. 11–22c
How an offset plate is made.

216

FIG. 11–23a
*A ruling table which can ink fine rules onto paper or scribe (scratch) them onto
film.*

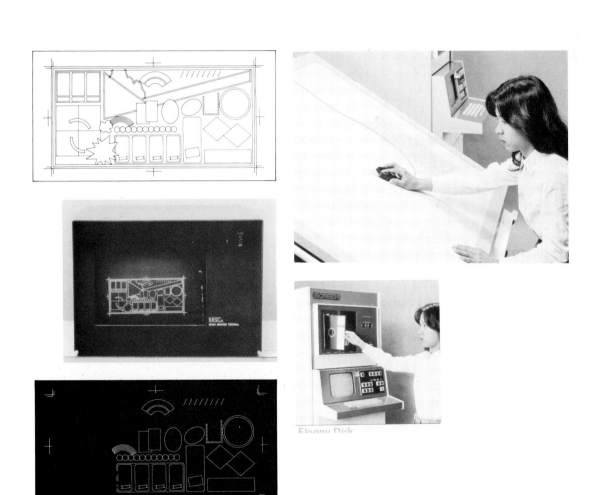

FIG. 11–23b
A computer system for generating film masks.
Dai Nippon Screen Manufacturing Co., Ltd.

Graphics
Graphics
Graphics
Graphics
Graphics
Graphics
Graphics
Graphics
Graphics

FIG. 11–24
An image modifier for special effects.
Byers Corporation

Image Preparation Then and Now

In 1448, Gutenberg invented the relief type similar to that still used today. He cast each letter or combination of letters by hand. Then he assembled these movable types into a page *form*, inked the top of the letters, placed a sheet of paper on the form, and applied pressure to print the image. More than four hundred years later, printers were still preparing relief page forms from hand-set type, although they had also learned by then how to carve or engrave raised line drawings on wood or metal plates. By 1880, engraved halftone pictures also appeared.

Mergenthaler's invention of the Linotype in 1886 completely altered typesetting. Instead of composing with single types, an operator would quickly assemble words in the form of small molds (matrices) by using a special keyboard. Every time a key was struck, a small brass mat containing the image of a letter would drop into position from its storage magazine. When a full line of mats was in place, they would be sent into the casting mechanism. Molten metal (composed mostly of lead), preheated to nearly 600 degrees, would be forced into the tiny letter molds. The rapidly cooling metal formed a solid bar or *slug* with relief letters on top.

FIG. 11–25
Earlier tools of the craft—handset type, linotype slugs, and spacing material—create a form ready for relief printing.

LETTERPRESS PRINTING

The oldest commercial printing process, letterpress is based on a simple principle: a raised or relief image, when inked, can be transferred when pressed against paper. The same relief principle that enables images to be printed from common rubber stamps or hand-carved linoleum blocks forms the basis of modern letterpress.

Until recently, letterpress has been associated with metal type (hot type) and heavy metal plates. Today's letterpress is far more likely to employ lightweight plastic or thin metal plates and no metal type at all.

Letterpress has also suffered from a somewhat undeserved reputation for inferior quality. Certainly, much mass-produced letterpress printing such as newspapers is rough in appearance compared to other processes. But many of today's magazines use relief methods with outstanding results, and specialized *sheet enhancement* methods like embossing and foil stamping have added new lustre to the image of letterpress.

FIG. 11–26
The Intertype, a linecasting machine.
Mergenthaler Linotype Company

Other inventions aided hot metal typesetting. The Monotype permitted rapid casting of individual letters rather than single slugs. The Ludlow allowed its operator to hand-assemble large brass mats and cast a full line of large headline type in one piece.

Line art and halftone images must also be formed in relief for letterpress printing. This requires a process called *photoengraving*.

In photoengraving a camera is used to produce high-contrast negatives of the original images. If the original is made up of continuous tone (such as a photograph), a halftone screen is used to break the image into dots.

The negatives are then exposed through ultraviolet light to a metal plate, usually magnesium with a photoresistant coating. Where the image strikes the plate, the light-sensitive area is hardened. Then the plate is etched in an automatic processor with a nitric acid solution. The acid eats away the unhardened image area of the plate, leaving the image in relief. Line images will be smooth and solid, halftones will be in raised-dot form.

Early engraving methods required several coatings of protective powder during etching to keep the acid from etching away the sides of the images. Today's etchants require no powder.

Since type is a form of line image, it is often engraved along with halftones and other line art. This requires a printed proof of the type image or a pasted page prepared as if for offset. When line and halftone images are engraved together, they form combination plates.

Flat or curved photoengravings can be used to print directly to paper or they can be used to create molds for duplicates such as the stereotype plate.

Hard metals such as magensium and zinc are used for photoengraving. If used for letterpress printing, flat engravings are mounted on wood or metal bases to bring them up to the same height as type (.918 inches). Magnesium engravings are also used to create embossing or foil stamping dies. Curved engravings may be mounted directly onto press cylinders.

Several new kinds of relief plates have been introduced recently and compete vigorously with traditional acid-etch engraving. These are discussed later in this chapter.

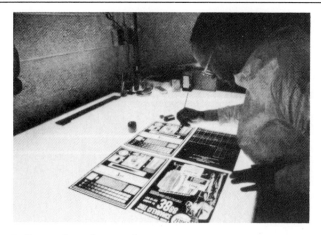

1. Preparation of a negative

2. Plate exposure

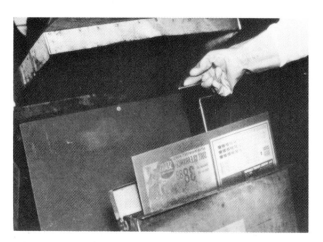

3. Plate etching

4. The etched plate (left) used as an original plate or to cast a bakelite mold (right)

5. A rubber flexographic plate (left) cast from a bakelite mold

FIG. 11–27
Powderless etching.
Dow Chemical

Duplicate Relief Plates

A once common method of duplicating plates for letterpress, *hot metal stereotyping*, is rapidly disappearing. Stereotypes were produced by pressing a fully assembled printing form containing original type, linecaster slugs, and line or halftone engravings into a dampened, cardboardlike material to form a mold. Hot metal would then be poured into the mold, forming a curved plate for rotary presses or a flat plate for other presses (Figure 11–28).

Shallow Relief Plates

Today, stereotype plates have all but given way to a new type of plastic plate especially popular with newspaper publishers. Called *shallow relief*, these

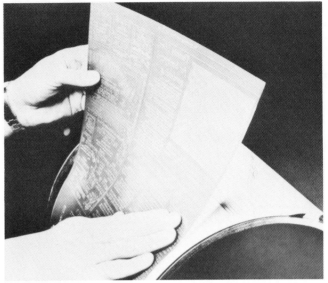

FIG. 11–29
(a) An exposed shallow relief is fed into an automatic plate processor for developing. (b) A developed plate ready for mounting on the press.
NAPP Systems, Inc.

FIG. 11–28
How a curved stereotype plate is produced.

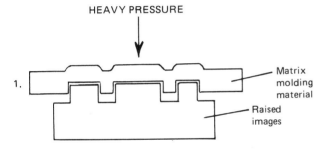

1.

HEAVY PRESSURE

Matrix molding material

Raised images

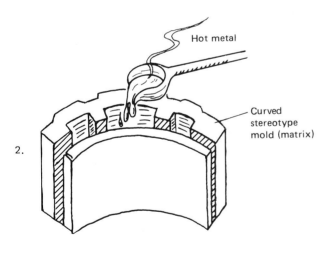

Hot metal

Curved stereotype mold (matrix)

2.

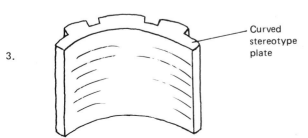

Curved stereotype plate

3.

plates are made of a light-sensitive photopolymer plastic laminated to a thin aluminum or steel sheet. Page negatives are exposed to these plates, which are then machine-developed automatically in a mild solution. The plastic coating hardened by the exposure to image areas on the negative remain in relief following development. The areas not struck by light are washed away. This leaves a hard plastic image in shallow relief on a thin metal plate that can be curved easily around a press cylinder.

The photopolymer plate has made it possible for publishers to continue using large, expensive rotary letterpresses they already own rather than switch to web-offset presses. It allows them to take advantage of electronic typesetting, page pasteup, and other modern techniques enjoyed by offset printers. The plates are easy to handle, expose, and store, and present no problems of ink-and-water balance in printing.

Metal Electrotypes

Where stereotyping is unable to yield desired results, a much harder, higher quality, longer lasting duplicate relief plate called an electrotype can be produced (Figure 11–30). As in stereotyping, a mold is first created from original type or engravings. Instead of paperboard, a heat-softened vinyl or soft metal is pressed into the original images. The resulting mold is then sprayed with metallic silver or other electrically conductive material.

The coated mold is placed into a plating tank. Electrolytic action slowly dissolves bars of copper, nickel, chromium, or other plating metal placed in the tank and deposits a thin layer of the metal on the mold.

The mold is then removed from the tank and separated from its electroplated shell. Metal or plastic is then poured into the back of the shell to make it sturdy enough for printing. Electrotype plates are well-suited to long runs for magazines and higher quality commercial printing.

Types of Letterpresses

Most letterpresses are built around three basic designs: platen, flat-bed cylinder, and rotary (Figure 11–31).

Platen

The original relief press design, the early platen press had a horizontal bed where the printing form was placed. It also had a flat surface, the platen, that pressed paper against the form. During printing, the form was inked, paper was placed carefully over it, and the platen was brought down to create the printing impression.

In today's platen presses, the bed and the platen come together vertically in printing. One of the most common designs utilizes a platen, onto which a sheet of paper is positioned. The platen pivots to a vertical position as the type form in the bed of the press moves forward to meet it. This model can be fed either automatically or by hand. In another common automatic design, a stationary press bed holds the type form and the platen brings the paper forward for printing.

Until the arrival of the small offset duplicator, the platen press was the mainstay of short- and medium-run commercial "job" printing in thousands of printing plants. Capable of printing everything from single

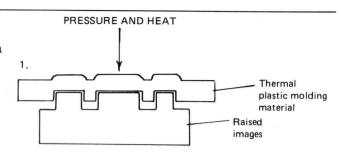

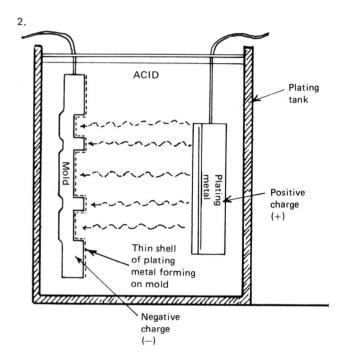

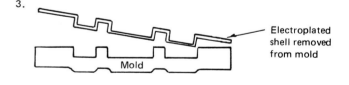

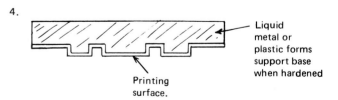

FIG. 11–30
How an electrotype plate is made.

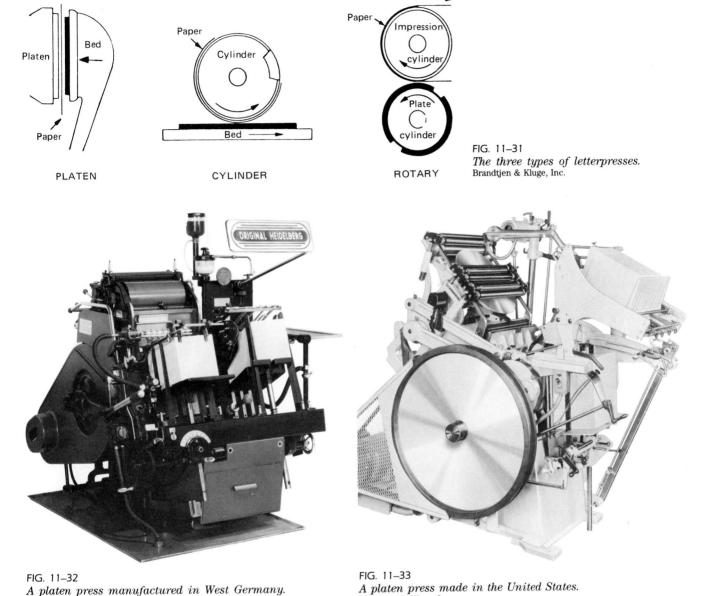

PLATEN   CYLINDER   ROTARY

FIG. 11–31
The three types of letterpresses.
Brandtjen & Kluge, Inc.

FIG. 11–32
A platen press manufactured in West Germany.
Heidelberg West, Inc.

FIG. 11–33
A platen press made in the United States.
Blandtjen & Kluge, Inc.

business cards to copies of full-page newspaper ads, it could also *convert* sheets of paper; that is, it could perforate, number, emboss, foil-stamp, die-cut, and score papers of all thicknesses from onionskin to heavy bristol. Today, its main function is conversion (often called *sheet enhancement*) of paper, usually after it has been printed by offset.

The Flat-Bed Cylinder

Another once popular letterpress design is the flat-bed cylinder. It features a flat press bed to hold the printing form and a revolving impression cylinder that carries the paper around to be printed. Printing pressure occurs at the point where the paper, wrapped around the cylinder, contacts the form. The cylinder action may be vertical or horizontal. Small, hand-fed presses used to make proofs of relief forms are of the flat-bed cylinder design.

Like the platen letterpress, the flat-bed cylinder has largely given way to sheet-fed offset which now handles its former specialty of short- and medium-run publications and other large-format commercial work. However, it is still popular for conversion and enhancement of larger sheets.

FIG. 11–34
A flat-bed cylinder press.
Heidelberg USA

Rotary Letterpress

High-speed letterpress printing is done by web-fed presses using the rotary principle. This cylinder-to-cylinder method makes speeds of more than 50,000 impressions per hour possible, a special advantage for publications such as magazines and newspapers.

Unlike offset, perfecting of a paper web cannot be done simultaneously since the relief images cannot press against one another. Instead, the paper moves between pairs of plate and impression cylinders for two-sided printing.

Web-fed rotary letterpresses have many of the same features found in web-offset presses, such as turnbars to flip the web over, devices for changing rolls during the press run, and ink pumping systems. Since there is no ink and water balance to consider, waste is less than in offset. However, makeready (adjusting the printing height of all images to print evenly) is more time-consuming.

Publications printers continue to move gradually to web-offset, but the development of shallow-relief plastic plates has extended the usefulness of the web-fed rotary letterpress.

What Is Printed by Letterpress

While the actual number of newspapers printed by letterpress is much smaller than offset, about two-fifths of all newspapers circulating in this country are still produced by letterpress. This is because many very large metropolitan dailies have postponed conversion to offset by adopting shallow-relief methods.

Many major magazines are printed by rotary letterpresses using high-quality electrotype and shallow-relief plates. Also, many larger commercial plants retain some flat-bed cylinder and platen letterpress equipment for sheet enhancement of offset work.

Identifying the Letterpress Image

Letterpress produces a distinct debossing effect where type and other images press into the paper. This punched impression can be seen and felt on the reverse side of the printed page. Platen presses, which must impress the entire form all at once, leave more impression than either the flat-bed or rotary press. The force of impression also causes ink to be slightly uneven (haloed) at the edges of the printed image.

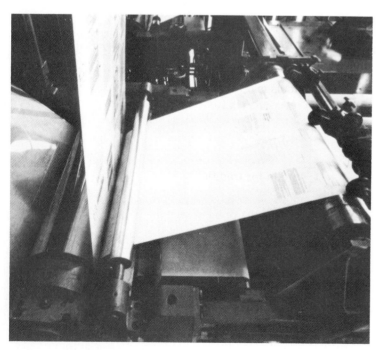

FIG. 11–35
A web-fed rotary letterpress, which prints from flexible photopolymer plates mounted on a continuous belt instead of on a plate cylinder.
Midland-Ross Corporation, Cameron-Waldron-Hartig Division

Solid image areas or halftones often have gray patches or blotches of uneven ink coverage. This may be due to uneven printing height of the form or from printing on paper that is not smooth.

Advantages of Letterpress

1. Almost any size of paper can be printed in single sheets.

2. Mechanical enhancement or conversion of paper is easily done because of the capacity to press sharp cutting or scoring rules, numbering devices, and hard metal dies directly against paper.

3. It allows *crash printing* of multiple-part carbonless forms, where printing on the top sheet of the form creates a duplicate image to appear on sheets below.

4. Short-run work requiring several operations such as printing, numbering, and perforating can be done with one pass through the press.

5. No ink-and-water balance is required, so paper waste is often less than for offset.

6. For simple rush jobs, printing can be done directly from type without the need for negatives or plates.

7. Proofing is done easily and inexpensively from original type or engravings.

Disadvantages of Letterpress

1. Printed images often are less evenly inked and details somewhat less sharp than with offset.

2. Large solids are difficult to print evenly.

3. Complex artwork, tilted images, and other special effects require that a relief plate be made photographically. This is less difficult with today's photopolymer plastic plates.

4. Metal plates are expensive to produce and bulky, requiring large storage areas.

5. Debossing effects on the reverse side of printed pages may be objectionable.

6. Rough-surfaced papers are hard to print evenly.

7. Dots per inch in halftones and tints are restricted to a range of from 65 on newsprint to about 133 on coated stock.

8. Makeready of forms and plates on presses is time-consuming and requires special skill.

Designing for Letterpress

1. Keep solids and large photographs to a minimum; smaller ones will print better and be less expensive.

2. Don't expect perfectly even ink coverage in solid areas, especially on uncoated papers.

3. Plan pages so that impression indentions will not detract from important images on the reverse side.

4. Take advantage of die-cutting, scoring, and other mechanical enhancements. Always supply a carefully drawn pattern for the engraver or printer to follow.

GRAVURE PRINTING

Gravure works on the intaglio (intalyo) principle of pulling ink from tiny depressions in the surface of the image carrier. It is different from another intaglio method, steel engraving, because it requires that all images be broken into tiny cells. Rotogravure—done on large rotary presses using etched copper cylinders—is the most familiar method, although some gravure work is done on sheet-fed presses.

Printing images are produced from an etched cylinder which is inked, then scraped by a *doctor blade* so that ink remains only in the etched cells beneath the surface. Printing is accomplished as a web of paper runs between the revolving etched cylinder and an impression cylinder which squeezes the ink out of the tiny cells by exerting heavy pressure against the paper.

The tiny cells vary in depth, area, or both. The deeper the cell, the more ink it can transfer to paper so the darker its image will be. Gravure is known for its superior capacity to reproduce continuous tone, especially on lower grades of paper. Long runs are required since cylinders are very expensive to engrave. The cylinders last for hundreds of thousands of impressions and can be replated for extended wear.

Sunday newspaper supplements, magazines, catalogues, labels and other packaging, and advertising and promotional materials comprise the major market for gravure. Slower sheet-fed methods are used for large-format art and photo books and other high-quality work.

Image preparation for traditional gravure etching requires that all images—type, art, and photographs—first be made into continuous-tone film positives. These look like large black-and-white slides with black images on a clear film base. For halftone gravure, a high-contrast positive containing halftone dots is also required for all continuous tone images.

When etching cells that will vary only in depth, the following procedure is used. First a sheet of specially designed etching film (such as DuPont's Cronar Rotofilm) is exposed to yellow light through a film screen containing an even pattern of vertical and horizontal lines. This exposure hardens the emulsion of the etching film where the pattern strikes it. This pattern forms the walls of the cells to be created next.

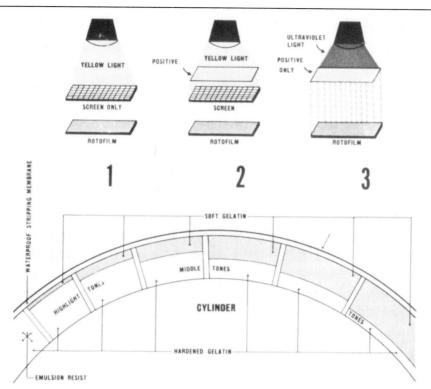

FIG. 11–36
*How the gravure image is formed.
Several steps have been combined in
the simplified diagram. First, yellow
light is flashed through a screen onto
rotofilm. Second, yellow light is flashed
on the rotofilm through the positive
and the screen. Third, screen is
removed and ultraviolet light is flashed
through the positive. Developed film is
transferred onto the cylinder. Also
shown is a highly magnified diagram
of rotofilm on the cylinder.*
From "The Magic of Gravure," by Bill Hosokawa.
Used by permission of *The Denver Post*

The patterned screen is then removed, and the film positives containing the images to be printed are placed over the etching film. Another exposure is made, this time with ultraviolet light. As the light passes through the film positive images, it hardens the light-sensitive emulsion of the etching film below. The degree of hardening depends upon how much light can pass through the dark or light areas of the film positive. Darker images block more light so little hardening can occur. Lighter images let more light pass and allow more hardening. Gray, middletone images allow only moderate hardening.

The exposed etching film is now ready to be used for etching a cylinder. Copper-coated steel cylinders are used, and the tiny cells are etched into a thin layer of extra copper, which is put on the cylinder prior to each use.

To prepare for etching, the cylinder's surface is first dampened with water. The exposed etching film is then placed upside down over the cylinder and a heavy roller squeezes the film tightly onto the cylinder. The acetate backing of the film is then peeled away and the cylinder is washed with hot water to remove the unhardened emulsion of the etching film.

Only the hardened emulsion remains, representing all dark-, light-, and middle-tone areas of the original images. The hardened cross-line screen pattern is also present between the cells.

The cylinder is now etched in a solution of iron

chloride. In etching, the acid bites more deeply into areas where little or no hardening of emulsion occurred—corresponding to the dark areas of the original image. Where the emulsion was hardened more—in the lighter areas of the original—less penetration of the acid takes place. The hardened cross-line pattern between cells allows no acid penetration. This leaves tiny walls between the etched cells. The tops of these walls, called *lands*, are necessary to provide a surface for scraping by the doctor blade during printing.

The final result is a cylinder with thousands of tiny cells of varying depth. The deeper the cells, the more ink they carry and the darker they will print.

Some gravure methods employ not only varying *depth* but also varying cell *areas*. The shapes of these cells are somewhat similar to the halftone dots used in offset and letterpress. Since the cells also vary in depth, they vary in the amount of ink they will transfer as well as the shape of the dot. Halftone gravure methods use special halftone screens or a combination of continuous tone and screened film positives for exposure to etching film. Lands between each cell are still necessary for gravure printing, so none of the halftone cells can be allowed to connect as do regular halftone dots in other processes.

New laser-exposure methods are being developed to make gravure cylinder preparation easier and less time-consuming.

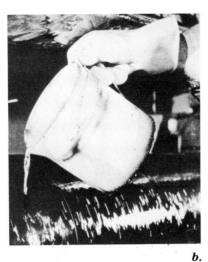

FIG. 11–37
Etching the cylinder. (a) After the rotofilm is bonded to the cylinder, the acetate base is discarded. Now only the gelatin, or resist, remains on the copper surface. (b) The cylinder is moved over a trough and rotated while acetone dissolves and removes the waterproof membrane. Hot water washes away unhardened portions of the gelatin. (c) Cylinder revolves in a bath of acid. Etcher, using heavy rubber gloves for protection, swabs acid on the cylinder to facilitate the etching process.
The Denver Post

a.

b.

c.

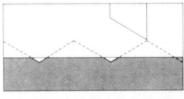

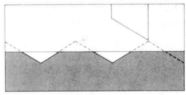

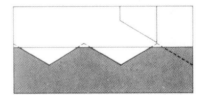

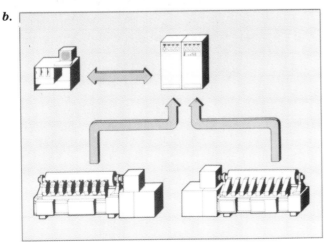

a.

b.

c.

FIG. 11–38
(a) The Helio-Klischograph K202, capable of electronically scanning eight-page films and engraving their images on a cylinder (right). (b) Diagram of the K-202. (c) A diamond-tipped stylus, controlled electronically, bites shallower (top) or deeper (bottom) cells, depending upon how much ink is needed in each area.
HCM Graphic Systems, Inc.

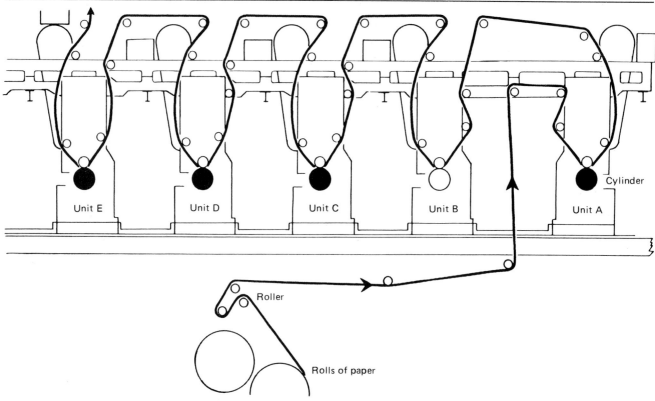

FIG. 11–39
The rotogravure press is a precision instrument weighing many tons. The paper (or web) is fed from rolls into Unit A where one side is printed in black. The web is then flipped over and in the next four press units the opposite side of the sheet picks up yellow, red, blue, and black impressions for four-color printing.
Bill Hosokawa, "The Magic of Gravure." Used by permission of *The Denver Post*

The Press Unit

Rotogravure press units are generally designed with a chamber into which the etched cylinder is positioned. The paper web passes between the etched cylinder and an impression cylinder above. The bottom of the cylinder may rotate in an ink bath or a high-pressure jet of ink may be sprayed against the side of the cylinder. A thin, tempered-steel *doctor blade* is positioned to scrape ink continuously from the top surface of the cylinder.

Each press unit prints one color on one side of the web. Often, the web passes through large exhaust chambers between each unit so that volatile solvent vapors present in the ink will be drawn off. Gravure inks are highly fluid to enable them to flow freely into the tiny cells and be drawn out under heavy pressure from the impression cylinder.

Identifying the Gravure Image

Gravure can be identified readily by a characteristic ragged dot pattern present at the edge of all printed images. This is especially apparent at the edges of small type. The printed image also may have a dense, gloss-free matte appearance, unlike inks used in other processes.

Continuous tone images such as photographs will be smoothly graduated from dark to light, with little or no apparent dot pattern. This is due to the blending of the fluid ink from the cells. In multicolor printing, the blending of dots is especially pleasing.

Advantages of Gravure

1. Good overall image quality, even on thinner and lower quality papers.

2. High speed and exceptionally long runs, beyond one million. Cylinders may be replated several times, if necessary.

3. Generally simulates continuous tone better than other commercial processes, especially when printing on lower paper grades. The increasing capacity of offset to reproduce very fine dot patterns has minimized this traditional advantage somewhat where coated stocks are concerned.

Disadvantages of Gravure

1. Most expensive and complex image carrier preparation.

2. Cylinders are heavy and require large storage areas.

3. Generally limited to longer runs where the cost of cylinders can be justified.

4. Use of volatile ink solvents requires generally more careful storage and handling than materials used in most other processes.

5. All images must be broken into tiny cells for printing. While this is no problem for continuous tone areas, fine type and rules may not reproduce clearly.

6. Proofing is more cumbersome since images must be printed directly from etched cylinders to show the final image accurately. However, specially designed proof presses eliminate the need for proofing on a large rotary press.

Designing for Gravure

1. Take advantage of the excellent photographic reproduction of gravure by using large images with full tonal ranges in both black-and-white and color.

2. Select only medium-weight body types of no less than 10-point. Avoid light sans serif or fine-serifed roman typefaces. Keep rules and fine detail in line art to at least a one-point thickness.

SPECIALIZED PROCESSES

In addition to the three major processes—offset, letterpress, and gravure—we should know about a few specialized processes sometimes considered to be on the fringes of the commercial printing and publishing industry. These include screen printing, flexography, and indirect relief. While less familiar, these processes have very important roles to play in printed communication.

Screen Printing

The oldest and most versatile of these processes is screen printing. It is based on the stencil principle, whereby ink, paint, or dye is forced through an image carrier onto the paper or other substrate to be printed.

Because it requires little pressure to transfer ink, screen printing can print on almost anything. Common commercial and industrial applications include T-shirts, billboards, posters, decals, signs of all types, point-of-purchase displays, glassware, textiles, wallpaper, greeting cards, electronic chips and circuits, and product faceplates, to name but a few. Artists also use the process to produce original color prints called serigraphs.

In the past, this process was limited to printing shorter runs due to the heavy, slow-drying ink films applied and predominantly hand-operated equipment. Fine details also were difficult to produce, especially in continuous tone. Recent developments in inks, solvents, screens, and press technology are eliminating many of the traditional disadvantages of screen printing and opening up vast new markets for this process.

Web-fed rotary speed and good halftone detail are now possible, adding to the inherent advantage of heavy ink coverage and the capacity to print on virtually any shape or surface.

Image Preparation

Screen printing techniques vary widely; therefore, the following is a somewhat generalized description of important steps required.

While some stencils are still cut by hand or painted directly on the screen, photographic stencil preparation is now standard (Figure 11–40). First, high-contrast film positives—for either line or halftone images—are prepared on a process camera. In a contact frame, these positives are then placed over a presensitized photostencil film attached to a screen or a screen coated with liquid photoemulsion. High-intensity light is used to expose the stencil. This hardens the photoemulsion where light passes through the clear base of the film positives.

Following exposure, the stencil and screen are removed and washed with water or solvent to remove the unhardened emulsion—the image area that was blocked by the dark areas of the film positives. This leaves the stencil open in the image areas so that ink

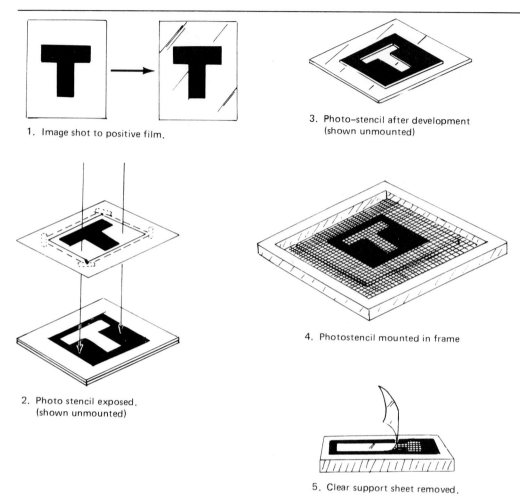

1. Image shot to positive film.

3. Photo–stencil after development (shown unmounted)

4. Photostencil mounted in frame

2. Photo stencil exposed. (shown unmounted)

5. Clear support sheet removed.

FIG. 11–40
How a screen printing stencil is made (photographic method).

can pass through it to the paper below. If photostencil film is used, the support sheet on which the emulsion was carried is peeled off after the stencil dries.

The developed stencil and screen is then mounted onto a frame, or *chase*, for printing. Sometimes the screen and stencil are mounted to a frame prior to exposure.

Printing Equipment

Several kinds of screen printing equipment are common: the hand-operated frame, the semi-automatic flat-bed or cylinder press, the automatic sheet-fed press, and the web-fed rotary press.

Manual and Semi-Automatic Operations

The manual frame is well-suited to short-run work. It may consist of little more than a solid base with a

hinged wooden frame to hold the stencil. The frame can be lifted and a sheet of paper placed beneath it. Ink placed on the stencil is then squeegeed through the unblocked openings of the screen onto the paper.

Sturdier, more sophisticated frames use a vacuum table to hold the paper into position and pedals or levers to position the screen. Semi-automatic presses pull the squeegee across the stencil automatically, but the operator still must position and remove the paper by hand.

Fully automatic presses are designed to feed, print, and deliver the sheets without manual operation. Many of these are equipped with sheet separating devices and dryers.

New web-fed rotary designs utilize curved screens and stencils wrapped around printing cylinders. One major innovator, Screen Systems, Inc., markets such a press capable of multicolor work at speeds up to 300 feet per minute. They have developed an all-metal screen/

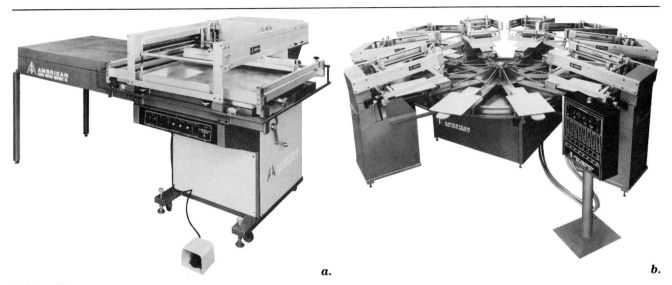

a.

b.

FIG. 11–41
(a) Semi-automatic screen press. (b) Computer-driven screen press capable of applying light colors in rapid sequence.
Advance Process Supply Co.

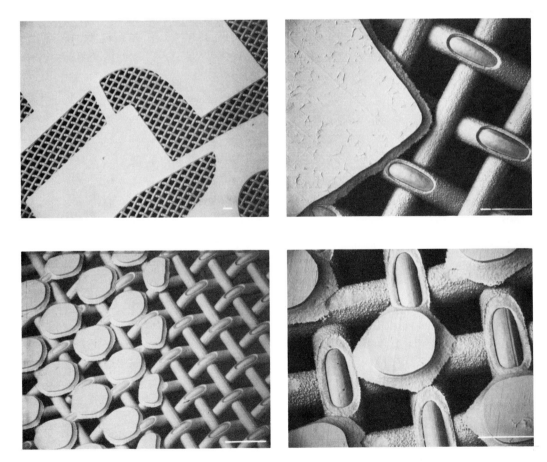

FIG. 11–42
Extreme close-ups of a 133-line halftone on a nickel foil screen, as seen by an electron microscope.
Screen Printing Systems, Inc.

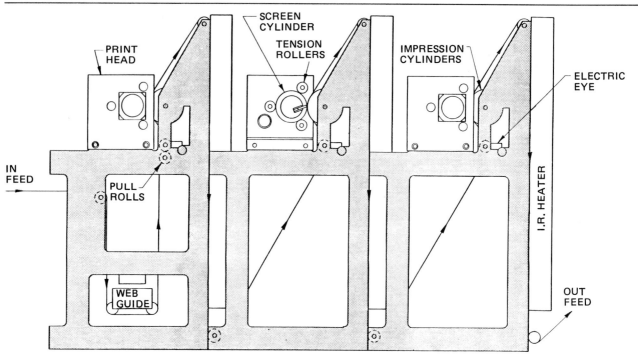

FIG. 11–43
Three rotary screen print stations in tandem.
Screen Printing Systems, Inc.

stencil capable of long runs and exceptional image detail. In this system, an ultrathin, precoated nickel foil is exposed, etched, and mounted to stainless steel mesh screen. Fine line images and 133-line halftone screen dots are possible, with simple exposure and etching procedures (Figure 11–42).

Of all processes, screen printing is capable of laying down the heaviest thickness of ink. This is a special advantage where light colors must be placed over

other inks or on a dark paper or other substrate. Rich fluorescent and metallic colors, difficult or impossible to print with other processes, are easily applied with stencil methods. New clear, water-soluble, and acrylic inks are also popular. Ink drying, a traditional problem for screen printing, is being resolved. For example, special opaque inks developed for Reinike rotary presses dry in from one to three seconds when passed through a small infrared dryer.

FIG. 11–44
Diagram showing web travel of a four-color rotary screen.
Screen Printing Systems, Inc.

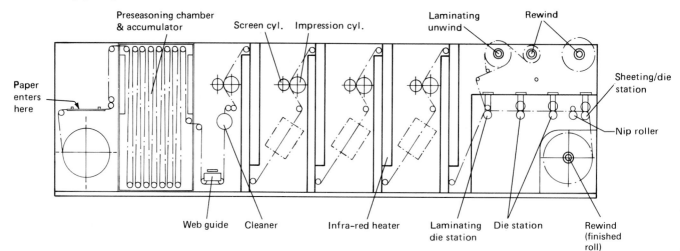

Screen printing in tandem with other processes is becoming common. Figure 11–45 shows a combination of web-fed rotary screen and letterpress, capable of multicolor screen and rotary letterpress printing, die-cutting, laminating, perforating, punching, and slitting. It can print on regular or laminated paper, pressure-sensitive materials, plastic film, cloth, or foil.

Summary of Screen Printing Characteristics

Screen printing can print virtually any color on any surface with little or no pressure. It can be identified by its heavy, very even ink coverage. Photographic stencils are most common, and can be used with a wide variety of screen materials. Traditional problems

FIG. 11–45
A screen/letterpress tandem printing device.
Mark Andy Inc.

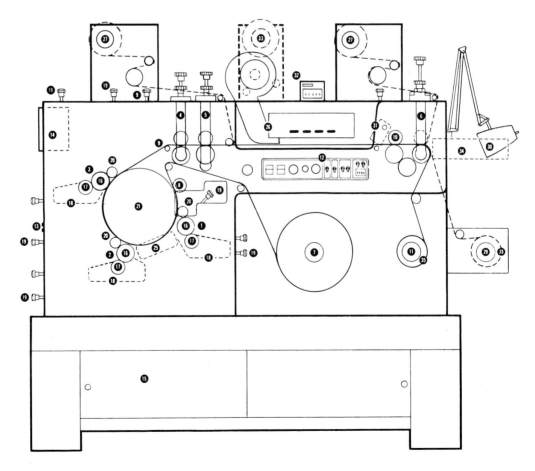

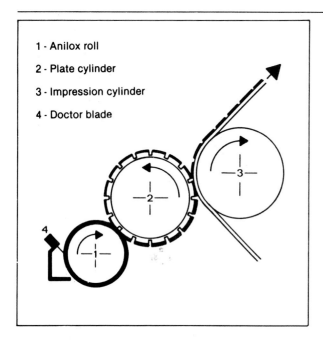

1 - Anilox roll

2 - Plate cylinder

3 - Impression cylinder

4 - Doctor blade

FIG. 11–46a
Principle of flexographic printing.
Industrial Specialties Division/3M

Cushion Mount

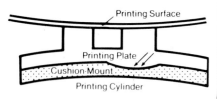

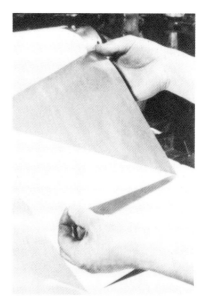

FIG. 11–46b
Mounting a flexographic plate.
BASF Systems Corporation/Printing Products

of speed, drying, and fine detail are being eliminated. However, manual printing methods are still slower than other processes.

Flexography

Another important process that has outgrown its second-class stature in the printing industry is flexography (flexo), a near relative of letterpress. It is a direct relief process which prints from resilient rubber or photopolymer plates on web presses. This versatile printing method is capable of high speeds and long runs. Plates made of rubber are especially good for printing on flexible packaging material and lower grades of paper where wrinkling and paper imperfections can damage harder plates.

Highly fluid aniline inks made from organic dyes are used, and only light printing pressures are required. Press designs are simplified since the inking systems require only a few transfer (metering) rollers and a single, specially textured "analox" roller to contact the plate. A doctor blade is used to scrape excess ink off the roller that contacts the plate, not off the plate itself.

Flexography, used extensively by the packaging and labelling industry, is now vying for new publication markets. While it is generally not capable of producing a sharp enough image for magazine or high-quality commercial printing, newspaper publishers are testing it as a possible alternative to offset.

Although a few flexographic plates are still cut by hand, most are either molded or photographically exposed and etched. If molded, original relief type or magnesium engravings are pressed into a heat-sensitive resin molding board. When cooled, this matrix mold hardens and can be used to cast rubber or plastic duplicate plates. Photopolymer plastic plates are often used, but the photopolymer is formulated to be softer and more flexible than that used for letterpress. Their preparation is described in the section on shallow relief (page 221).

The flexibility of the plates gives the process its name. Plates can be wrapped around a cylinder easily and attached with double-coated foam tape which cushions the impression. This eliminates much of the makeready common to letterpress.

Indirect Relief (Dry Offset or Letterset)

This process is essentially a modification of offset printing, but instead of a flat plate, a relief plate prints against the rubber blanket. This eliminates the need for water, and allows printing from the soft rubber blanket instead of a plate.

The flexible photopolymer plastic plate must be etched in *right-reading* form, since its image will be printed against the intermediate rubber blanket, then transferred to the printed item.

Popular in the packaging industry, indirect relief is used to print against metal, aluminum, or heavy paperboards. It lays down a heavy film of ink. Specially adapted presses are capable of printing two-piece aluminum cans individually in several colors. During the printing cycle, each separate color is transferred to a master rubber blanket, one at a time, to form the entire multicolor image. Then the can rotates into position to pick up the complete, multicolor image.

Indirect relief uses the best characteristics of both letterpress and offset.

FIG. 11–47
Flexomaster II with male/female punch and fan folder.
Allied Gear and Machine Co.

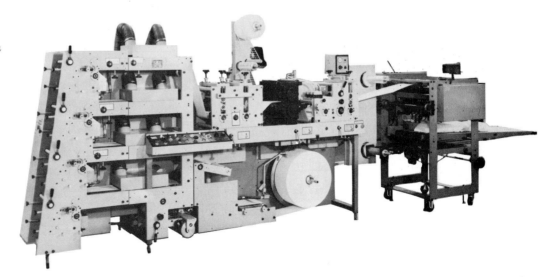

THREE-DIMENSIONAL PRINTING AND CONVERSION PROCESSES

Steel Engraving

An intaglio process, steel engraving is perhaps the most elegant form of printing. This process is widely used for producing expensive social invitations, announcements, and business printing.

Beginning as a flat plate of steel (or sometimes copper), an image is engraved into the surface of the *die*, creating a depression. When the die is inked, its engraved image holds ink while its top surface is wiped clean. In printing, paper is placed over the die and an impression roller passes over it. The pressure squeezes the paper into the die, drawing out the engraving ink. The combination of a heavy ink film and the pressing of the paper into the die creates an embossed or raised-image effect.

Although it is an intaglio process, steel engraving differs from gravure in a significant way. It does not require that all images be broken into tiny cells for printing.

While dies are sometimes etched by hand or by a special engraving device called a *pantagraph machine*, most are etched by methods similar to those used in regular photoengraving. Steel dies are sometimes hardened after engraving.

Because of its unique ability to print ultrafine lines that are difficult to counterfeit, steel engraving is used to print valuable items such as currency, stamps, and bonds.

Thermography

Actually more a finishing technique than a printing process, thermography is a specialized method of producing raised images in printing through the application of heat. Often used to simulate steel engraving, it has become highly popular for business cards and letterheads.

The paper is printed normally, usually by offset, and is transferred to a conveyor belt. Then a thin stream of thermographic powder containing resin is sprinkled onto the paper while the ink is still wet. Powder that does not adhere to the image is vacuumed away and the sheet is sent into a heating chamber or "tunnel." The heat causes the powdered ink to rise, forming raised images.

When compared with engraving, thermography can be identified easily since no embossing of the sheet toward the printed image occurs. In fact, if printed by letterpress the image may be somewhat debossed.

Once done only by a few large specialty houses, thermography is now a common service offered by many smaller printers.

Embossing

Embossing is the creation of an inkless raised image on paper or other substrate. It is usually performed on a platen press or a specialized embossing machine by pressing the paper between a heated die and a counter die attached to the platen.

Numerous effects are possible through variation in dies and temperature. *Scorch* embossing is possible by increasing the heat of the die. This creates a pleasing discoloration of the paper in the image area. *Glazing* of textured papers, also done through increasing die temperature, creates a smooth, shiny image for contrast.

Embossing dies, about 1/4-inch in thickness, come in five types: single level, multilevel, bevel edge, sculptured, and stamp.

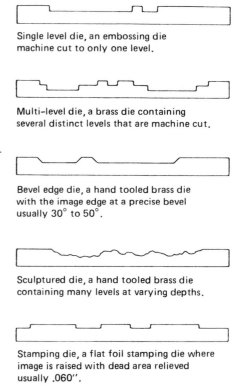

Single level die, an embossing die machine cut to only one level.

Multi-level die, a brass die containing several distinct levels that are machine cut.

Bevel edge die, a hand tooled brass die with the image edge at a precise bevel usually 30° to 50°.

Sculptured die, a hand tooled brass die containing many levels at varying depths.

Stamping die, a flat foil stamping die where image is raised with dead area relieved usually .060''.

Source: Brandtjen & Kluge, Inc.

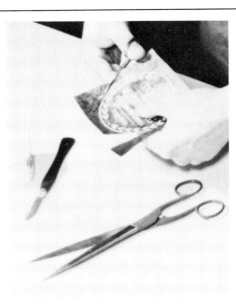

FIG. 11–48
Preparing an embossing die.
Heidelberg West, Inc.

Dies are made either of magnesium or brass. Magnesium dies are easily produced by photographic engraving methods, described previously. They are used for quick, single-level embossing. Brass dies, used where several distinct levels or sloped edges are desired, require extensive hand work and machining. They are more expensive, but last longer.

A raised counter die, the exact shape of the die's engraved image, must be attached to the press platen. As the press closes, the counter die forces the paper into the depression of the heated die, embossing it. Die manufacturers provide counter dies, but many printers mold their own on the press using counter-cast resin or embossing board.

One of the most dramatic sheet enhancements available, embossing can be expensive but is highly popular for special effects.

Embossing is usually *blind;* that is, it creates a raised image in the paper alone. When embossing is done over a previously printed image, it is called *registered* embossing.

Foil Stamping

When a thin layer of colored foil is applied to the printed material with a heated stamp or die, it is known as *foil stamping* or *foil embossing.*

Foil stamping requires only a flat, raised-image stamping die similar to a letterpress engraving. Foil embossing requires both a die and counter die similar to those used in regular embossing.

FIG. 11–49
A hot foil stamping press.
Heidelberg West, Inc.

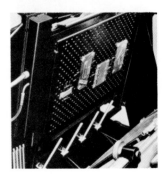

FIG. 11–50
A web-fed foiling operation for printing continuous forms.
Brandtjen & Kluge, Inc.

Foils come in rolls of various widths and in a wide range of metallics, gloss and dull colors, and clear films on a cellophane or polyester base. Press attachments feed the foil strips continuously over the heated die during printing. Pressure and heat release the foil from its base and cause it to adhere to the paper.

Foil application is another popular way to create unusual effects not possible with traditional printing processes.

Die Cutting

Another noninking enhancement of paper and other materials, die cutting is the punching out of special shapes through the use of sharp, tempered steel rule dies. The effect is similar to that produed by a common cookie cutter. Although heavy die-cutting machines are commonly employed, platen, flat-bed cylinder, and rotary letterpresses may all be adapted for such work.

Dies may be heated for *thermal* die cutting of hard, nonporous, or pressure-sensitive materials, but unheated dies are often used on papers.

Manufacturing of die-cutting forms is highly specialized and can be quite expensive, depending upon the intricacy of the shapes required.

Tiny nicks in the sharp rules prevent the die-cut shape from separating from the larger sheet after it has been cut. This keeps paper from falling into and damaging the press.

FIG. 11–51a
The die-cutting form, with steel cutting rules, can be locked easily and quickly.
Heidelberg West, Inc.

FIG. 11–51b
A precision modular rotary die-cutting unit.
Bernal Rotary Systems, Inc.

FIG. 11–52
A high die. Photographed by permission of Smith Printers and Lithographers, Tustin, CA.

On-Press Scoring, Perforating, and Numbering

Scoring, the creasing of heavy cover papers and other materials so they will fold without cracking, is often accomplished on the printing press. Raised rules are locked into the printing form and used to print against the paper without ink, creating a crease for subsequent folding. The sheet is folded *away* from the creased image, not toward it.

Perforating is similar to scoring, except that a sharper, serrated rule is used. It also requires a steel backing strip on the platen under the paper so that the rule will produce a satisfactory perforation.

Machine numbering is possible on virtually all types of printing presses today. It requires small machines containing numbering wheels that rotate after each

FIG. 11–53

impression and provide sequential numbers. In printing forms such as multiple checks, where several duplicate images are printed on the same sheet, *skip wheels* allow numbers to jump sequence. For example, the top numbering machine on a three-check page would skip from number 1 to number 4, since the two other machines would already have printed numbers 2 and 3 on the same page. The second and third numbering machines would skip to numbers 5 and 6 for printing the next sheet.

Scoring, perforating, and numbering may all be done off-press with specially designed equipment.

Color in Graphics

Bringing Realism and Excitement to the Printed Page

Nothing brings the printed page to life better than color. It can soothe, shock, stimulate, or seduce. It can make an image all too real or delightfully unreal. In print, as with the other media we enjoy today, color has an unmistakable power to make us sit up and take notice.

Introduction

For many people, color and printing are synonymous. So many direct-mail pieces, magazines, books, and brochures we see today utilize full-color illustrations that we can easily take their beauty and impact for granted. High quality color in commercial and publications printing is a recent phenomenon, however. Emerging in the late 1950s, it helped set the stage for color television, whetting the appetites of advertisers and viewers alike for the realism of full color. Today, color is so dominant in both print and broadcast media that occasionally black-and-white is used to set a message apart rather than simply to save money.

Little doubt remains about the general effectiveness of color in print. Virtually every study conducted to date indicates that color sells and communicates well. Newspaper readers prefer full-color pictures on page one. Magazine advertisers are told that color enhances the attention-getting power of their message by up to 40 percent. Direct mailers know that colored paper alone may help double their returns if used intelligently with an appropriate message. While color doesn't guarantee sales, it seems to ensure that a message will at least get noticed—a prerequisite to communication in print.

Color can be used effectively by anyone today. Both the mysteries surrounding the preparation of original color materials and the chances of a printer failing to reproduce color acceptably have been greatly reduced. Like any other printing technique, though, the more we know about the process the better our results will be.

The Meaning of Color

The physiological and psychological impact of color stimuli and their effects on perception are only beginning to be understood. Designers usually employ the warm end of the spectrum—reds, oranges, and yellows—to show action. Cooler hues—blues, greens, violets—depict calmness and serenity. Here are some of the other connotations usually assigned colors:

- greens and browns—colors in nature, trees, grass, earth.
- blues—also natural, as in water, sky; also shows seriousness or sincerity (true blue).
- violets and purples—royalty, elegance.
- gold—value, rarity, stability.
- reds—passion, intensity, warmth, sunshine, protection, life (blood).
- white—cleanliness, purity.

The use of behavior-influencing colors in today's retailing, marketing, and advertising is pervasive. Fast-food restaurants are designed in warm or hot colors to depict speed of service, warmth, and an exciting atmosphere. Hospitals use whites and restful blues and greens. Exclusive restaurants and department stores often choose rich purples, golds, and earth tones to convey a sense of quiet elegance.

Reds, whites, yellows, and browns line the shelves of grocery stores, along with full-color product illustrations. Pure-looking primary hues and realistic colors are especially essential for food products. Blue, less suitable for food, still remains highly favored by both sexes in most other product promotion and packaging. His-and-her marketers usually offer a navy blue for men and a blue-green for women, reflecting a belief that males prefer darker tones.

Notions about color effects and preferences should be weighed carefully against common sense. Color fads in cars, clothing, furniture, and other everyday items often temporarily distort our views about color preference. This year's high fashion pink and gray pullover may well be next year's discount store reject. Designers are rarely surprised when formerly taboo colors in clothing, cars, and packaging make a comeback, however brief. They remain ready to take advantage of, and often influence, the seasonal popularity of color.

The effectiveness of a color in print depends on many things—its contrast with other colors, its illumination, how well it matches the mood of the viewer, how accurately it is reproduced, how much of it is used, and how well it complements an overall color theme. One thing is sure—nearly everyone believes that color is worth the price and the planning.

Color Basics

People who use color every day—art directors, graphic designers, painters, interior decorators—usually think of colors according to their relationships on a color wheel. The wheel contains a triad of *primary* colors: red, yellow, and blue. They are called primary because all other colors can be created from them, at least in theory. *Secondary* colors: orange, green, and violet, appear between the primaries on the wheel and represent mixtures of two primaries. Finally, *intermediate* colors are those created by mixing a primary color with an adjacent secondary color. Figure 12–1 shows the standard color wheel.

The color wheel indiates which colors will harmonize or balance one another when used together. Figure

12–2 shows three common color strategies: complementary (colors opposite each other on the wheel); split complementary (colors on both sides of a color balancing the complement of that color); and analogous (neighboring colors on the wheel).

Three terms are used to describe the nature of color: *hue, intensity,* and *value.*

Hue is the common *name* of a color, such as *red.* Often we use terms such as "reddish" to describe the most prominent hue we see in a color.

Intensity, sometimes called *chroma,* refers to the relative saturation or *purity* of the hue. The intensity of a hue can be reduced by mixing it with varying amounts of its complement. For example, the intensity of red can be reduced by adding small amounts of green. Red and green, as with all opposites on the wheel, are approximately the same intensity until mixed with one another.

Value is the relationship of the color to black or white (usually thought of as its approximate placement on a gray scale). In painting, lighter color values, called *tints,* are created by mixing white pigment with a color. In printing, the same effect is usually created by breaking the color into small dots and allowing the white paper to show through. To the eye, this is the same as mixing white ink with the color. The smaller the color dots the lighter the tint or value. Of course, it is also possible to mix white ink with a color to form a tint, similar to the way painting tints are created.

Shades are mixed by adding either black or a darker neighbor on the color wheel to a hue. Small black or dark color dots printed over a lighter color achieves roughly the same effect in printing.

Using Color Theory

Understanding what color is and how it works helps us use it more effectively. The objects we see around us every day—green trees, yellow road signs, red sweaters—would be colorless without light. White sunlight contains all the rainbow colors of the visible light spectrum. With the help of a prism, we can see the individual colors of each light wavelength.

Ours is a world of color only because the objects in it absorb and reflect light in different ways. In practice, we consider white light to be a combination of red, green, and blue primaries. The light color reflected from an object determines the color we see. For example, a red sweater is simply one that contains dyes that absorb blue and green light and reflect only red.

The light primaries, red, green, and blue, are known as *additive* primaries; that is, they form color effects by being overlapped or merged. Anyone who has observed the multicolored spotlights at an auditorium show has noticed the almost magical formation of white light when red, green, and blue beams converge on the performer in center stage. This is the additive principle in action.

When two additive light primaries combine, they form what is known as a *subtractive* primary. For instance, a red spotlight combined with a green one would produce yellow. One can think of the yellow as *minus blue*—it contains only red and green light but no blue.

Printers use the subtractive (pigment) primaries in full-color (four-color) printing to create the light primaries present in our world of color. By using transparent inks with color pigments that match the subtractive primaries, they can control the color of light that is reflected from white paper. This technique reconstructs the color of an original image or object. Here's how it works:

• Cyan ink absorbs (subtracts) the red waves of white light and reflects green and blue.

• Magenta ink absorbs green light waves and reflects red and blue.

• Yellow ink absorbs blue light and reflects red and green.

Each one of the inks absorbs a different light primary and reflects the two remaining.

When all three subtractive pigment primaries are printed in controlled proportions, the full range of light waves is reflected back to us from the paper. An image of a lemon, with its predominantly yellow process ink dots, reflects mostly green and red light to our eye. This combination of light waves appears yellow to us.

Of course, most color images we see printed contain a virtual rainbow of color. This is made possible by varying the *size* of halftone dots used to print the colors in each area of the illustration. Figure 12–3 shows an enlarged area of a full-color halftone. The halftone dots used to print color are similar to those used for black-and-white, except that four negatives are required, one for each primary ink and another for black. Black provides better definition and shadow intensity than the three primaries alone can yield. It is also used to print black type. The process required for separating the component colors for printing is discussed later in this chapter.

Types of Color Printing

We have seen how transparent pigment primaries are used to reflect mixtures of light back to our eye to recreate full-color images. Much of the color printing we see, however, is not so complex.

Color printing is easier to understand if we think of the two basic kinds of inks used—process and nonprocess. *Process* inks are transparent so they can present different color mixtures to our eye when printed in the same spot on paper. *Nonprocess* (flat) inks are largely opaque. They reflect only a single hue. The more opaque the ink, the less the paper color beneath can influence the color of the printed image.

Three categories of color printing are produced from the two basic types of ink:

Flat color—images printed from a *single*, nonprocess or process ink. These are usually line images such as type, screen tints, or borders, but line and halftone illustrations also can be reproduced. Flat color is also referred to as single color, spot color, applied color, or second color.

Full color—images made from the overprinting of halftone dots, usually in yellow, magenta, cyan, and black—all at different screen angles—so that the subject colors appear natural. This type of color is often called four-color or four-color process, although it is possible to use more than four inks, as well as nonprimaries such as fluorescents, to enhance an image. Full color requires photographic or electronic separation of all continuous tones into four negatives, each to be printed in a separate color.

Manufactured color—images produced through the overlapping of solid blocks or screen-tint dots of more than one flat or process color. This creates the appearance of having used a flat color. Another name used for this is mechanical color.

Flat Color

Flat colors are prepared either by the ink manufacturer or by the printer. Printers can buy the particular Passionate Persimmon hue they need, already premixed and ready for the press. They may choose, however, to use a system of mixing colors, somewhat like a paint store, to create a needed color themselves. Either way, the color they place into the ink fountain of the press will be the color they see coming out on the printed sheets.

Flat colors produced from a mixing system are specified by number, with each number representing a

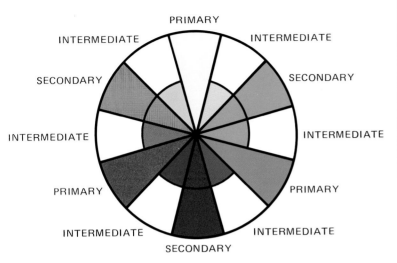

FIG. 12–1
A standard color wheel.

FIG. 12–3
An enlargement of a four-color image.
Courtesy of 3M, St. Paul, MN

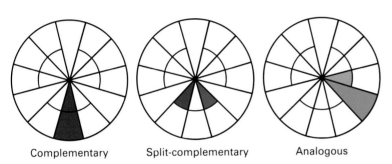

Complementary   Split-complementary   Analogous

FIG. 12–2
Common color strategies.

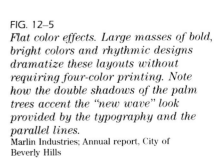

FIG. 12–4
*Pantone Matching System color
formula guide and color specifier book
containing tear-out swatch "chips" for
attaching to artwork.*

FIG. 12–5
*Flat color effects. Large masses of bold,
bright colors and rhythmic designs
dramatize these layouts without
requiring four-color printing. Note
how the double shadows of the palm
trees accent the "new wave" look
provided by the typography and the
parallel lines.*
Marlin Industries; Annual report, City of
Beverly Hills

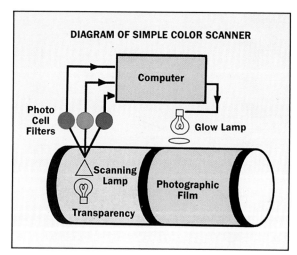

DIAGRAM OF SIMPLE COLOR SCANNER

Computer

Photo Cell Filters

Glow Lamp

Scanning Lamp

Transparency

Photographic Film

FIG. 12–6
Diagram of a simple color scanner.
Courtesy of Eastman Kodak Company

FIG. 12–7
Light path for transparent copy. After passing through the analyzing drum, the light passes through filters and photomultipliers, where it is converted into electronic signals.
Courtesy of Eastman Kodak Company

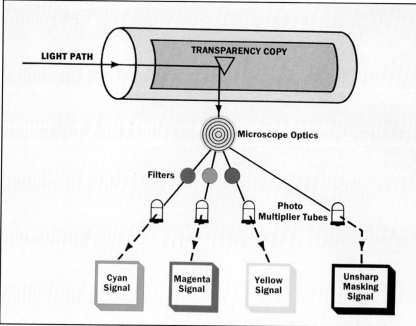

LIGHT PATH

TRANSPARENCY COPY

Microscope Optics

Filters

Photo Multiplier Tubes

Cyan Signal

Magenta Signal

Yellow Signal

Unsharp Masking Signal

FIG. 12–9
Using a reflection color densitometer.
Cosar Corporation

FIG. 12–10
Standard Offset Color Control Bars available from the Graphic Arts Technical Foundation.

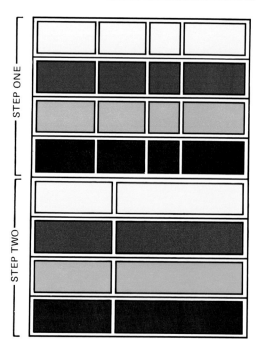

STEP ONE

STEP TWO

FIG. 12–8
How seven sets of color separations can be "ganged" onto a single piece of scanner film in two separate steps. The actual film is black—the colors in the diagram indicate the positions of each color.
Dai Nippon Screen Mfg. Co., Ltd.

FIG. 12–11
Dramatic color impact is achieved by: blue image posterization and silver type on a black background; black-and-white versus color for contrast; and a double-page spread title ending in a color "starburst."
Annual report, Orange County, CA, Performing Arts Center; *Vectors* Magazine, Hughes Aircraft Company; *Marathon World* magazine, Marathon Oil Company

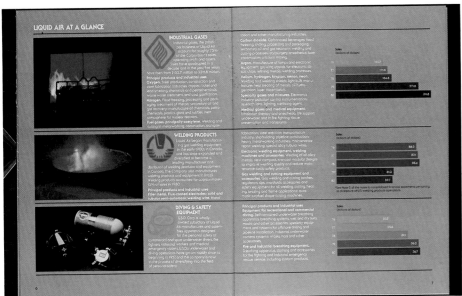

FIG. 12–12
Using a fifth color for layout harmony. A silver background enhances the richness of the four-color photographs; symbols and bar graphs derived from process colors add to the effect.
Annual report, Liquid Air Corporation

FIG. 12–13
Creative ink and paper combinations: red-and-black line art printed on silver cover paper; two-color duotones that give a four-color effect; and a four-color image printed on the gold side of a gold/white duplex cover paper.
Kurtland Community College, Consolidated Papers, Inc. and Appleton Papers Inc.

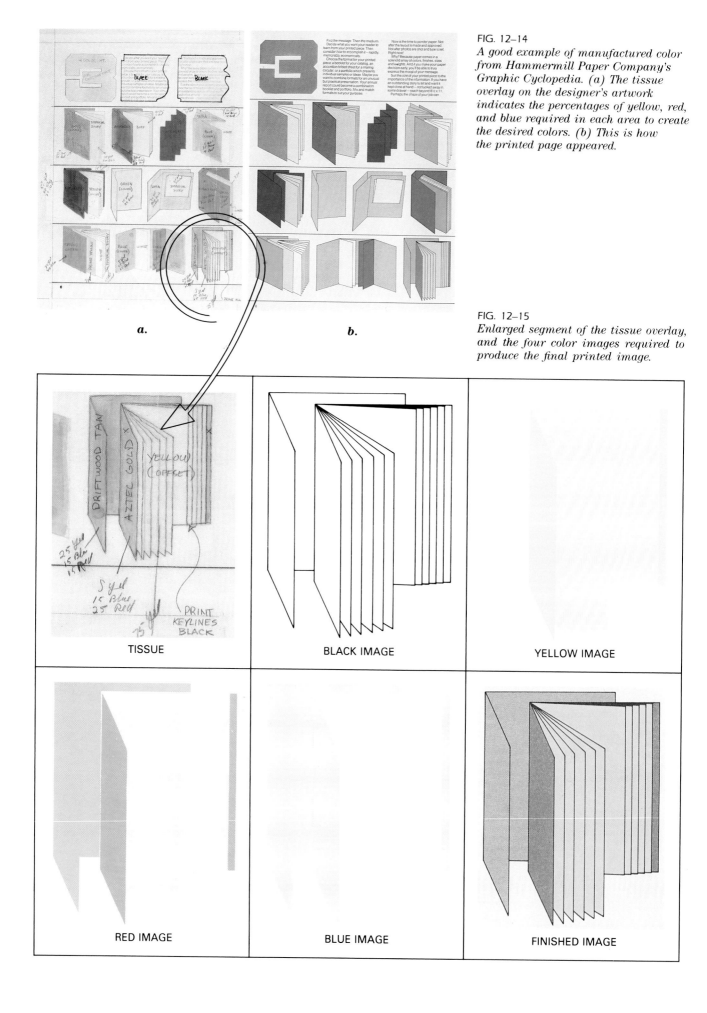

FIG. 12–14
A good example of manufactured color from Hammermill Paper Company's Graphic Cyclopedia. (a) The tissue overlay on the designer's artwork indicates the percentages of yellow, red, and blue required in each area to create the desired colors. (b) This is how the printed page appeared.

a.

b.

FIG. 12–15
Enlarged segment of the tissue overlay, and the four color images required to produce the final printed image.

TISSUE

BLACK IMAGE

YELLOW IMAGE

RED IMAGE

BLUE IMAGE

FINISHED IMAGE

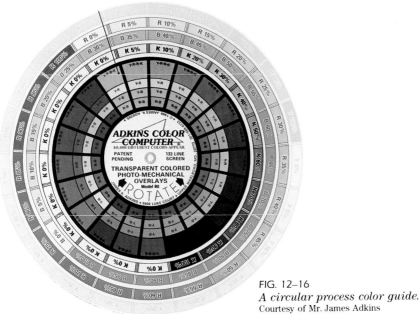

FIG. 12–16
A circular process color guide.
Courtesy of Mr. James Adkins

a. *Negative image of circle line conversion overprinting 50% yellow, 10% red, and 30% blue.*

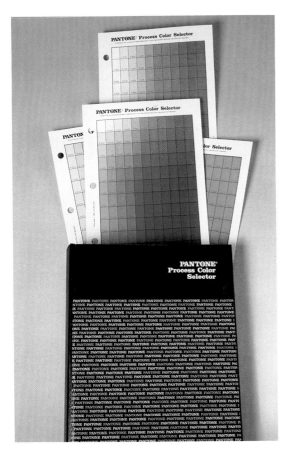

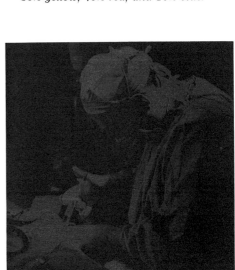

FIG. 12–17
The Pantone Process Color Selector. Nearly 9,000 color combinations in 10 percent tint steps are indicated for both coated and uncoated papers.

b. *Black etchline conversion overprinting 50% yellow, 40% red, and 10% blue.*

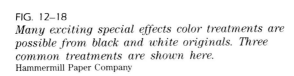

FIG. 12–18
Many exciting special effects color treatments are possible from black and white originals. Three common treatments are shown here.
Hammermill Paper Company

c. *Blue halftone overprinting solid (100%) red.*

FIG. 12–19
The Hell Chromacon electronic page makeup and retouching system employs four major elements: a programming station (right); an input station with control computer, VDT and scanners; an image processing station for page makeup and image manipulation; and an output station.
HCM Graphic Systems, Inc.

a.

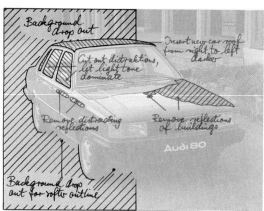

b.

c.

FIG. 12–20.
Imperfections in original images are no problem for electronic color systems. Here we see: (a) the original film transparency, (b) a tissue overlay describing the retouching work to be done, and (c) the final image.
HCM Graphic Systems, Inc.

a.

FIG. 12–21
*Here we see the transformation of (a)
an original image into (b) an exciting
color cover. Note the specific markings
for graduating the color tints from
light to dark, as marked on the
instruction tissue (c).*
HCM Graphic Systems, Inc.

b.

FIG. 12–22
*An early snowfall descends on this
pastoral mountain scene (center), as
photos of late summer (top) and
winter (bottom) are merged
electronically.*
HCM Graphic Systems, Inc.

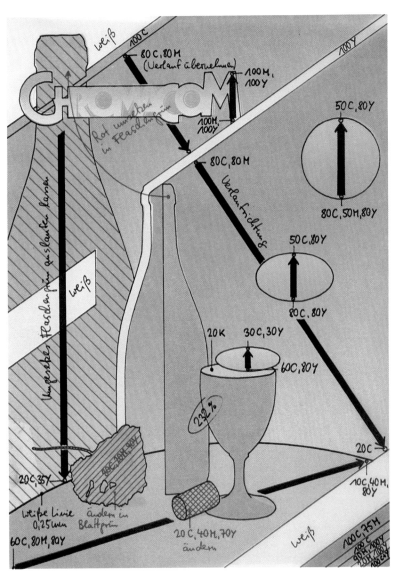

c.

FIG. 12–23
Applying hand-cut, self-adhesive color film to original art.
Chartpak

FIG. 12–24
An overlay produced by exposing a negative of the original art to 3M Color Key film. The overlay is then developed, rinsed, and dried. Nine colors and black are available.
Courtesy of 3M, St. Paul, MN

FIG. 12–25
Custom color transfers made from negatives of type and other original images. This system allows a wide range of colors, including fluorescents and metallics, to be produced.
Chromatec Scientific Corporation

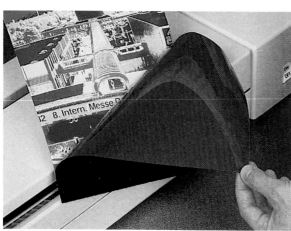

FIG. 12–26
The Agfa-Gevaert Copycolor system produces peel-apart color print through the use of a process camera and diffusion transfer processor.
Agfa-Gevaert, Inc. and *Graphic Arts Monthly*

specific mixing formula. One well-known system, the PANTONE MATCHING SYSTEM®*, utilizes color swatch books which show inks printed on both coated and uncoated paper.

Flat inks may be used full strength or mixed with additives like varnish, drier, or even transparent white mixing ink for increased volume.

Solid coverage, or 100 percent of color in a printed area, requires no dot pattern. Reducing the value of a flat color requires the use of screen tints. Just as black appears as increasingly lighter gray as the dots used to print it are decreased in size, colors appear lighter in value as dots get smaller—the white paper showing between the printed dots causes this effect.

Screen tints give a multicolor appearance with only a single ink color. We can consult a standard color tint guide to specify which dot percentage will yield the desired value.

A fan-out guide for selecting PANTONE Color numbers and formulas is shown in Figure 12–4. Also shown is a booklet containing matching color "chips" that can be removed and attached to art boards, and a guide which shows how screen tints, halftones, and type reproduce in each color.

Figure 12–5 shows examples of flat or single color effects.

One color and black may be used to create a *duotone, tint block,* or other color effect in illustrations. Duotones are two-color images made from black-and-white continuous tone originals. Two halftone negatives of the same image are used, one usually printed in black, the other in red, brown, blue, or other color.

To produce the two duotone negatives, the camera operator typically shoots one with halftone dots at a 45-degree angle. Then the halftone screen is rotated to a 75-degree angle and another shot of the same image is made on another sheet of film. Turning the screen between shots prevents the dots from the two separate colors from printing exactly over one another and hiding the lighter color. With the screen angle 30 degrees apart, the two sets of halftone dots—black and color—print side by side in tight patterns, providing color enhancement and richer detail to what otherwise would have been a single-color image.

Either negative in a duotone can be adjusted for a longer tonal range from light to dark. Larger or smaller dots (meaning more or less color) in the highlights, middletones, or shadows for either black or color can be produced by adjusting camera and flash exposures.

A dark color such as brown is sometimes used for duotones rather than black. One of the most striking duotone effects is the black/black *double print,* which renders striking black-and-white detail, especially in shadow areas.

Fake duotones are created by printing a single halftone—usually in black—over a flat color or a tint block. A disadvantage of fakes is that their highlight areas can be no brighter than the background color on which they are printed; therefore, they have less contrast. Also, they still require the printing of two ink colors, even though only one halftone is needed.

An advantage of fake duotones is that they require less precision in printing. Usually, however, true duotones are seen to be worth the small extra expense.

Color Separation by Scanner

Electronic scanners have largely replaced process cameras and color enlargers as the primary means for separating and recording on film the color elements of originals for full color printing. Costs of scanners have plummeted in recent years from more than half a million dollars to less than a fifth of that, depending upon the type of equipment.

What does a scanner do? It accomplishes electronically what other separation methods do photographically. "In the scanning process, a picture is examined bit-by-bit by an optical device, and this information is electronically modified and reproduced as pictures. Like the color camera, the scanner separates the color copy into negative or positive records of the standard four images—cyan, magenta, yellow and black" (*The Color Separation Camera,* Kodak, p. 2).

The simplified diagrams in Figures 12–6 and 12–7 depict the basic operating elements of typical scanners. A color original, usually a film transparency (slide) is mounted on a transparent drum. Light from an analyzing lamp illuminates the transparency from beneath. An optical device above the revolving transparency splits the transmitted light into four beams. Three of the beams pass through different color separation filters—red, green, or blue—and into a photo multiplier tube. A fourth beam, used for unsharp masking, passes directly into a similar tube without filtration.

---

*Pantone, Inc.'s check-standard trademark for color reproduction and color reproduction materials.

Separation signals from the three color-filtered beams and the masking beam are transmitted to a computer which modifies them. Modifications are done according to pre-set instructions or they can be controlled by the scanner operator. A fourth signal—for black—is created by a combination of the three color signals.

As the exposure unit of the scanner receives the modified separation signals from the computer, it exposes the film accordingly. Argon-ion or helium-neon lasers, as well as gas discharge lamps, may be used for exposing film.

How do the color filters break down the original into its component colors for printing? Recalling our earlier discussion of light and pigment primaries, we remember that *cyan ink absorbs (subtracts) red light* and reflects green and blue; *magenta absorbs green light* and reflects red and blue; and *yellow absorbs blue light* and reflects red and green. The red, green, and blue filters in the scanner produce *negative* exposures, on film, of the light *reflected* (transmitted) from the original. The clear (unexposed) areas of the negative are records of light *absorbed* by the original. When a printing plate is made, light passes through these clear areas of the negative to create dots on the color plate.

Now we can see that the red filter transmits only red light signals from the original to the film. The unexposed parts of the film (the clear areas that will create the image on the printing plate) represent a record of the green and blue light the red filter absorbed. A plate made from this negative will be printed with cyan ink, the pigment that reflects green and blue light. The result? *We capture a record of the blue and green light the filter absorbed, using cyan ink to print this record.*

Today's scanners are often classified according to their method of output: continuous tone, contact screen, or electronic-dot-generating. *Continuous tone scanners* provide panchromatic film negatives that can be screened following separation. *Contact screen scanners* use dot-producing screens somewhat similar to those used in the process cameras to produce regular halftone dots. *Electronic-dot-generating scanners* use a multiple array of tiny exposure lamps or lasers to produce micro dots. These ultra-fine dots, as many as 144, may be used to create a single halftone dot. As in black-and-white halftoning methods, electronically generated dots may be round, square, chain, elliptical, or any other shape the computer program calls for.

Up to 2,000 scans per inch can be made on an original, but when the image is screened the output is converted to standard rulings—lines (dots) per inch. Typically, this would be 133-line for web offset or 150- to 200-line for sheet-fed. This means that several scans are made for each tiny strip of halftone dots produced.

An enlargement/reduction ratio of 10 to 3,000 percent is possible on one manufacturer's machine. Output speed varies with the type of scanner and the manufacturer. It is given as a fraction that lists seconds per scanned inch and number of scans in that inch. Example: the Crosfield Magnascan 570 lists a rate of 30/400, which means that it takes thirty seconds to cover one inch with 400 scan revolutions.

Color originals are *transmission* copy (slides) or *reflection* copy (photographs or other illustrations on an opaque base). Slides, usually called *transparencies*, may be any size that will fit on the scanning drum, but are usually 35 mm., 2¼ in. square, or 4×5 in. formats. Often, duplicates are made of originals. This makes color densities more similar if several transparencies are to be *ganged* (grouped) together for separating, as shown in Figure 12–8. It also allows emulsion stripping, whereby the base of the film can be removed and the color emulsions assembled into a single piece for scanning. This saves scanning and image assembly later on.

Reflection copy must be flexible enough to wrap around the scanning drum, and should not have tape, labels, or other significant buildup on the back. Several manufacturers' drums will accommodate up to a 20×25 inch original. Glossy surfaces which reflect evenly are preferred, although any continuous tone color original can be used. Valuable paintings or other hard-mounted originals are usually copied onto slides or separated by a color camera.

A film output size of about 20×24 inch is common for scanners, many of which can produce images one-up, two-up, or four-up, depending upon desired sizes.

Image Control

Significant changes in original colors, such as flesh tones and intensity of red roses or blue skies, are possible with scanning. It is important that the machine operator know the color effects desired ahead of time to avoid remakes. Also, since the printing process cannot produce the entire tonal range of the original, the operator must *compress* the image into fewer printable tones. If the operator is told which details in the shadows, middle tones or highlights are most important, dot sizes and tonal contrast can be adjusted to achieve the best results.

Color Control in Printing

Process color requires careful monitoring by press operators during printing if color effects are to match those originally obtained in the separations. After color proofs have been approved by the customers, the printer is responsible to see that the finished color matches them as closely as possible. Since all required color corrections will have been made on the separation films, the printing plates will be exposed carefully so that they contain the correct dot sizes, in correct position, for each ink color. It then becomes the press operator's job to see that the right amount (density) of each ink is printed and that dot positions (registration) are maintained.

As each plate transfers its image to the paper, it also prints a *color bar* along the edge of the sheet. When all colors have been printed, a full color bar will appear containing yellow, magenta, cyan, and black printed together in solids, screen tints, and several overlapped combinations. Special symbols and patterns to indicate dot gain, trapping (how well one color accepts a color printed over it), neutral grays, slurring, and registration of dots, also appear in the bar.

Figure 12–9 depicts a modern reflection densitometer and its use. Density is determined by the amount of reflection on a scale of 0.0 (maximum reflection) to 2.0 (least reflection). For example, a densitometer reading of 1.05 yellow, 1.26 magenta, 1.34 cyan, and 1.50 black might be obtained for a given set of adjacent color bars. Colors not within acceptable limits—about ± 0.02 for colors or ± 0.04 for black—would be adjusted.

Figure 12–10 shows a color bar available from the Graphic Arts Technical Foundation.

Press operators use a color densitometer to determine the density of ink being applied to the sheet. If too much or too little of one pigment is applied, the hues of the entire image will be affected. Each color on the printed bar is viewed with the reflection densitometer, and digital density readings are compared to pre-established values. If the color density of a certain bar is not within an established tolerance, all images printed beneath that bar will be incorrect in color. To correct the color, the press operators manually adjust the keys that regulate the flow of ink from the fountain until the right amount is transferred to each section of the plate.

Popular Color Effects

Full-color printing offers nearly unlimited opportunities for editors and designers to add visual impact to their messages. The pages shown in Figure 12–11 illustrate how dramatic color use can lend excitement to a printed piece.

Many creative new color enhancements—both striking and subtle—have become commonplace today as more people explore the potential of full-color printing. Some of them are described below.

Fifth-Color Backgrounds. Usually printed in light, flat colors such as warm gray or cream, these unscreened hues appear lighter than colors manufactured from process dots. Such backgrounds are used in annual reports, brochures, and other high-quality pieces. Sometimes a faint watermark-like pattern depicting an organization's identity symbol is used.

If a fifth-color background is printed over an entire page, bleeding to the edges, it can create the illusion that four-color images have been printed on a colored paper. Actually, it is necessary for the printer to provide *windows* in the background color so that four-color illustrations can print on white paper, not on the background color. Otherwise, the color in the illustrations would be distorted.

Figure 12–12 shows an example of fifth-color use.

Full Color on Colored Paper. Although white, coated book paper is usually preferred for full-color printing, colored papers can produce strikingly beautiful effects when used sensibly. The main thing to remember is that transparent process inks allow paper color to show through. This changes the hues in the printed image. For example, full-color images on cream-colored paper will become much warmer—their blues will become blue-green, and their reds will turn to red-orange.

Color shift due to printing on colored papers can be controlled in two ways. The color separation negatives—discussed in the next section—can be corrected to compensate for paper color. For example, reds and yellows can be reduced for printing on cream-colored paper. Another method is to print opaque white windows on the paper, then print full-color images in the windows. This creates an effect somewhat similar to having printed the images on white paper.

Figure 12–13 shows the effects of paper color on images printed with color inks. Shown too are examples of two-color duotones, also produced by scanning methods, but printed with nonprocess inks.

Special Inks.   Corporate trademark and identity colors are often printed as flat, fifth colors rather than as manufactured color. The basic mixing colors of the Pantone Matching System (or other color system) can be used to produce custom color formulas, or the system manufacturer will provide special mixtures for a fee. Ink manufacturers also offer custom color services.

Unlike manufactured colors, flat colors are unlikely to shift in hue during printing. Such a shift is a real possibility in manufactured color, where even the slightest changes in size of the overlapped color dots during printing can be significant. A good rule to remember: if color match is critical, stay with flat color.

Fluorescent inks are popular as flat colors, especially since they retain their intensity in the high-moisture environment of offset printing. Dramatic full-color printing is also possible with fluorescent pigments. Producing effective color separation negatives for multiple fluorescents is a highly specialized task.

Varnish Overcoats.   Full-color images, especially on solid-coverage covers, often require a protective coat of printed varnish to prevent fingerprints from showing. Photographs and other areas can be spot varnished, or the entire sheet can be treated. Gloss varnish enhances the richness of full color. Sometimes a dull varnish is placed around photographs to bring out the natural gloss of the full-color inks. Spot varnish can produce dramatic black-on-black or other dull/gloss contrasts of the same color.

Manufactured Color

Although overlapping of flat colors can produce manufactured color, this on-press mixing is usually done with standard process primaries—yellow, magenta, cyan, and black. From these virtually all others except fluorescents can be duplicated reasonably well. It is especially important to realize that *process inks cannot exactly duplicate flat colors; they only approximate them by overlapping dots and/or solids in varying proportions.* Manufactured

color should be selected from a process color guide, not from a flat color swatch book.

In full-color printing, photographs and other illustrations require that all three primary colors plus black be used. Taking advantage of the presence of all color primaries, printers overlap two or more of them to create color effects in headlines, borders, and other elements. If a headline is printed in both yellow and blue, it will become green. A border printed in yellow and red will be orange. Treatments such as these are known as *using process color to simulate flat color.* The color is "manufactured" in the sense that no ink of that color is placed in the press—the color appears only on the printed sheets.

The examples shown in Figures 12–14 and 12–15 depict how manufactured color is specified by the designer and how the printer uses the four standard process colors to simulate the printing of numerous flat colors.

Many types of process color guides have been devised for making color selection more convenient and accurate when process colors must be combined to manufacture other colors during printing. Two such guides are shown in Figures 12–16 and 12–17.

While line art is commonly reproduced with manufactured colors, special effects halftones, discussed in Chapter 6, also lend themselves especially well to such color. Figure 12–18 suggests the variety of color/image combinations possible with special effects coloring of black and white originals.

Electronic Color Preparation

The many recent technological advances in image creation and preparation discussed in Chapter 7 have been especially beneficial in color printing. Color electronic prepress systems (CEPS) allow virtually unlimited creation, storage, and manipulation of color images. All color elements can be merged with text and other graphics electronically. Then fully color-separated films containing all the halftone, tint, line, and text images for entire pages can be produced.

Figures 12–19 through 12–22 depict the almost unlimited control such systems provide.

Previsualizing Color

Long before electronic color systems provided ways of previewing color on-screen before it is printed, other color visualization systems had been widely used and still enjoy great popularity. These color "comprehensive" systems or techniques are generally low in cost and easy to use. They may be used in conjunction with electronic systems, where they are less expensive and faster in certain applications.

The simplest system of "comping" color is that of using markers, pencils, or self-adhesive color films to fill in desired effects. Other systems require the use of a process camera to produce a negative of the original art for use in creating a color overlay or custom color transfer. Still other systems use the process camera to produce either a diffusion transfer print or a direct positive print.

Figures 12–23 through 12–26 illustrate these systems.

Color Proofing Methods

Because of the great expense involved with separating and printing color, it is necessary that the final images and color effects be proofed and approved or corrected *before* the press run begins. Many hard copy proofing methods are possible, depending upon budget restrictions and how well the proof must reflect the final printing. These methods include:

Prepress proofing—loose overlays, overlays laminated on a base sheet, overlays laminated on the desired paper or substrate;

Press proofing—proof press sets, and production press sheets.

Prepress methods are least expensive, and are generally considered acceptable for most standard web and sheet-fed color work. Recent developments in ink transfer to paper and control of color dot *gain* to simulate press effects have made prepress proofing an even more attractive alternative to expensive press proofing.

Overlay methods such as 3M's popular Color Key proof employs clear polyester sheets precoated with a single color pigment. Each of the negatives to be used for full color is exposed to the appropriate color-coated sheet. As the sheets are developed, only the exposed dot areas remain, and four overlays—each with a different color image—are produced.

When overlayed on a white base sheet, in perfect register, the four clear sheets containing their yellow, magenta, cyan and black images form a good representation of the final full-color image. A big advantage of loose overlays is that they can be lifted and each set of dots inspected individually. One disadvantage is that inexperienced viewers of the proof may have difficulty grasping the final printed effects from this three-dimensional set of shiny overlays.

Laminated overlays on white base sheets have become popular. Instead of forming loose overlays, the individually exposed and developed polyester sheets are laminated to a white base sheet to form a single piece. 3M's Matchprint system uses precoated color sheets, while Du Pont's Chromalin process employs dry, pigmented toners to add color to the sheets following exposure. Laminated proofs are more rugged and can be finish-coated to look more like a printed sheet.

In the Spectra Color proofing system each negative is exposed to a sheet of clear imaging film and toner processed to the proper color. The color images are then transferred, one at a time, to a clear receptor sheet. The resulting transparency proof can be laminated directly to any paper or other substrate a customer specifies. The advantages of one's seeing an image on the exact paper stock, perhaps on both sides like the final printing, make this a popular system (Figure 12–27).

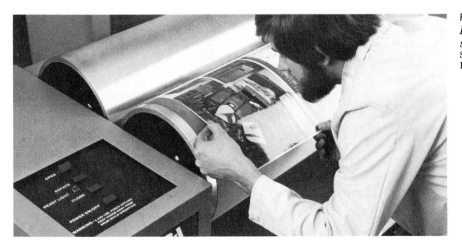

FIG. 12–27
Laminating a set of imaging films to a sheet of printing paper.
Spectra Group-EAC Graphics, Div. of Heidelberg Eastern, Inc.

A recent development by Spectra is the transfer of ink pigment directly to paper without the need for a receptor sheet.

Press proofing is still the favored hard copy proofing method for full color when optimum quality, precise color matches, or multiple-location printing is required.

Proofing press designs vary from single-color, flat-bed, to fully automated four-color versions that operate very much like standard production presses. Proofing devices such as the HCM Hell Chromaprint 4074 can be programmed to produce a full set of *progressive proofs*—different color combinations printed on separate sheets. These proof sets, sometimes called *progs*, show numerous combinations of the yellow, magenta, cyan and black images. This can be a great aid to printers in duplicating the ink densities for each color. Figure 12–28 shows a flat-bed offset proofing press.

Some buyers of printing insist upon seeing an actual sheet that has been printed on the press that will be used for the job. If a customer can arrange to approve such *press sheets* as the color is being adjusted—without causing undue press delays—proof press proofs may be avoided, and a good check of final color can be made. However, not all color adjustments can be made by simply fine-tuning the amount of ink that contacts the plate, and a lot of expensive time can be wasted at this late stage. For this reason, prepress proofs should be made and carefully examined if a set of proof press progressives is not made.

A full set of proofs made on a production press can cost roughly ten times as much as an overlaid or laminated prepress proof.

Electronic soft copy *previewing* is now being used today to avoid costly remaking of separations. Video analyzers with display screens allow the operator to manipulate the color image of a transparency and predict printing effects produced by different papers

and dot size fluctuations. Data from video analysis can be fed into the scanner's computer during the separation process to modify the negatives for improved color reproduction. Some systems, such as the Hazeltine Separation Previewer, read and display an image showing a composite of the values from completed color separation films. The operator can adjust the color shown on the screen and determine what film corrections are required or whether a rescan is needed.

Major strides have been made recently in automating color adjustments. Today, large printers utilize computers and full sheet scanners to analyze the color bars across the entire width of a press sheet in just a few seconds. A video monitor displays all the color readings, shows which bars are in or out of tolerance, and provides data for adjusting the press fountains. Some systems allow the operator to send signals to individually motorized sliding units in the ink fountain. This eliminates manual turning of fountain keys. Figure 12–29 shows such a system.

Sheet-scanning systems generally allow presetting of all ink fountains to the values used previously on a printing job. Other systems scan an entire plate to determine proper ink settings if the job has not been run before. Both procedures greatly speed initial ink adjustments and press makeready.

Even small duplicating presses today are often equipped with an additional printing unit to allow two-color printing in one pass. Some of these units can be removed easily and replaced with a similar unit containing a different ink color. This procedure eliminates the need for washing different ink colors off the press rollers after printing.

Split-fountain color printing allows more than one flat color to be printed by the same plate and printing unit. Dividers in the ink fountain keep the colors separated. Figure 12–30 shows how both single and two-color presses can be set up to produce multiple color splits.

FIG. 12–28
Automatic flatbed offset proof press.
Consolidated International Corporation

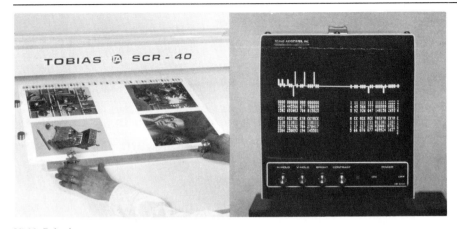

Miehle-Roland

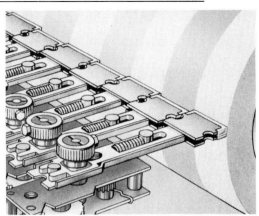

Rockwell International

FIG. 12–29
Sheet-scanning device reads color densities (left), displays read-out of color values on monitor (right).
Motorized blade segments on inking fountain are controlled by computer. Operator can control movement of the segments from a remote console, adjusting ink distribution.

FIG. 12–30
Split-fountain color effects, shown here with black representing one color and gray representing another color.
Hammermill Paper Company

TWO COLORS FROM A ONE COLOR PRESS

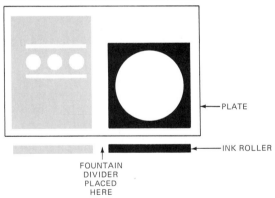

PLATE

INK ROLLER

FOUNTAIN
DIVIDER
PLACED
HERE

MULTIPLE SPLITS ON A ONE COLOR PRESS

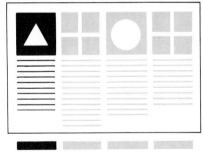

MULTIPLE SPLITS ON TWO COLOR PRESSES

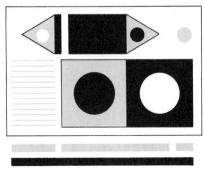

Industry Standards

Precise color evaluations depend upon proper lighting and viewing. The American National Standards Institute, Inc., in 1972, approved ANSI Standard PH2.32 which states conditions for appraising color quality and color uniformity in printing. Among its recommendations is the establishment of a very white 5000-degree Kelvin color temperature for light falling *on the copy* when color quality such as proof versus press sheet is evaluated. A somewhat more blue 7,500-degree Kelvin temperature is used for comparing the *consistency* of sheets from the same press run. The presence of a color-viewing station such as the one depicted in Figure 12–31 by Graphic Technology, Inc., is a good sign that a printer is concerned with accurate color reproduction. Printing buyers who use color frequently would be wise to view color materials under the same conditions as the printer.

In 1976, color standards were developed for publications printing by web offset. The "Recommended Standard Specifications for Advertising Reproduction Material for Magazine Web Offset Printing" were endorsed by the American Business Press, the Magazine Publishers Association, the American Association of Advertising Agencies, and six other industry associations. Here are a few of the many recommendations:

• All images of a given color should be assembled on a single piece of film.

• Standard inks should be used, and metallic or fluorescent inks should not be screened.

• Register marks should be at least ½ inch long and not thicker than .003 inch.

• Density of tones (dot percentages of all combined colors in a given area) should be limited to 160 percent for two colors, and 260 percent for four colors.

FIG. 12–31
A color viewing station.
Graphic Technology, Inc.

• Ink colors should be screened to a maximum of 85 percent or made completely solid.

• Reverse type should be done with as few colors printing together as possible, and small, thin type should be avoided.

13 Paper and Ink

Choosing the Appropriate Message Bearers

Picture the Mona Lisa painted on a garbage sack. Imagine the Declaration of Independence printed in fluorescent orange. The medium in which we view an image is all important to its meaning. And the ability to bring together an ideal blend of aesthetic appeal, reproduction quality, and practicality is one of the most valuable attributes a graphic communicator can have.

Introduction

Few of humankind's achievements have had a more profound effect on human development than the invention of paper. Indeed, the recording of civilization without paper is inconceivable today. Despite years of predictions from futurists that we are becoming a paperless society. the world is more dependent than ever upon the convenience and flexibility of paper and ink as our information messengers.

Computer systems are not complete without the capacity to produce data printouts—on paper. Tiny hand calculators now provide thermal images on specially treated paper.

Try to image a world without print—without books, newspapers, magazines, posters, billboards, labels, resumes, grocery bags, coupons, stationery, instruction manuals, or party decorations. According to recent estimates by the American Paper Institute, twenty billion copies of newspapers and two billion books are printed annually in the United States alone.

With apologies to Mark Twain, rumors of the demise of print are greatly exaggerated. Printing, on paper, is here to stay.

How Paper Is Made

Modern paper manufacturing bears little resemblance to the methods of Ts'ai Lun, the Chinese court official credited with inventing papermaking about 105 A.D. The handcrafted writing substrate he made from mulberry bark, hemp, rags, and water was first compressed by hand into a pulp, then dried on a mat. Today, electronically controlled machines longer than a football field spew forth rolls of paper 30 feet wide at a rate of more than 3,000 feet per minute.

Paper is made from cellulose fibers, largely obtained from trees, but also from cotton, flax, sugar cane, and other fibrous plants. Cone-bearing (coniferous) softwood trees such as pine, fir, and cedar produce long, strong fibers. Hardwood (deciduous) trees such as beech, maple, oak, and poplar produce shorter but more even fibers.

NOTE: The author wishes to thank the Graphic Arts Technical Foundation, particularly Raymond N. Blair, Editor-in-Chief, for assistance with the paper section of this chapter. Technical information and manufacturing descriptions are abstracted from their excellent book *What the Printer Should Know About Paper*, by William H. Bureau.

Major manufacturers like International Paper Company have now developed so-called "super trees" through selective seeding, thinning, and fertilizing. These trees grow faster and produce one-fifth more pulp.

Papermaking is made possible by the special molecular structure of wet cellulose fibers, which form a tight bond when dried in close proximity to one another. These cellulose fibers are the primary ingredient in pulp.

Pulp (or pulpwood) can be produced by mechanical or chemical means, or by a combination of the two. More than three-fourths of the wood pulp made in the United States today is chemically derived.

The Mechanical Process

Trees, harvested as commercial crops by paper companies, are first sheared at ground level in preparation for pulping. They are then cut into smaller roundwood logs, which are forced against huge grindstones or drums to reduce the fibers to groundwood pulp.

Newer methods utilize revolving disks to refine the fibers of wood chips ground from the harvested trees or obtained from secondary sources such as sawmills. This method produces *refiner mechanical pulp* (RMP). This process has been further enhanced today by the addition of steam pressure which softens the wood chips. Pulp made in this fashion is known as *thermomechanical pulp* (TMP). Since less fiber damage occurs, thermomechanical pulp can be mixed with other, weaker pulps to strengthen the paper.

Paper made from mechanical pulp is used for making newsprint and coated papers for publications needing only a brief useful life. Lignin, the glue that holds the cellulose fibers together in nature, is not removed in mechanical processing. As a result, paper discoloration occurs after a short time.

The Chemical Process

The most prominent methods for producing pulp rely upon chemical or semichemical processes. Wood chips are fed continuously through huge cooking vats (digesters) where high pressure dissolves the lignin and separates the wood fibers. The resulting pulp is next washed to remove impurities. Bleaching follows, but is done carefully so that the fibers are not excessively weakened. Further refining is done with

rotating disks or a cone-shaped device called the "jordan." Coloring dyes and pigments, starchlike sizing (to strengthen and seal the fibers), and waterproofing resins all may be added. Nonfibrous fillers, or "loading," also are added. These fillers, largely clay, titanium dioxide, and calcium carbonate, improve opacity, brightness, and other desirable characteristics.

Today's paper mills are equipped to control precise blending of fibers, fillers, and other elements in the stock preparation stage.

The Paper Machine

The prepared pulp stock is now highly diluted with water—about one part fiber to 200 parts water. This solution, called "furnish," is conveyed to the headbox of the papermaking machine.

Paper machines perform three essential tasks to convert pulp to paper: forming, pressing, and drying.

The papermaking cycle begins at the wet (forming) end of the machine. A carefully regulated bead of diluted pulp flows from the headbox onto a horizontally moving endless wire screen. A major innovation today is the twin-wire machine which uses two continuous wire screens to "sandwich" the pulp vertically at the forming stage. This prevents the weight of the wet pulp from forming a wire pattern on the underside of the fiber mass. As water drains through the wire, the cellulose fibers begin to interlock, and more water is vacuumed away.

A "dandy roll" riding atop the moving mat of fibers is used to produce a distinctive finish or *watermark* pattern on certain papers. When wet fibers contact the design on the dandy roll, they are displaced, leaving a distinct pattern without adversely affecting the thickness of the paper.

Entering the press section, the moving (web) still consists of more than four-fifths water. Multiple groups of pressing cylinders compress the fibers and reduce the water content an additional 20 percent. Still more water by weight than fiber, the web now enters the drying section. Here, numerous heated rolls drive out all but about 5 percent of remaining moisture, while additional surface sizing or coating pigments also may be applied.

Next, calendering stacks containing heated metal rollers then smooth the sheet, depending upon the type of paper being produced. Extra smoothing or polishing, called supercalendering, is needed on many high-quality papers. This is done on a separate machine. Embossing of pebble, linen, and other textured patterns also may be done on a machine equipped with engraved metal cylinders. The finished paper is wound into massive rolls, which are then slit and wound into smaller rolls. Additional slitting, sheeting, and packaging, as required, completes the papermaking process.

One of two types of coatings may be applied as a final finish to many papers: (1) *pigments* (usually clay) which provide a smooth, filled surface; and (2) *binders*, either natural, such as starch, or synthetic, such as rubber or polymers. These provide extra smoothness, absorbency, and other desirable characteristics.

Coatings are usually applied by allowing the paper to come into contact with a cylinder which contains a constantly replenished supply of coating material. Excess coating is scraped away evenly by a blade. This is called blade coating. Another process, air knife coating, utilizes a high-pressure jet of air to spread the coating material to an even thickness.

Cast coated papers are produced by contacting their fresh, wet coatings to a smooth, heated drum which "bakes" them to an extremely high gloss without the need for supercalendering. These surfaces are ideal for printing.

Figures 13–1 through 13–13 show the major steps in paper manufacturing.

Characteristics of Paper

Virtually all of the important properties of paper are defined scientifically by the paper industry through the use of highly precise measuring instruments. Our descriptions will employ more general terms.

Basic Size is the size in inches adopted over time as standard for each type or grade of paper—for example, bond: 17×22, book: 25×28, cover: 20×26, and newsprint: 24×36.

Basic Weight is the weight in pounds of a ream (500 sheets) of a given basic size. For example, a ream of 60-pound book paper, in the basic book paper size of 25×38, weighs 60 pounds. Twenty-pound bond paper is paper that weighs 20 pounds per ream in the basic size for bond paper—17×22.

Many office workers who use bond paper regularly may be confused about ream weights. They see the 20-lb. label on an $8\frac{1}{2} \times 11$ ream and know that this small package cannot possibly weigh 20 pounds. The basic size of bond—17×22—is what determines the basis weight, not the sizes of the reams they use, which are only one-fourth of the basic size. The actual weight of the $8\frac{1}{2} \times 11$ bond ream is only five pounds.

Most papers are readily available in nonbasic sizes. S. D. Warren Paper Company lists 29 regular sizes of book paper other than the basic 25×38. However, not all are regularly stocked. It is wise to consult the catalogs of each paper distributor to determine availability when selecting paper.

Sometimes paper weight is given according to a 1,000 (M) sheet quantity. The 60-lb. book paper used in the example above would be designated as "25 by 38—120M" if sold in this manner.

Today, we may also specify paper metrically, by grammage. This is the weight in grams for a single sheet one square meter in size. This is abbreviated g/M^2.

Finish. This is the tactile nature of a paper's surface, primarily thought of as its texture, smoothness, or degree of gloss. Uncoated papers are finished on the papermaking machine, where variations in the dandy roll, pressing, and drying determine surface characteristics. Coated papers often receive off-machine coating, supercalendering, and/or embossing in addition to on-machine finishing, depending upon the type of paper desired.

Color. This is the degree to which pigmentation, dyes, or impurities cause a paper to vary from white, as indicated by unequal reflectance of light wavelengths. A wide variety of colors are available in most paper stocks today. Generally, colored papers cost more than white. Because of the expensive pigments used, darker colors usually are more expensive than lighter ones. Color tinting of continuous-web papers on the printing press has become popular in plants which run business forms in numerous colors. This technique allows them to stock only white paper and tint it to suit any need.

Brightness. Often used interchangeably with the term whiteness, brightness is the percentage of a standard blue light wavelength a given paper reflects. Printing papers fall within a range of about 60 to almost 100 percent reflectance. Newsprint measures in the high 50s to low 60s. Most printing papers fall between 75 and 85. Premium coated papers are rated at 85 and above.

Fluorescent dyes and optical brighteners can be added to enhance paper brightness and apparent whiteness.

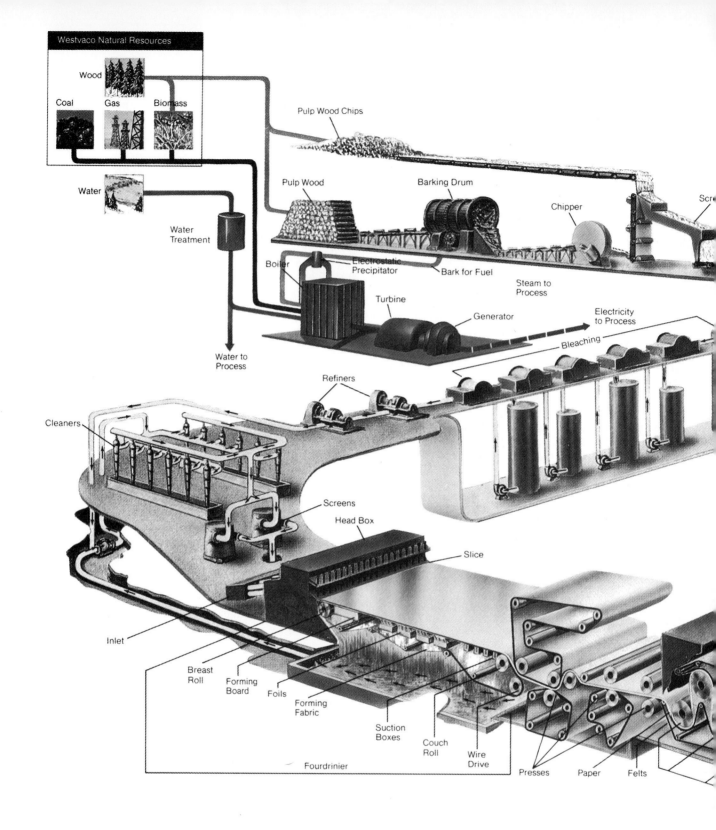

FIG. 13–1
Flow diagram of the papermaking process.
© Westvaco Corporation

254

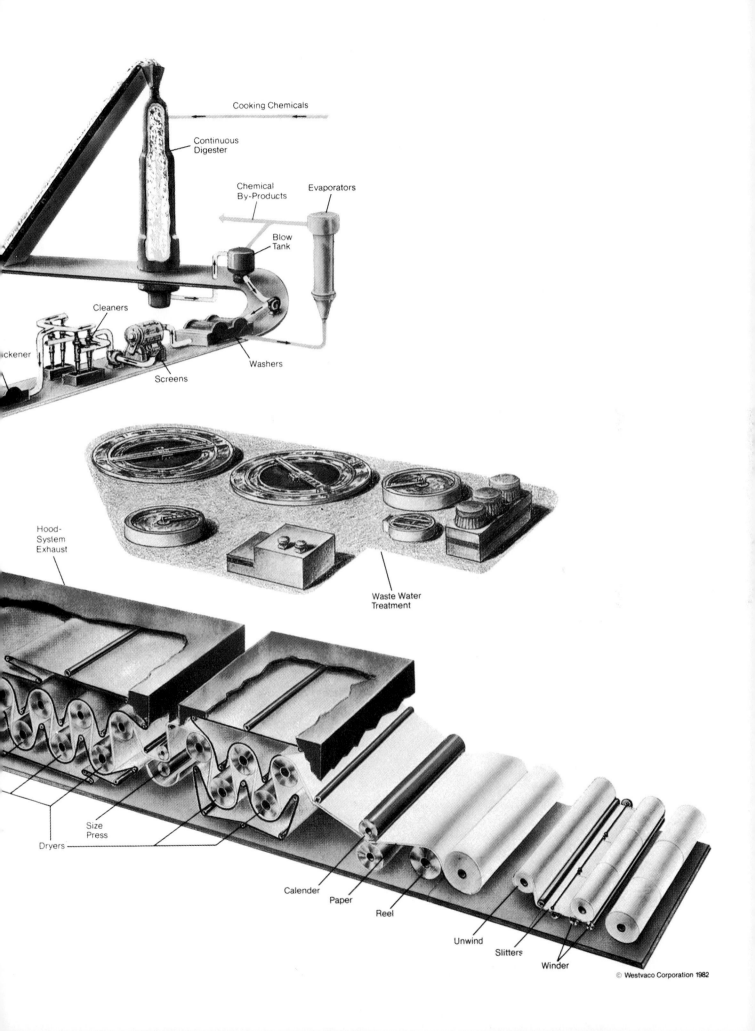

Cooking Chemicals

Continuous Digester

Chemical By-Products

Evaporators

Blow Tank

Cleaners

ckener

Screens

Washers

Hood-System Exhaust

Waste Water Treatment

Size Press

Dryers

Calender

Paper

Reel

Unwind

Slitters

Winder

© Westvaco Corporation 1982

FIG. 13-2
Ground-level hydraulic shear.
Graphic Arts Technical Foundation

FIG. 13-3
Debarking drum.
Hammermill Paper Company

FIG. 13-4
Chips ready for conversion into chemical pulp.
Hammermill Paper Company

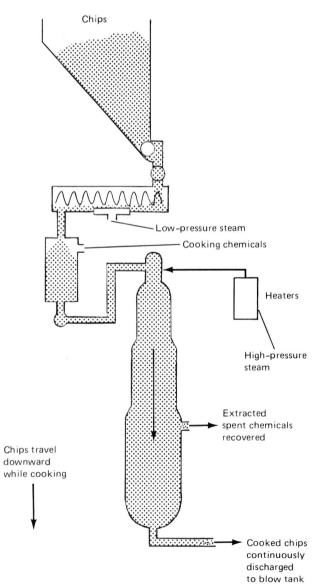

Chips

Low-pressure steam

Cooking chemicals

Heaters

High-pressure steam

Extracted spent chemicals recovered

Chips travel downward while cooking

Cooked chips continuously discharged to blow tank

FIG. 13-5
Continuous-pulping system.
Graphic Arts Technical Foundation

256

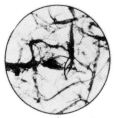 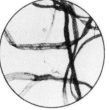  

groundwood softwood sulfate hardwood sulfite hardwood sulfate semi-chemical

FIG. 13–6
Five types of wood fiber used in paper making.
Kimberly Clark Corporation

FIG. 13–8
Twin-disk refiner.
Graphic Arts Technical Foundation

FIG. 13–7
Following washing and bleaching, pulp is blended with precise amounts of sizing, fillers, dyes, and other materials.
Hammermill Paper Company

FIG. 13–9
The Fourdrinier paper machine, with the starting (wet) end at the left.
Hammermill Paper Company

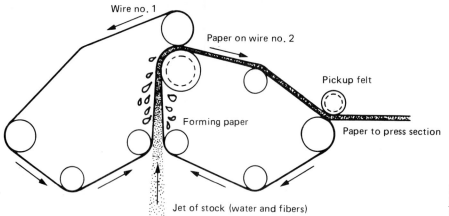

Wire no. 1

Paper on wire no. 2

Pickup felt

Forming paper

Paper to press section

Jet of stock (water and fibers)

FIG. 13–10
How a twin-wire former works.
Graphic Arts Technical Foundation

FIG. 13–11
A supercalender, which irons a smooth finish on the surface of the paper, consists of a stack of alternately soft and hard rolls.

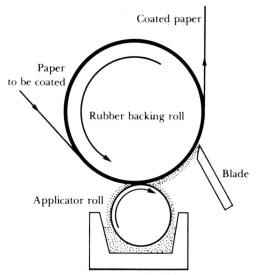

Coated paper

Paper to be coated

Rubber backing roll

Blade

Applicator roll

FIG. 13–12
How blade coating is performed.
Graphic Arts Technical Foundation

FIG. 13–13
(a) Finished rolls must be slit, cut, counted, wrapped, and labelled. (b) Testing stations throughout the paper mill ensure quality control. Some thirty-eight separate tests are performed.
Hammermill Paper Company.

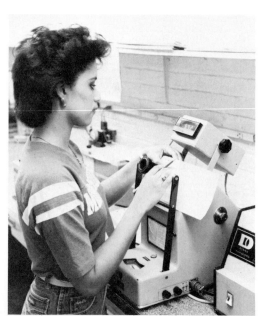

a.

b.

Whiteness. This is the degree to which all light rays are reflected evenly from a paper's surface. It varies from warm (cream) to cool (blue). Neutral white is preferred for most color printing since it is less likely to affect the colors of images printed in transparent process inks. Because of wide variations in white papers, it is important that we view actual paper samples under balanced white light in selecting a white paper.

A block of prepared magnezium oxide is used as a 100 percent standard for ideal whiteness, although no paper achieves this ideal. This substance is also used to measure paper color and brightness.

Grain. The alignment of cellulose fibers constitutes a paper's grain. Fibers normally align parallel to the direction the wet pulp is moving on the wire of the paper machine. Grain direction can be seen in two common ways: a sheet of paper, when wet on one side, will curl parallel to its grain direction. A sheet of paper tears more evenly with the grain than against it.

Knowledge of grain direction is very important to a printer. Sheets with a grain running parallel to the axis of press cylinders bend around the cylinders more readily. Large sheets folded down to signatures for binding lie flatter when the grain is running in the direction of the publication's bound edge.

Grain direction is identified as long or short in paper catalogs by a horizontal line placed under one dimension. For example, $20 \times \underline{26}$ means that the grain of this particular cover paper is running the long way on the sheet.

Formation is an expression of how evenly the fibers are spread within a sheet of paper. The more evenly they are distributed, the better the formation. "Wild" or uneven formation means that ink absorption will be uneven as well, and blotchy solids or halftones may result during printing.

Caliper refers to a paper's thickness and is expressed in points, or thousandths of an inch. A sheet that measures .003 is 3 points or 3-thousandths of an inch thick. Bulk, a term used to describe the cumulative effect of sheet thickness in combination with compressibility, is described in publications work as pages per inch (PPI). The number of paper pages per inch is multiplied by two in order to yield the actual number of printed pages per inch, since each sheet contains two printed pages, one on either side. Degree of bulking is important when we want a publication to look thicker or thinner.

New manufacturing methods such as Warren's Warrenflo process eliminate supercalendering and provide papers with higher bulk and less weight. This is important for publications which must be mailed. Uncoated papers usually provide higher relative bulk than coated varieties.

Caliper is especially important where return-mail printing is required. The United States Postal Service applies a heavy surcharge to pieces of mail that fail to meet the .007 minimum thickness its automated equipment requires.

Opacity is the capacity of a paper to block light. Unprinted opacity relates to the degree to which we can see through a sheet to printed matter on the sheet below. Printed opacity is how well the paper prevents the printing on one of its sides from showing through from the opposite side. Unbleached groundwood papers, with their many impurities, have greater opacity. Generally, opacity increases with sheet thickness in all paper grades.

There are exceptions, but as a rule of thumb, book papers of less than 60-lb. basis weight should be avoided where opacity is a critical design factor. Many new lightweight papers, such as Zellerbach Paper Company's Zellopaque, have been developed recently. They offer good opacity and more flexibility in the selection of thinner papers for mailing economy.

Ink Absorbency. The ability of a paper to absorb ink is important, since excessive ink spread in printing can inhibit image quality and affect proper drying of certain inks. A sheet also should be able to "hold out" ink to prevent it from soaking in too much ("striking through") and becoming visible from the reverse side. Holdout is not the same as printed opacity, but the effects of both may appear to be the same.

Permanence is the ability of paper to withstand changes over time, primarily yellowing and cracking. Papers with high cotton fiber content and low levels of acid and impurities offer the highest degrees of permanence. Unfortunately for important documents, the same acid sizing that has been applied to most papers for the past century to enhance printing characteristics eventually will decompose paper unless it is specially treated or restored.

Printing Paper Grades

The American Paper Institute divides the manufacturing of paper into two main groups—printing paper grades and packaging/industrial grades. We are concerned here only with the printing grades.

Paper can be classified in many ways. The categories

below reflect a grouping by common end use of each grade, not necessarily by similarities in manufacturing.

Simply describing the unique look and feel of printing papers here doesn't do them justice. Try to examine actual paper samples as you study this section. Obtain sample books from paper distributors, printers, or your instructor. You'll note that brand names and classification methods vary somewhat between companies. However, you should have no difficulty matching the samples to the descriptions below.

Newsprint. Made for short useful life, this familiar, inexpensive paper is made primarily from groundwood pulp. Fiber strength and opacity are positive attributes, but impurities cause it to discolor rapidly. Most newsprint today is shipped in rolls for use on web presses.

A significant new source of pulp, a rapidly growing, fibrous plant named Kenaf offers great promise for moderating newsprint costs. It produces almost nine times more pulp per acre than softwood trees.

The basic size for newsprint is 24×36.

Bonds and Writing Papers. These uncoated papers are made either from chemical wood pulp or cotton fibers. Their surfaces are designed to accept handwriting in ink and pencil, as well as typing and printing. They must withstand the folding, erasing, filing, and handling that is common in business environments.

Better quality watermarked bonds contain all chemical pulp. Very high-quality watermarked bonds are distinguished by watermarks that indicate their percentage of cotton fiber—from 25 to 100 percent.

The basic size for bond and writing papers is 17×22.

Book. The graphics professional's mainstay, this group of papers is used not only for books but also for flyers, brochures, booklets, magazines, promotion pieces, annual reports, and a host of other printed materials. Book papers come in a wide variety of finishes, weights, and colors. Basic size is 25×38. The main divisions of book papers are coated, uncoated, and text.

Uncoated. These papers range in cost from very low to very high, and reflect wide differences in manufacturing techniques and appropriate printing applications.

Woven or antique finish papers offer high bulk, but, because of their relatively rough surface, they do not reproduce halftone detail as well as other papers.

Machine finish (MF) papers are smoother and less bulky. Offset papers, especially designed to meet the specifications of the offset printing process are smooth, lint free, and offer good moisture resistance. They are available in many colors.

The fiber formation of lower grade uncoated books may be poor. Because printing ink is absorbed readily by most uncoated papers, poor fiber formation results in uneven absorption and blotchy solid and halftone images. The good fiber formation present in high-grade uncoated papers, plus extra bleaching and whitening in many cases, provides a much better printing surface.

Text. While this unique group often is considered to be in a separate class from regular book papers, the function of these beautiful papers is most closely related to book stocks. These high-quality uncoated papers have either natural or embossed textures which can range from subtle to highly pronounced. They provide a "designer" look to booklets, brochures, invitations, and special publications of all kinds. Available in a variety of both light and deep hues, most texts also have cover weights and envelopes to match. Deckled (feathered) edges may also be specified. Text papers are among the most expensive uncoated papers.

Coated. Coated papers offer the highest quality printing surfaces because they prevent ink from penetrating into the paper fibers. Therefore, fiber formation is less critical. There are four main types of coated papers:

1. *Matte.* A thin layer of coating pigments is applied to these papers by a blade-coating device. This coating fills in the spaces between the fibers and provides a surface that is much smoother than that of uncoated papers. These papers are not supercalendered. They are a top value for users such as editors of company publications who need the better quality of coated papers but are also faced with a limited budget. Ink holdout is moderate, and large solids may tend to be a bit drab, but the quality of both black-and-white and color images is usually quite acceptable.

2. *Gloss coated.* These quality papers receive one or more enamel coatings, usually by blade, and much thicker than the pigment coatings of matte papers. The paper then is supercalendered to various degrees, depending upon the type of paper being manufactured. Since these papers control ink penetration, their smooth, reflective surfaces yield maximum sharpness and realistic color reproduction. These papers are selected where clear, high-quality images and a sense of professionalism must be conveyed. Annual reports, corporate magazines, promotional mailings, and brochures are common end products.

3. *Dull coated.* These papers are enamel coated in much the same way as gloss papers, but supercalendering is modified to produce little or no gloss. Dull-coated papers are preferred where the printing advantage of a heavy coating is desired without the distracting glare often present with gloss coating. Publications containing a high proportion of reading matter utilize dull-coated stock extensively. These papers are used for all types of quality magazines, brochures, and other fine publications.

4. *Cast coated.* A special coating process distinguishes these papers from all others. Freshly coated paper is pressed against a hot, polished drum, causing the coating to dry to a hard gloss. This mirrorlike finish provides an ideal surface. Figure 13–14 illustrates how coatings affect the surface of paper.

While coated papers are available mostly in variations of white, certain companies, such as Appleton Papers, Inc., specialize in coated colors. Appleton offers blue, canary, cream, goldenrod,

FIG. 13–14

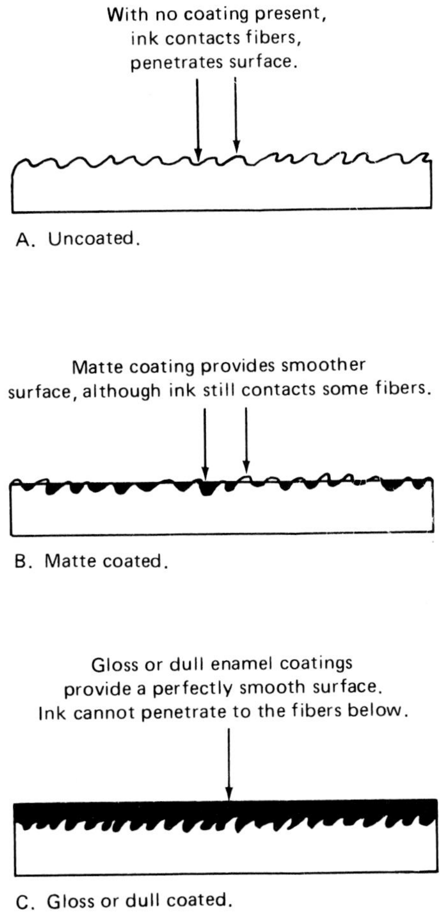

With no coating present, ink contacts fibers, penetrates surface.

A. Uncoated.

Matte coating provides smoother surface, although ink still contacts some fibers.

B. Matte coated.

Gloss or dull enamel coatings provide a perfectly smooth surface. Ink cannot penetrate to the fibers below.

C. Gloss or dull coated.

green, india, ivory, pink, and tan. If possible color distortion in four-color printing is taken into account, colored enamels offer an unusually beautiful printed effect.

Cover. These thicker papers, designed to produce the outer wrap of a publication, come in many coated and uncoated varieties, many to match book and text finishes. Many specialized covers—leather-embossed, fluorescent, metallic, even two-sided duplex styles—are available. Cover papers are manufactured to withstand embossing, scoring, die-cutting, varnishing, and other finishing processes which add dramatic impact.

The basic size of cover paper is 20×26. Caliper often is referred to in points, for example, 10-point Chromecoat Cover.

Bristol. These heavier "card stocks" come in several common types:

Index Bristol—used for filing and record keeping, is smooth finish and comes in numerous colors for easy color coding. It utilizes a combination of chemical pulp and cotton fiber for strength. The basic size is $22\frac{1}{2} \times 28\frac{1}{2}$.

Printing Bristol—is made from chemical pulp and offers bulk and ease of printing on its smooth, absorptive finishes. Basic size is $25\frac{1}{2} \times 30\frac{1}{2}$.

Other bristols include postcard, tag, file folder, and coated versions.

Blanks—commonly called posterboard, these are a type of paperboard used for posters of all kinds, car signs, and point-of-purchase displays. They may be coated on one or both sides, and consist of laminated layers or plies.

Specialty Papers. Many papers are produced for highly specialized uses. These include translucents for fly leaves in publications, safety papers for checks and other documents which need protection from forgery, and lightweight papers (formerly known as Bible papers) for economy in mailing. Also included are parchment papers for documents, wedding papers for formal announcements and invitations, carbonless papers for business forms, and gummed label papers of all types.

Relating Paper Size to Job Format

The chart in Figure 13–15 indicates the major grades of paper, standard sizes, and common basic weights. It also shows which weights are equivalent between different grades of paper. Figure 13–16 lists the thirty regular sizes of book paper offered by one paper manufacturer, along with thousand-sheet weights for each size.

To avoid paper waste, the designer and printer work together to determine which standard paper size will produce the job most efficiently. The relationships of sizes and quantities of paper to price is depicted in Chapter 15.

FIG. 13–15
Standard Sizes and Basis Weights for Major Paper Grades

	BOOK 25 × 38	BOND AND LEDGER 17 × 22	COVER 20 × 26	PRINTING BRISTOL 22½ × 28½	INDEX 25½ × 30½	TAG 24 × 36
Book	30	12	16	20	25	27
	40	16	22	27	33	36
	45	18	25	30	37	41
	50	20	27	34	41	45
	60	24	33	40	49	55
	70	28	38	47	57	64
	80	31	44	54	65	73
	90	35	49	60	74	82
	100	39	55	67	82	91
	120	47	66	80	98	109
Bond and Ledger	33	13	18	22	27	30
	41	16	22	27	33	37
	51	20	28	34	42	46
	61	24	33	41	50	56
	71	28	39	48	58	64
	81	32	45	55	67	74
	91	36	50	62	75	83
	102	40	56	69	83	93
Cover	100	40	55	68	82	91
	110	43	60	74	90	100
	115	45	63	78	94	105
	119	47	65	80	97	108
	124	50	68	85	103	115
	146	58	80	99	120	134
	164	65	90	111	135	149
	183	72	100	124	150	166
	201	79	110	136	165	188
	219	86	120	148	179	199
Printing Bristol	100	39	54	67	81	91
	120	47	65	80	98	109
	148	58	81	100	121	135
	176	70	97	120	146	162
	207	82	114	140	170	189
	237	93	130	160	194	216
Index	110	43	60	74	90	100
	135	53	74	91	110	122
	170	67	93	115	140	156
	208	82	114	140	170	189
Tag	110	43	60	74	90	100
	137	54	75	93	113	125
	165	65	90	111	135	150
	192	76	105	130	158	175
	220	87	120	148	180	200
	275	109	151	186	225	250

NOTE: Equivalent weights from grade to grade are shown within vertical rules.

Source: S. D. Warren Paper Company

FIG. 13–16

Thirty Regular Sizes and Equivalent M Weights of Book Papers

BASIS	45	50	60	70	80	90	100	120
17 × 22	35	39	47	55	63	71	79	94
17½ × 22½	37	41	50	58	66	75	83	99
18 × 24	41	45	55	64	73	82	91	109
19 × 25	45	50	60	70	80	90	100	120
20 × 26	49	55	66	77	88	99	109	131
20 × 35	66	74	88	103	118	133	147	177
23 × 26½	58	64	77	90	103	115	128	154
23 × 29	63	70	84	98	112	126	140	169
23 × 35	76	85	102	119	136	153	169	203
23½ × 35	78	87	104	121	139	156	173	208
24 × 36	82	90	110	128	146	164	182	218
25 × 38	90	100	120	140	160	180	200	240
26 × 40	98	110	132	154	176	198	218	262
28 × 40	106	118	142	166	188	212	236	282
35 × 45	150	166	198	232	266	298	332	398
35 × 46	152	170	204	238	272	306	338	406
36 × 46	156	174	210	244	278	314	348	418
36 × 48	164	182	218	254	292	328	364	436
37 × 49	172	190	230	268	306	344	382	460
38 × 50	180	200	240	280	320	360	400	480
41 × 54	210	234	280	326	372	420	466	560
41 × 61	236	264	316	368	422	474	526	632
42 × 58	230	256	308	358	410	462	512	616
44 × 64	266	296	356	414	474	534	592	712
44 × 66	276	306	366	428	490	550	612	734
45 × 68	290	322	388	452	516	580	644	774
46 × 69	300	334	400	468	534	602	668	802
49 × 74	344	382	460	536	612	688	764	920
50 × 76	360	400	480	560	640	720	800	960
52 × 76	374	416	500	582	666	748	832	998

Source: S. D. Warren Paper Company

Choosing an Appropriate Paper

Today, as we are so often required to communicate through cold and impersonal electronic gadgetry, we are refreshed by the unhurried pace and unobtrusive way fine printing brings us information. Who among us has not sensed the quality of a special greeting card or scenic calendar just by touching its surface? The tactile qualities of fine papers are indeed some of their strongest selling points.

Picking an appropriate paper is one of the most important decisions we can make in designing an effective printed message. In a recent study sponsored by the Direct Mail/Marketing Association, where colored and/or textured papers were used, responses were improved significantly in ten out of twelve direct mail tests.

The National Paper Trade Association notes that many factors affect paper choice—reproduction quality, color, weight, price, size, availability, folding quality, and psychological factors. They also say that price sometimes may be the least important of all. They report that when we consider the costs of research and copy writing, art and composition, printing, binding, and mailing of a typical promotion piece, the difference between the highest and lowest priced coated paper may be only four percent of the total outlay. For uncoated papers, the figure is about seven percent.

Most printers are well qualified to assist you in selecting the right paper for your jobs. While many customers have enough experience to specify accurately the type of paper they want, including a particular manufacturer's grade name, they are wise to allow intelligent substitutions when a printer can supply equivalent paper for a more favorable price or suggest a paper that will give a better effect. The most important thing is that you are satisfied with both the cost and the quality of the finished work.

Printers deal with paper suppliers every day. As

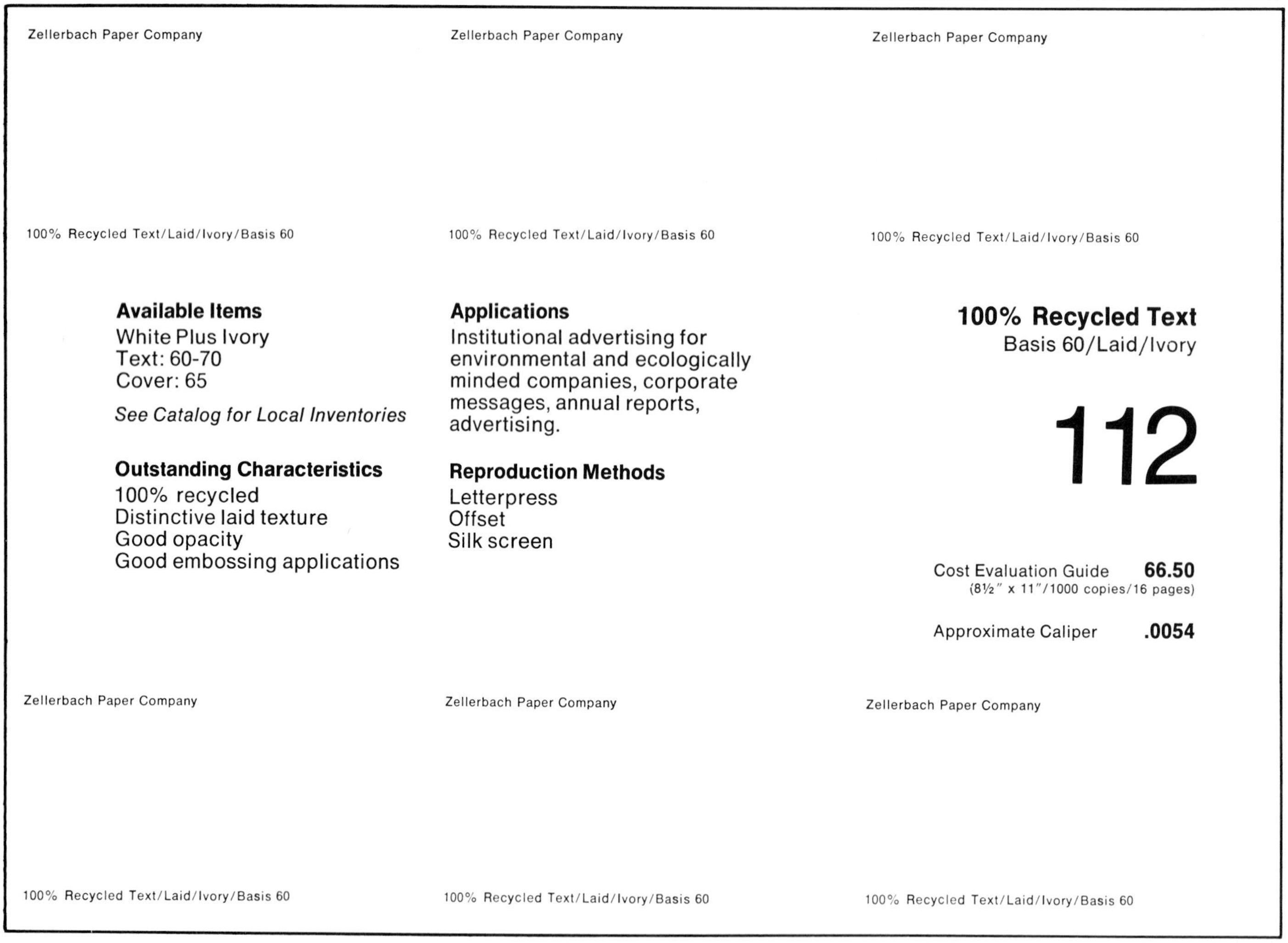

Zellerbach Paper Company

Zellerbach Paper Company

Zellerbach Paper Company

100% Recycled Text/Laid/Ivory/Basis 60

100% Recycled Text/Laid/Ivory/Basis 60

100% Recycled Text/Laid/Ivory/Basis 60

Available Items
White Plus Ivory
Text: 60-70
Cover: 65

See Catalog for Local Inventories

Outstanding Characteristics
100% recycled
Distinctive laid texture
Good opacity
Good embossing applications

Applications
Institutional advertising for environmental and ecologically minded companies, corporate messages, annual reports, advertising.

Reproduction Methods
Letterpress
Offset
Silk screen

100% Recycled Text
Basis 60/Laid/Ivory

112

Cost Evaluation Guide   **66.50**
(8½" x 11"/1000 copies/16 pages)

Approximate Caliper   **.0054**

Zellerbach Paper Company

Zellerbach Paper Company

Zellerbach Paper Company

100% Recycled Text/Laid/Ivory/Basis 60

100% Recycled Text/Laid/Ivory/Basis 60

100% Recycled Text/Laid/Ivory/Basis 60

FIG. 13–17
Page from a book of paper samples.
Zellerbach Paper Company

valued customers of paper merchants, they naturally receive quantity discounts, special services, and sound advice on how best to buy and use printing papers. They are a valuable resource when you need help, as are the graphic arts consultants employed by paper companies.

A single paper company may distribute paper products made by several hundred different suppliers. However, printers normally use a variety of paper houses, since no distributor handles all available papers.

When you design a printing job, use paper sample books to compare and select paper. Figure 13–17 shows a page from a typical paper sample book. Exact samples of each type of paper in all available

variations are bound together for convenience. Each page contains the manufacturer's name and grade for the paper, its range of weights and finishes, characteristics and suggested uses. Customers simply select the paper they want, then tear out a perforated sample to attach to the job.

A chart of standard page sizes is depicted in Figure 13–18. It shows how many standard page size booklets, folders, etc. may be obtained from stock size papers.

Printers know that paper spoilage is a normal part of production, so they are careful to have extra paper on hand. Figure 13–19 shows how much allowance must be made for spoiled paper.

FIG. 13–18

Quick Reference Chart Showing Standard Page Sizes
for Sheet-Fed Booklets and Folders

SIZE OF PAGE	NO. OF PAGES	SIZE OF PAPER	NO. OUT OF SHEET	PRESS SIZE SHEET	SIZE OF PAGE	NO. OF PAGES	SIZE OF PAPER	NO. OUT OF SHEET	PRESS SIZE SHEET
3 × 6	4	25 × 38	24	6¼ × 12½	5 × 8	4	35 × 45	16	10¼ × 16½
	8	25 × 38	12	12½ × 12½		8	35 × 45	8	16½ × 20½
	12	25 × 38	8	12½ × 18¾		12	28 × 42	4	20½ × 24¾
	16	25 × 38	6	12½ × 25		16	35 × 45	4	22¼ × 35
	24	25 × 38	4	18¾ × 25		24	28 × 42	2	24¾ × 41
						32	35 × 45	2	35 × 45
3½ × 6¼	4	28 × 44	24	7¼ × 13	5½ × 7½	4	35 × 45	16	11¼ × 15½
	8	28 × 44	12	13 × 14½		8	35 × 45	8	15½ × 22½
	12	28 × 44	8	13 × 21¾		12	38 × 50	6	15½ × 33¾
	16	28 × 44	6	14½ × 26		16	35 × 45	4	22½ × 35
						32	35 × 45	2	35 × 45
3¾ × 6⅞	4	32 × 44	24	7¾ × 14¼	5½ × 8½	4	35 × 45	16	11¼ × 17½
	8	32 × 44	12	14¼ × 15½		8	35 × 45	8	17½ × 22½
	12	32 × 44	8	15½ × 21⅜		16	35 × 45	4	22½ × 35
	16	32 × 44	6	14¼ × 31		32	35 × 45	2	35 × 45
	24	32 × 44	4	21⅜ × 31	6 × 9	4	25 × 38	8	12½ × 19
4 × 5½	4	25 × 38	16	8¼ × 11½		8	25 × 38	4	19 × 25
	8	25 × 38	8	11½ × 16½		16	25 × 38	2	25 × 38
	12	38 × 50	12	16½ × 17¼		32	38 × 50	2	25 × 38
	16	25 × 38	4	16½ × 23	7 × 10	4	32 × 44	8	16 × 22
	24	38 × 50	6	16½ × 34½		8	32 × 44	4	22 × 32
	32	25 × 38	2	23 × 33		16	32 × 44	2	32 × 44
4 × 6	4	25 × 38	18	8¼ × 12½	7½ × 10	4	32 × 44	8	16 × 22
	8	38 × 50	18	12½ × 16½		8	32 × 44	4	22 × 32
	12	38 × 50	12	16½ × 18¾		16	32 × 44	2	32 × 44
	16	25 × 38	4	16½ × 25	8 × 10	4	35 × 45	8	17½ × 22½
	24	38 × 50	6	16½ × 37½		8	35 × 45	4	22½ × 35
	32	25 × 38	2	25 × 33		16	35 × 45	2	35 × 45
4 × 9	4	25 × 38	12	8¼ × 18½	8½ × 11	4	35 × 45	8	17½ × 22½
	8	38 × 50	12	16½ × 18½		8	35 × 45	4	22½ × 35
	12	25 × 38	4	18½ × 24¾		12	35 × 45	2	35 × 35
	16	38 × 50	6	16½ × 37		16	35 × 45	2	35 × 45
	24	25 × 38	2	24¾ × 37	9 × 12	4	25 × 38	4	19 × 25
	32	35 × 45	2	33 × 37		8	25 × 38	2	25 × 38
4¾ × 6¼	4	28 × 42	16	9¾ × 13		16	38 × 50	2	38 × 50
	8	28 × 42	8	13 × 19½					
	12	32 × 44	6	13 × 29¼					
	16	28 × 42	4	19½ × 26					
	32	28 × 42	2	26 × 39					
5 × 7	4	32 × 44	18	10¼ × 14½					
	8	32 × 44	8	14½ × 20½					
	12	32 × 44	6	14½ × 30¾					
	16	32 × 44	4	20½ × 29					
	24	28 × 42	2	21¾ × 41					
	32	32 × 44	2	29 × 41					

Source: Graphics Master, Dean Lem, Associates

FIG. 13–19
Spoilage Charts for Offset Printing

SHEET-FED OFFSET*	1,000	2,500	5,000	10,000	25,000 AND OVER
Single Color Equipment					
One color, one side	8%	6%	5%	4%	3%
One color, work and turn or work and tumble	13%	10%	8%	6%	5%
Each additional color (per side)	5%	4%	3%	2%	2%
Two Color Equipment					
Two colors, one side	—	—	5%	4%	3%
Two colors, work and turn or work and tumble	—	—	8%	6%	5%
Each additional two colors (per side)	—	—	3%	2%	2%
Four Color Equipment					
Four colors, one side only	—	—	—	6%	5%
Four colors, work and turn or work and tumble	—	—	—	8%	7%
Bindery Spoilage					
Folding, stitching, trimming	4%	3%	3%	2%	2%
Cutting, punching or drilling	2%	2%	2%	2%	2%
Varnishing and gumming	7%	5%	4%	3%	3%

*Percentage represents press size sheets, not impressions. Figures do not include waste sheets used to run up color as it is assumed that waste stock is used for this purpose.

WEB OFFSET	WASTE % OF TOTAL IMPRESSIONS*
Press Run	
Up to 25M	18
Over 25M to 50M	15
Over 50M to 100M	13
Over 100M to 200M	11
Over 200M	9
Penalties to be added:	
For each additional web over 1	1
For using 2 folders	1
For 3, 4 or 5 colors	1
For Coated Paper	5
For Light Papers under 40#	2
For Heavy Papers over 60#	2

*Includes waste for core, wrappers, and damaged paper, which is estimated at 2½%.
The chart is for blanket-to-blanket presses running two colors on two sides of the web, on uncoated paper 40 to 60 lbs., and using one folder. The chart includes makeready spoilage.

Source: S. D. Warren Paper Company

Tips on Selecting Paper

1. Match the paper to the printing process. Letterpress papers must be smooth so that uneven impression is prevented. They also require fairly good ink absorbency. Offset papers require moisture resistance and should be lint- and chemical-free. Fiber bonding in the paper must be strong to resist the pick of the tackier inks used. Gravure papers must be extremely smooth and soft, but not particularly clean. Screen printing papers require resistance to change in dimensions under heat drying conditions.

2. Avoid color problems. Normally, use neutral white papers for full-color printing unless a special effect (perhaps a historic or futuristic quality) is intended. Papers that deviate significantly from white may affect the balance of four-color images adversely unless picture areas are first printed with a white opaque background. Remember, too, that uncoated papers in dark colors may reproduce flat color effects much differently than white grades. Also check printed examples to see how an ink will change when printed on a particular stock. If possible, see that an exact sample of paper color and finish is compared.

3. Determine important fiber characteristics. Is good fiber formation important to the job in question? It makes quite a difference on uncoated papers, since printing inks come into direct contact with the fibers. Hold a sheet up to bright light. Do you notice a "wild," uneven pattern compared to other papers? If so, heavily inked areas or photographs may not reproduce well on the sheet.

Is grain direction a consideration? It may affect binding, especially if many folds are required. What about fiber strength for proper die-cutting, scoring, and other finish work? Discuss all these with the printer ahead of time, even if you are not furnishing your own paper. If you are, be certain to discuss these things before you order.

Figure 13–20 rates book and text papers, from high to low, on numerous printing and overall characteristics often looked for in paper. These comparisons apply only to average characteristics of such papers, not necessarily to any given brand or trade name.

A few of the more common problems printers encounter with printing papers due to a combination of factors, including, on rare occasions, defective paper, are:

Hickies—blemishes caused from lint, dust, or other materials in the paper.

Pilling—coatings, fillers, or paper dust adhering to printing blankets.

Ghosting—the appearance of unwanted images during and after printing. There are two kinds: mechanical—caused when ink is insufficient to cover adequately all images in the path of the ink rollers; and chemical—which occurs during ink drying, where uneven glossing or absorption creates unwanted images.

Setoff—the unintended transfer of ink from one paper surface to another.

Showthrough—printed images which can be seen from the reverse side of a sheet. This may be caused by ink penetration of uncoated papers or by insufficient opacity.

Printing Inks

However vital it may be to printing, paper is still only the stage upon which printed images perform. It is the combination of chemical compounds with substrate—ink on paper—that makes a printed message possible.

FIG. 13–20

General Ratings of Paper Types on Important Characteristics

TYPE OF PAPER			COST	BULK	QUALITY OF FORMATION	GENERAL REPROD. QUALITY	NUMBER OF HALFTONE DOTS PER INCH	DETAIL OF COLOR REPROD.	SUITABILITY FOR SCORING, EMBOSSING, DIE-CUTTING	GLARE POTENTIAL
Uncoated Book	Bottom-of-Line		Low	High	Low	Low-Medium	Low-Medium	Low	Low-Medium	Low
	Top-of-Line		Medium-High	High	Medium-High	Medium-High	Low-Medium	Low-Medium	Medium-High	Low-Medium
Coated Book	Matte		Medium	Medium	Medium	Medium-High	Medium-High	Medium-High	Medium-High	Low
	Dull		High	Low	High	High	High	High	High	Low
	Gloss		High	Low	High	High	High	High	High	High
Text	Bottom-of-Line		Low-Medium	Medium	Low-Medium	Low-Medium	Low-Medium	Low-Medium	Medium	Low
	Top-of-Line		High	High	High	Medium-High	Low-Medium	Low-Medium	High	Low

*This chart indicates *general* characteristics only. Paper performance can vary widely depending upon manufacturing technique and materials used. Use this chart as a starting point, but judge each paper on its own merit.

Ink manufacturing is one of the most technologically advanced areas of graphic communication. Like so many other segments of the printing and publishing industry, both its products and processes have advanced today from crude historical beginnings to amazing levels of complexity and importance.

Technical assistance, such as the recent Standards for Web Offset Publications (SWOP), supported by the National Association of Ink Manufacturers and others in the industry, helps keep professional agencies, publishers, and other ink users up to date.

Detailed discussions of formulations and testing procedures used in ink manufacture are not necessary for communicators to appreciate how inks affect their printed creations. Discussed below are some of the most important types of ink, their characteristics, printing applications, and some problems printers may encounter with them.

Basic Ingredients of Ink

Three main ingredients are used in the manufacture of printing inks—pigments, vehicles, and additives.

Pigments

Pigments are the solid elements that are suspended in the mixture and which give the ink its primary color. They also determine its degree of opacity, its permanence, and its reactions to temperature and chemical variations.

Black pigments, once primarily lampblack, are now mostly produced as carbon from an oil burning "furnace."

White pigments are useful for blending with other inks or for serving as an underlay for other inks printed on colored paper. They are made from titanium dioxide, zinc oxides, calcium carbonate, clay, and other materials.

Pigments for colored inks may be inorganic, organic, and metallic. Dyes are also used for coloring. Unlike pigments, they are liquid and dissolve in the ink solution.

Inorganic pigments include lead compounds for colorfast, opaque "chrome" colors; iron compounds for transparent blues, and cadmium mercury for reds.

Metallic pigments are made from powdered metals—aluminum for silver and brass, and copper for gold.

Organic pigments are more predominent today, and have helped replace more environmentally dangerous materials such as lead. Examples of organic pigments include Yellow Lake, Persian Orange, and Fire Red. Organic formulations sometimes include dyes.

Fluorescent pigments also are now available in many brilliant colors. Printed pieces require bright, balanced light in order to emit maximum radiant energy or fluorescence. The inks are especially effective for screen printing, where a heavy film or ink can be applied.

Vehicles

The vehicle is the fluid portion of ink in which pigments are suspended or dyes dissolved. Vehicles may be of several types. These include nondrying and drying oil, solvent resin, glycol, resin-oil, resin-wax, and photoreactive vehicles.

Inks containing nondrying oils set to a state suitable for handling by absorption. Their vehicles soak into newsprint or other soft papers, leaving the pigment at the surface. You probably have noticed a slight film of ink on your hands after reading a newspaper. This is because the news ink used does not fully dry.

Drying oil vehicles are common today for most offset and letterpress printing other than newspapers. Such vehicles dry by oxidation—that is, they react with oxygen and eventually solidify. Heat-treated linseed oil is the most common drying vehicle, although cottonseed, soybean, and other oils may be used.

Both gravure and flexographic printing depend upon the evaporative properties of solvent-resin inks. These processes often require printing on substrates such as plastic or aluminum which have little or no absorptive qualities. Inks used in both processes are highly liquid, and contain resins and gums. Flexographic inks also may contain alcohol or water. During printing, the solvents evaporate rapidly leaving pigments and resins behind.

Glycol vehicles are used where chemical separation ("precipitation") is needed for drying. Some ink formulations require steam or other moisture to separate the vehicle from the pigment and resin.

So-called "quick-set" inks with resin-oil vehicles contain solvents that are absorbed rapidly into paper. This leaves an ink film that, while still drying by oxidation, is set well enough to prevent accidental ink smearing or setoff.

Cold-set inks require special press equipment to heat their resin-wax vehicles. Just as a hot mixture of paraffin and chocolate, when cooled, solidifies into a bar of candy, heated resin-wax vehicles harden quickly after being printed on a cool paper surface.

Recently, photoreactive vehicles have been introduced for use in inks that can be cured by ultraviolet (UV) radiation. However, their popularity has been tempered by their relatively poor running characteristics on presses and their much greater cost.

Inks with vehicles designed for infrared (IR) drying are also available. However, most of the benefit derived from the radiation in short wave (or "cold") infrared driers is achieved for standard inks as well, so IR ink popularity remains moderate.

Both UV and IR drying systems offer the advantages of higher sheet stacking and the reduction or elimination of setoff-reducing sprays.

Figure 13–21 shows the relationship of printing processes, drying systems, and types of vehicles used in inks.

FIG. 13–21

Ink-drying Methods Used in Each Printing Process

Ink-drying Method	Letter-press	Flexo-graphic	Litho-graphic	Gravure	Silk Screen
Oxidation	X		X		X
Heat Set	X		X		
Moisture Set	X				
Polymerization			X		X
Evaporation		X		X	X
Absorption	X				
Fusion	X		X		
Hot Wax	X				
Filtration	X		X		
Radiation	X		X		X

Source: NAPIM

Additives

Many additives are used to alter the working or drying properties of inks. Cobalt driers hasten oxidation, waxes reduce tack and "shorten" an ink, thinning oils and solvents improve absorption and drying, and varnish adds viscosity. Other additives, including anti-drying agents, may also be needed.

Figure 13–22 shows some of the important equipment and procedures used in the manufacturing and testing of printing inks.

Problems Related to Ink

Every printing process has characteristics which interact with ink properties to create problems in production. Some of the most important ones are:

Setoff. In sheet-fed letterpress and offset printing, both dependent heavily upon oxidation for drying, inks can be problematic. Ink not properly set will transfer or "setoff" to the bottom of the sheet above when the pile of paper being printed gets too high. Anti-setoff powder or liquid is sprayed between each sheet during printing to reduce this effect. Also, cobalt driers may be added to the ink.

Stripping and Tinting. These problems are related primarily to the fountain solution used in offset printing. Stripping, the inability of ink to adhere to inking rollers, usually means that excessive moisture has crept into the inking system during printing. Unevenly inked, weak images will result.

When a thin layer of ink, a tint, appears in nonimage areas of a page during printing, ink contamination of the dampening system probably has occurred. Scumming, an effect which resembles tinting, is often traced to plate oxidation, where nonimage portions of a metal offset plate become ink receptive. This is usually due to improper plate densitization or strong fountain solutions.

Picking. When paper sticks to plates or blankets excessively, or ink is pulled away from a sheet or plate, the effect is called picking. This is caused when an ink is too sticky (tacky). Varnish can be added to reduce tack. Poor ink trapping, the inability of one ink color to adhere to another in multicolor printing, is a related problem.

Mottling. This is uneven, blotchy printing, especially in solid areas. This may be caused by improper viscosity of ink, but may also appear where plate impression is too heavy in letterpress printing.

Ghosting. A chemical or ink ghost usually appears as an unwanted dull image within an otherwise glossy area. This is the result of differing rates of ink absorption. Dull images result where more ink vehicle penetrates into the paper. It is thought that oxidation of large solid areas on surrounding sheets may rob oxygen from an inked surface. Slowing press runs and stacking freshly printed sheets in smaller lifts usually helps cure the problem. Proper selection of papers which allows less vehicle penetration also is helpful.

Hickies. Like ghosting, hickies are sometimes traced to paper imperfections as well as ink. These little troublemakers are the result of dried ink particles becoming attached to the plate or blanket on the press and preventing proper inking of the image. Particles may come from dried ink "skin" in the fountain or the ink container itself. Washup of affected surfaces is required, and sometimes a complete ink change is necessary.

Figure 13–23 illustrates various problems encountered with inks.

FIG. 13–22
*Equipment used in the manufacture
and testing of printing inks.*
NAPIM

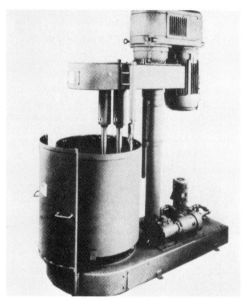

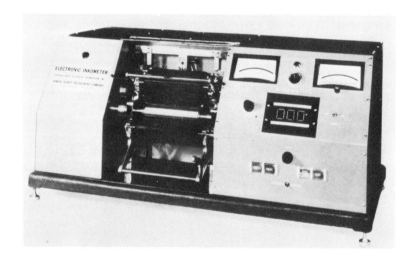

Tinting

Ink has emulsified into the water fountain. Pigment is being put on the plate from the water dampeners. Tint can easily be washed off plate.

Inadequate drying

Printed ink film is wet or tacky for an unreasonable length of time.

Misting or flying

Fine droplets or filaments of ink are formed on the roller train during film splitting. Ink may form a mist, or it may actually be sprayed or thrown off the press rollers.

Ink piling

Ink builds up on areas of the rollers, blankets, and/or plate, creating a dry accumulation known as "caking" or "piling".

Scumming

The inability of water to keep the non image area of the plate clean. Scum cannot be easily washed from the plate.

Poor trapping

Superimposed inks are being improperly laid down on the previously printed colors, causing poor color balance and poor overall appearance.

Blanket embossing & roller swelling

The blanket develops a relief image of the image on the plate, and/or the rollers swell so that they no longer stay within their normal settings.

Dot Spread

The halftone dots increase in size, causing the printed signature to lack sharpness.

Poor roller transfer and/or glazing

The ink appears to dry on the rollers or the inking system seems to be unable to adequately transport the proper ink/water emulsion down the roller train onto the plate.

Poor rub & scratch resistance

The printed ink film appears dry but exhibits poor rub and/or scratch resistance nonetheless.

Premature plate blinding

A strong image area of the plate is progressively losing its receptivity to the ink.

Linting/picking or paper "piling"

Paper surface is roughed or torn, hickies are visible. "Linting" refers to pulling of the fibers on uncoated stocks, "picking" to lifting of the coating on coated stocks onto blankets, plates and rollers.

FIG. 13–23
Common ink problems.
Sun Chemical Corporation,
General Printing Ink Division

New Developments

One of the most promising trends in printing today is heat-set web offset for publications. Inks designed for this process must provide acceptable gloss and ink holdout on newsprint and other uncoated papers previously expected to dry inks primarily by absorption.

Continuing developments in UV and electron beam curing are projected by leading industry experts like Sun Chemical Corporation's Terry Scarlett. The environmental benefit of solvent vapor reduction, he contends, coupled with high-glossing for record jackets, paperback book covers, and synthetic substrates, makes the future of these processes bright.

Scarlett also sees increased use of water-based inks to reduce vapor emission problems in gravure and flexographic processes, traditionally heavily dependent upon volatile solvents.

271

14 Bindery and Finishing
Customizing the Visual Presentation

The best format, the most creative writing, the most innovative graphics all will fall short unless they are packaged well. Bindery and finishing operations are the means to ensure that the finished piece or publication is properly "dressed" to meet its audience. Today's methods offer variety and speed undreamed of only a few years ago.

Introduction

Nowhere is the custom nature of producing printed products more apparent than in the bindery section of a major plant. Scores of printing jobs may be in the final manufacturing stages, all requiring entirely different steps and degrees of skill on the part of workers. Few industrial plants are prepared to deal with such diversity, but printers would have it no other way.

Bindery and finishing operations generally include anything that alters or adds significantly to paper or other substrate after it has been printed. Usually this includes common operations such as cutting, folding, stitching, and trimming pages or folded signatures. However, it can include any other operations that enhance the final product.

Most bindery operations can be thought of as *on-press*, performed on the press while it is operating, or *post-press*, performed following the press run. Operations that are handled by machines or attachments working in tandem with the press are described as *in-line* processing.

This chapter discusses a number of common operations, some of which we described in Chapter 11 as being related to major or specialized printing processes.

Prepress Operations

Printers usually think of *prepress* operations as the mechanical production of images—usually on film—and the proper assembly of these images for platemaking. The exposure and development of the plates themselves is also sometimes included under this catch-all term. Here, we use a broad definition of prepress. We include all the preparatory steps required on a job before the printer begins mechanical work, since all such steps affect the finished piece of printing.

Planning

The objectives, format, psychological appeal, distribution, production schedule, and budget should be carefully considered for each job before any production work begins. Since binding and finishing can have such a great influence on the effectiveness of a printed piece, these last few steps in the production cycle may be the most crucial. Job specifications that can affect bindery and finishing decisions are discussed in the next chapter.

Formats

The element of planning most related to bindery and finishing is the selection of a page format or finished size. Standard page formats that match the basic sheet sizes of paper are shown later in this chapter. The exact size of paper that will be run on the press and the press size itself depend on several things:

1. *The number of copies needed.* The larger the number, the more likely the job will be run with several copies to a sheet to make printing more efficient. Printers schedule work carefully to ensure that presses offering the most efficient sizes of plates and paper are available at the right time for each job.

2. *The printing quality desired.* Heavy ink coverage and precise registration of color images require larger presses, even for small runs.

3. *The nature of post-press operations.* If a printed sheet requires only that it be cut into equal pieces and/or folded without further trimming, press sheet sizes may be only exact multiples of the finished size ordered by the customer. Bleeds, gripper edges, and signature stitching all require oversized paper; therefore, press sizes needed may also increase.

The following examples show some of the ways presswork must be planned for proper bindery.

Example 1.

Suppose we select a standard 8½ × 11 format for printing a business letterhead on bond paper. If we order a small amount and print only a simple, one- or two-color design, the job probably will be done on an offset duplicator. The maximum size for most duplicators is 11 × 17; therefore, the job could be run in single sheets, 8½ × 11 or on 11 × 17 sheets, with two letterheads per sheet. Bond paper is readily available in reams of either size. The basic size of bond paper is 17 × 22, so *four* 8½ × 11 or *two* 11 × 17 sheets can be cut from one basic size sheet, if the printer does not buy precut reams.

If a long run, better inking, or bleeds are required, a larger press capable of printing a sheet 17 × 22 or larger probably would be used. Figure 14–1 shows the alternatives for one-up, two-up, or four-up printing of letterheads without trim.

Bleeding an image off the edge of the sheet requires paper to be cut larger than the finished size prior to printing because the sheets must be trimmed precisely at the edge of the image. A ⅛-inch extra overlap should be allowed; otherwise, a tiny white strip might appear at the trimmed edge if the cut is slightly off.

A. Short run, one or two colors — run on *duplicator*.

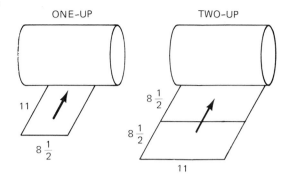

B. Long run, better inking, bleeds — run on *large press*.

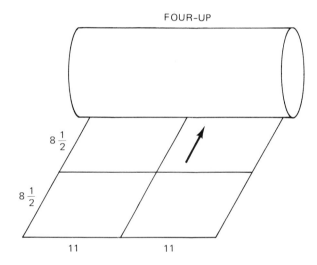

FIG. 14–1
Alternatives for running a simple 8½ by 11-inch form.
S. D. Warren Paper Company

Also, the edge of the sheet that is grasped by the small fingerlike *grippers* of the press must be free of printed image. From ¾-inch to ½-inch extra is required, depending on the type of press. If a sheet is to be folded into a signature, it must be trimmed after stitching so the pages will be separated. This requires at least an additional ⅛-inch around three sides of the folded paper.

Example 2.

To see the necessity of allowing for bleed, gripper, and trim, imagine we are printing a small number of 8-page, 5 × 8, self-cover booklets on an offset duplicator. A folded *dummy* of the booklet shows where allowance for basic trim is required if no bleed is used (Figure 14–2). The printed sheet should be at least 10¼ × 16½ to allow three-sided trim of the sheet when folded. Note that the grippers may extend into the page area if no image is present.

Figure 14–3 indicates the effects of bleed and gripper on the same booklet. Suppose we want the design to bleed on pages 1, 4, 5, and 8. The printed sheet should now be at *least* 10½ × 17 to allow for gripper area during printing and for bleed trim following folding.

These examples show how even simple forms printed on an offset duplicator must be carefully planned.

Work printed on large presses requires even more care since more pages with bleed and trim areas will be printed at one time. Regardless of press size, all bleed, trim, and gripper space must be determined and marked on masking sheets before any negatives can be positioned properly for plate burning. Those doing pasteup of booklets and other publications should always consult the printer to determine if bindery and finishing allowances must be made on the boards and, if so, where. Many jobs are pasted up with incorrect

FIG. 14–2
Figuring a no-bleed design. Printed sheet should be at least 10¼ by 16½ to allow three-sided trim of sheet when folded. Note that grippers may extend into page area if no image is present.
S. D. Warren Paper Company

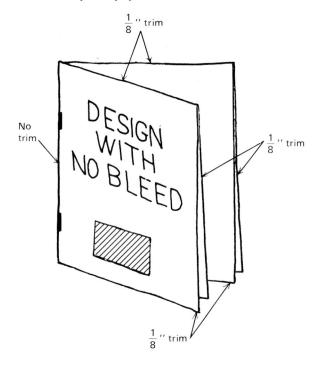

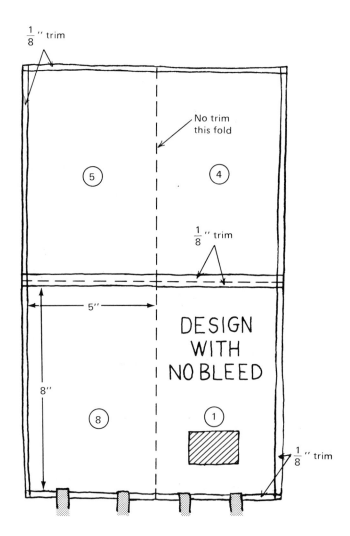

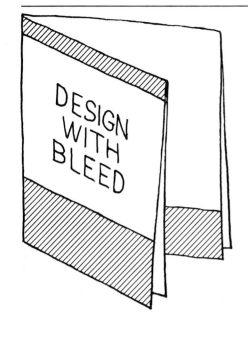

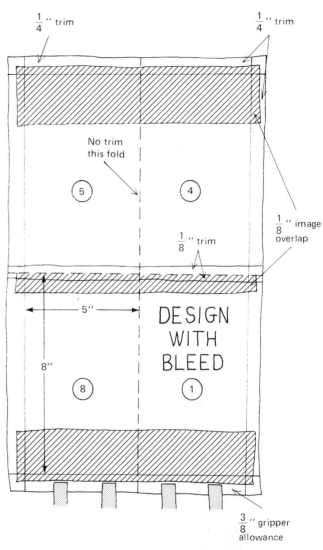

FIG. 14–3

For a bleed design, printed sheet should be at least 10½ by 17 to allow for
gripper area and bleed trim when the sheet is folded.
S. D. Warren Paper Company

spacing between pages, and simply create more work for the stripper who must cut pages or negatives apart and reposition them.

Standard Paper Sizes and Formats

Figures 14–4, 14–5, and 14–6 show many standard combinations of paper sizes and formats for both sheet-fed and web printing. Trim and gripper areas are not shown, but are taken into account for each size. In planning a printed piece, we can consult these charts and be assured that the finished size selected will cut and trim from standard size printing papers without excessive waste. Nonstandard sizes can be purchased but since they cost more, they are not cost effective unless the printing run is large.

The web diagrams in Figures 14–5 and 14–6 are for two common web sizes only—23½ × 35 *(full web)*, and 17¾ × 26 *(half web)*. Other press sizes such as 17¾ × 23 half web, a 23 × 18 *jobber web*, and a 45½ × 35 *double web* also are used. Anyone who is considering web printing should check exact sizes, formats, and page impositions before preparing any materials.

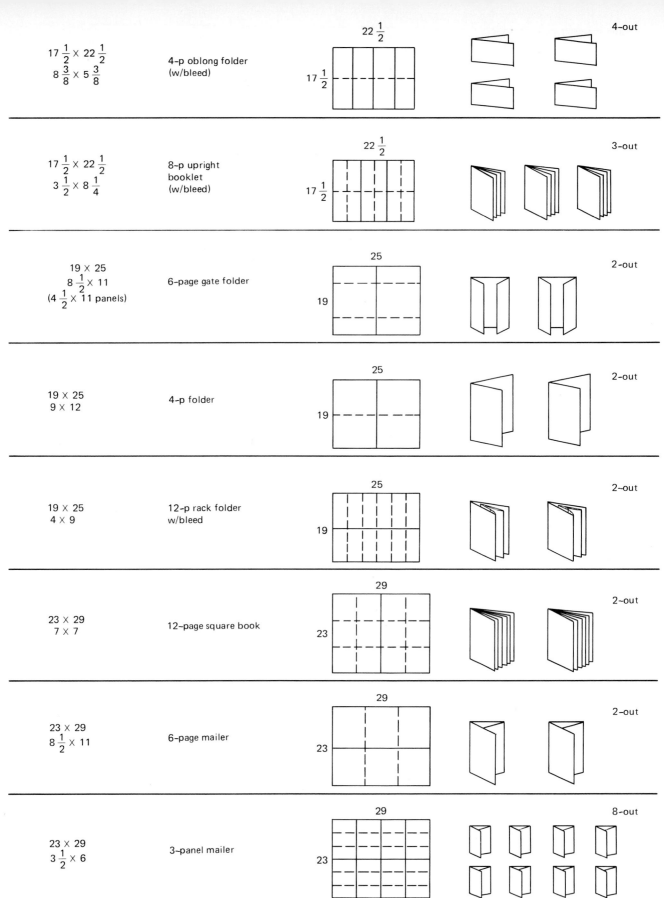

$17\frac{1}{2} \times 22\frac{1}{2}$
$8\frac{3}{8} \times 5\frac{3}{8}$

4-p oblong folder
(w/bleed)

$22\frac{1}{2}$

$17\frac{1}{2}$

4-out

$17\frac{1}{2} \times 22\frac{1}{2}$
$3\frac{1}{2} \times 8\frac{1}{4}$

8-p upright
booklet
(w/bleed)

$22\frac{1}{2}$

$17\frac{1}{2}$

3-out

19×25
$8\frac{1}{2} \times 11$
$(4\frac{1}{2} \times 11$ panels)

6-page gate folder

25

19

2-out

19×25
9×12

4-p folder

25

19

2-out

19×25
4×9

12-p rack folder
w/bleed

25

19

2-out

23×29
7×7

12-page square book

29

23

2-out

23×29
$8\frac{1}{2} \times 11$

6-page mailer

29

23

2-out

23×29
$3\frac{1}{2} \times 6$

3-panel mailer

29

23

8-out

FIG. 14-4
Sheet-fed formats.
S. D. Warren Paper Company

276

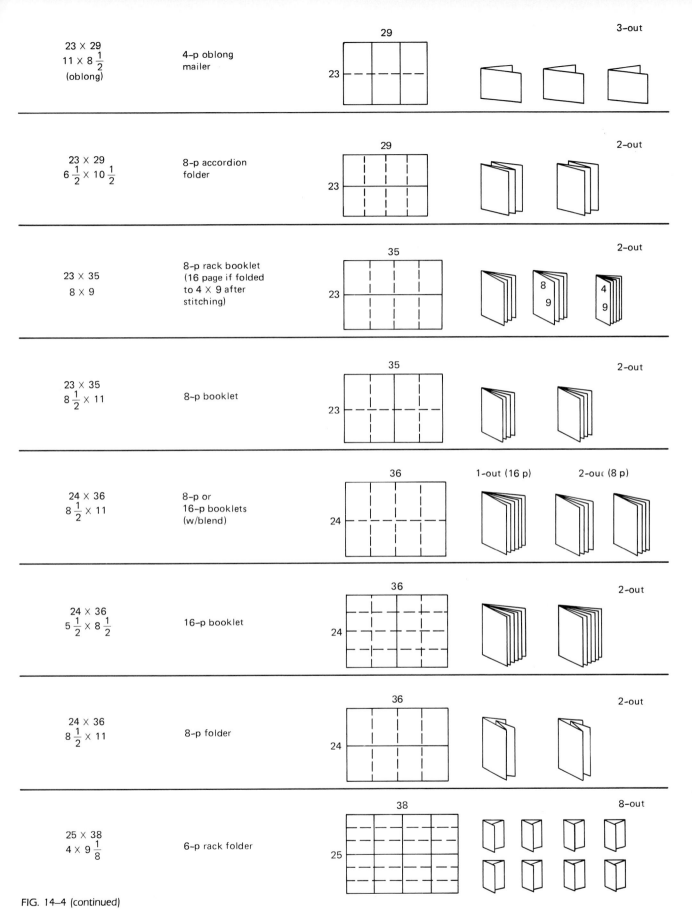

23 × 29 11 × 8 $\frac{1}{2}$ (oblong)	4–p oblong mailer	**3-out**
23 × 29 6 $\frac{1}{2}$ × 10 $\frac{1}{2}$	8–p accordion folder	**2-out**
23 × 35 8 × 9	8–p rack booklet (16 page if folded to 4 × 9 after stitching)	**2-out**
23 × 35 8 $\frac{1}{2}$ × 11	8–p booklet	**2-out**
24 × 36 8 $\frac{1}{2}$ × 11	8–p or 16–p booklets (w/blend)	**1-out (16 p) 2-out (8 p)**
24 × 36 5 $\frac{1}{2}$ × 8 $\frac{1}{2}$	16–p booklet	**2-out**
24 × 36 8 $\frac{1}{2}$ × 11	8–p folder	**2-out**
25 × 38 4 × 9 $\frac{1}{8}$	6–p rack folder	**8-out**

FIG. 14–4 (continued)

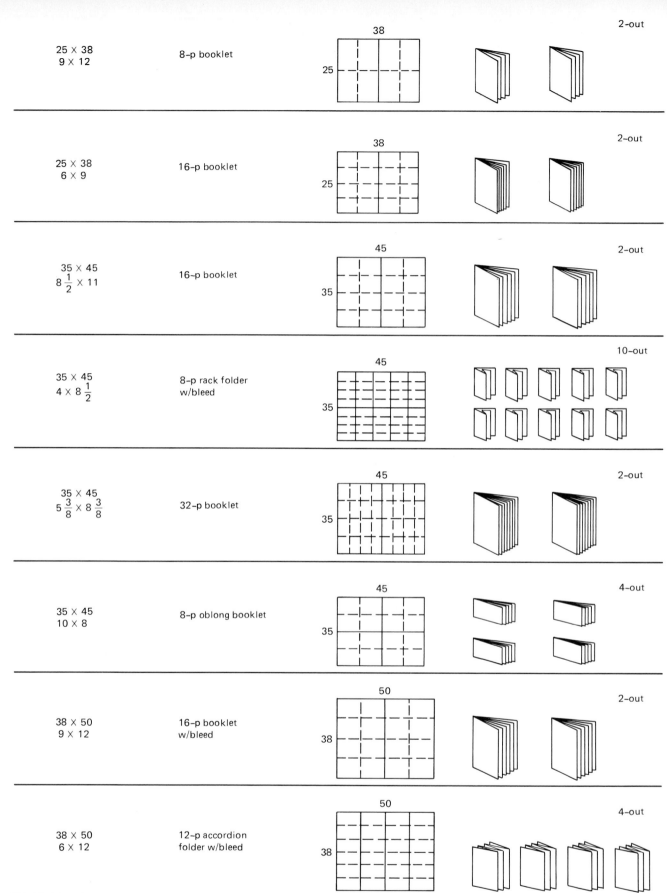

25 × 38 9 × 12	8-p booklet	2-out
25 × 38 6 × 9	16-p booklet	2-out
35 × 45 $8\frac{1}{2}$ × 11	16-p booklet	2-out
35 × 45 4 × $8\frac{1}{2}$	8-p rack folder w/bleed	10-out
35 × 45 $5\frac{3}{8}$ × $8\frac{3}{8}$	32-p booklet	2-out
35 × 45 10 × 8	8-p oblong booklet	4-out
38 × 50 9 × 12	16-p booklet w/bleed	2-out
38 × 50 6 × 12	12-p accordion folder w/bleed	4-out

FIG. 14–4 (continued)

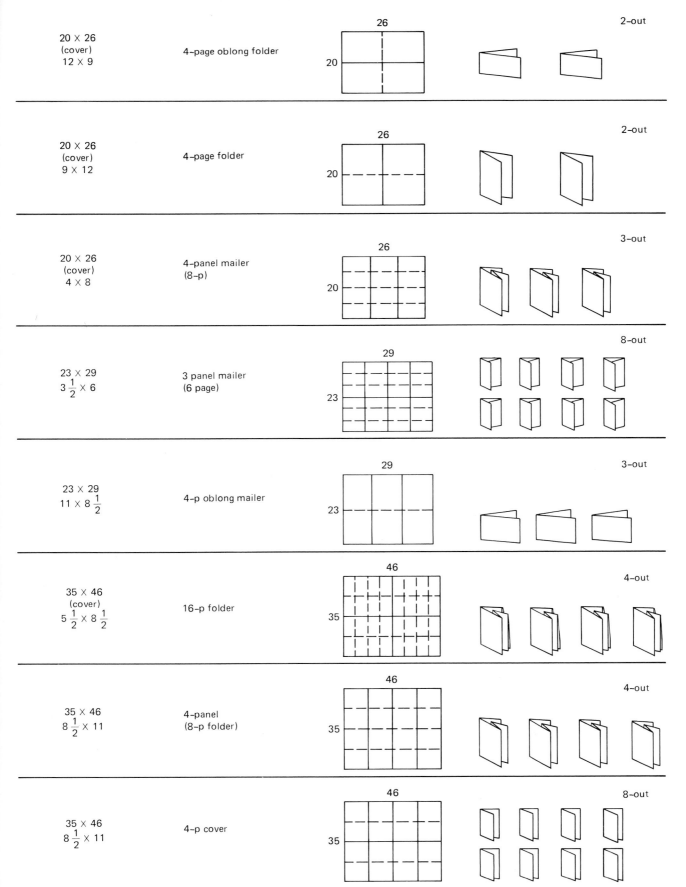

20 × 26 (cover) 12 × 9	4-page oblong folder	26 / 20	2-out
20 × 26 (cover) 9 × 12	4-page folder	26 / 20	2-out
20 × 26 (cover) 4 × 8	4-panel mailer (8-p)	26 / 20	3-out
23 × 29 $3\frac{1}{2}$ × 6	3 panel mailer (6 page)	29 / 23	8-out
23 × 29 11 × $8\frac{1}{2}$	4-p oblong mailer	29 / 23	3-out
35 × 46 (cover) $5\frac{1}{2}$ × $8\frac{1}{2}$	16-p folder	46 / 35	4-out
35 × 46 $8\frac{1}{2}$ × 11	4-panel (8-p folder)	46 / 35	4-out
35 × 46 $8\frac{1}{2}$ × 11	4-p cover	46 / 35	8-out

FIG. 14–4 (continued)

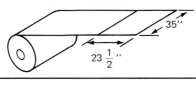

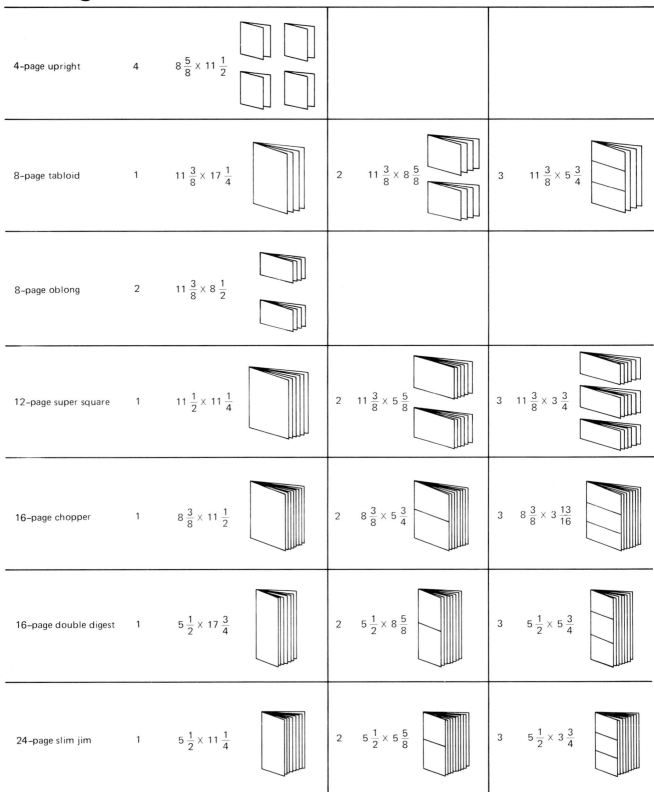

4-page upright	4	$8\frac{5}{8} \times 11\frac{1}{2}$					
8-page tabloid	1	$11\frac{3}{8} \times 17\frac{1}{4}$	2	$11\frac{3}{8} \times 8\frac{5}{8}$	3	$11\frac{3}{8} \times 5\frac{3}{4}$
8-page oblong	2	$11\frac{3}{8} \times 8\frac{1}{2}$				
12-page super square	1	$11\frac{1}{2} \times 11\frac{1}{4}$	2	$11\frac{3}{8} \times 5\frac{5}{8}$	3	$11\frac{3}{8} \times 3\frac{3}{4}$
16-page chopper	1	$8\frac{3}{8} \times 11\frac{1}{2}$	2	$8\frac{3}{8} \times 5\frac{3}{4}$	3	$8\frac{3}{8} \times 3\frac{13}{16}$
16-page double digest	1	$5\frac{1}{2} \times 17\frac{3}{4}$	2	$5\frac{1}{2} \times 8\frac{5}{8}$	3	$5\frac{1}{2} \times 5\frac{3}{4}$
24-page slim jim	1	$5\frac{1}{2} \times 11\frac{1}{4}$	2	$5\frac{1}{2} \times 5\frac{5}{8}$	3	$5\frac{1}{2} \times 3\frac{3}{4}$

FIG. 14–5
Full-size web formats.
S. D. Warren Paper Company

280

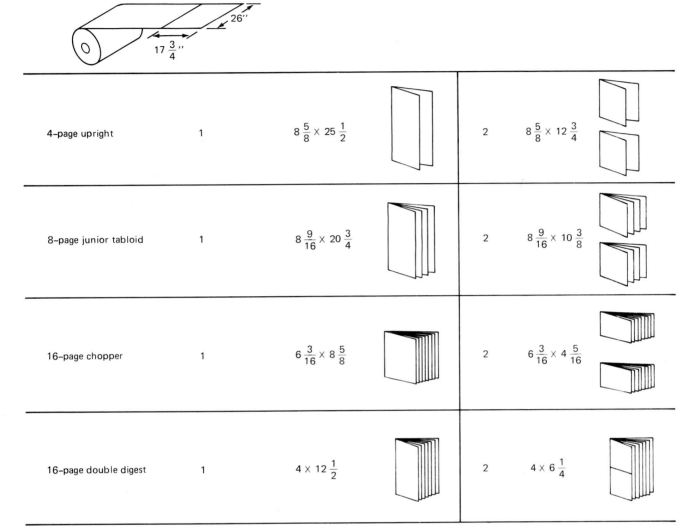

HALF-SIZE WEB (26″ width × 17 $\frac{3}{4}$ cutoff)

4–page upright	1	$8\frac{5}{8} \times 25\frac{1}{2}$		2	$8\frac{5}{8} \times 12\frac{3}{4}$
8–page junior tabloid	1	$8\frac{9}{16} \times 20\frac{3}{4}$		2	$8\frac{9}{16} \times 10\frac{3}{8}$
16–page chopper	1	$6\frac{3}{16} \times 8\frac{5}{8}$		2	$6\frac{3}{16} \times 4\frac{5}{16}$
16–page double digest	1	$4 \times 12\frac{1}{2}$		2	$4 \times 6\frac{1}{4}$

FIG. 14–6
Half-size web (26-inch width by 17-3/4 cutoff)
S. D. Warren Paper Company

Imposition

The diagrams presented thus far show only what format—page sizes and number of pages per piece—are possible from standard sheet and web sizes. We can see how important it is to have pages printed correctly on the paper so they come out in proper sequence after folding. This is called *imposition.* As we discussed in Chapter 10, pasteup of pages should never be done without consulting the printer, because imposition depends entirely upon the size of press and paper to be used.

Fundamentals of Imposition

Three basic types of imposition are common today: *work and back; work and turn;* and *work and tumble. Work and back* (also called sheetwise) is done where a different form or set of pages is to be printed on each side of a sheet. The sheet is not cut apart after printing; instead, it probably will be folded into a signature, unless no folding at all is desired.

Work and turn is a method of imposing and printing where all the pages of a job are printed in one large form on a double-size sheet. The sheet is then turned

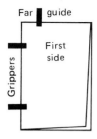

WORK AND BACK

Different form on each side of sheet.
Sheet is not cut or slit after printing.
Not necessary to trim or square sheet before printing.

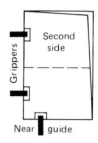

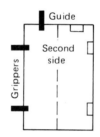

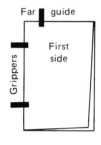

WORK AND TURN

Same form on both sides of sheet.
Sheet is cut or slit after printing.
Not necessary to trim or square sheet before printing.

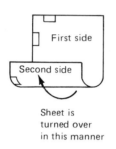

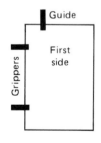

WORK AND TUMBLE

Same form on both sides of sheet.
Sheet is cut after printing.
One end of sheet may be left untrimmed, other sides must be squared.

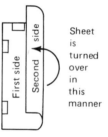

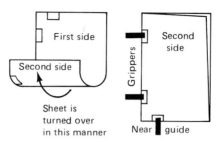

FIG. 14–7
Common imposition methods.
Franklin Offset Catalog, Porte Publishing Company

over end-for-end (usually on the long dimension of the sheet) and the same form or set of pages is printed on the back. The sheet is then cut apart, producing two copies of the job from each large sheet.

Work and tumble is similar to work and turn except that the printed sheet is turned over on the short dimension for printing on the reverse side. Two copies are produced per sheet, similar to work and turn. Figure 14–7 shows all three common imposition schemes for sheet-fed printing.

Figure 14–8 indicates a number of imposition diagrams for popular sizes of booklets and brochures, showing the positions of pages for each configuration. While these impositions are common, it is not wise to assume that a job will always be done according to "formula." Availability of presses and paper, bindery

considerations, special effects, and position of solids and other heavily inked areas can affect where pages will appear on the plate.

Figure 14–9 shows two plates, one with pages positioned so that proper ink coverage will be difficult, the other with more balanced placement of inking areas. The form on the right would be less troublesome to print because there is no long, uninterrupted solid in the direction the ink rollers must travel.

Another potential imposition problem is the *crossover,* an image, usually an illustration, photograph, or border, that is spread across facing pages of a booklet or folder. Crossovers work best in the natural (center) spread of a piece. There is no chance of misalignment in folding since the facing

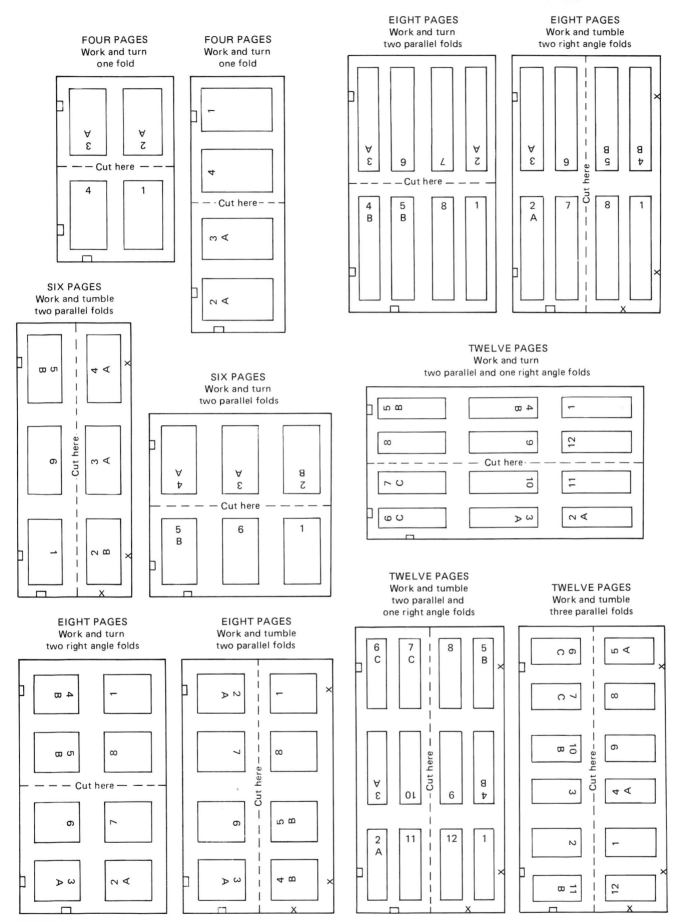

FIG. 14–8
Popular impositions.
Franklin Offset Catalog, Porte Publishing Company

SIXTEEN PAGES
Work and turn
three right angle folds

SIXTEEN PAGES
Work and turn
two parallel and one right angle folds

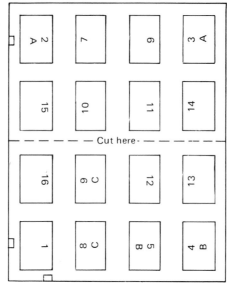

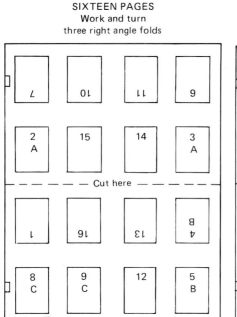

FIG. 14–8 (continued)

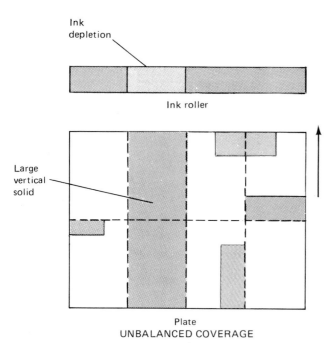

FIG. 14–9
Effects of page placement on inking.

pages are adjacent when printed. Also, there is no need for a duplicate negative to be made so that separate portions of the image can be stripped into different areas of the masking sheet. Note in Figure 14–10 that only one negative is required for the image on pages 8–9, whereas two negatives are required for the crossover on pages 4–5.

Postpress Operations

The operations of binding and finishing that occur after planning, imposition, and printing take place can be divided generally into seven groups: sheet enhancing, cutting, folding, assembling, mechanical binding, adhesive binding, and packaging.

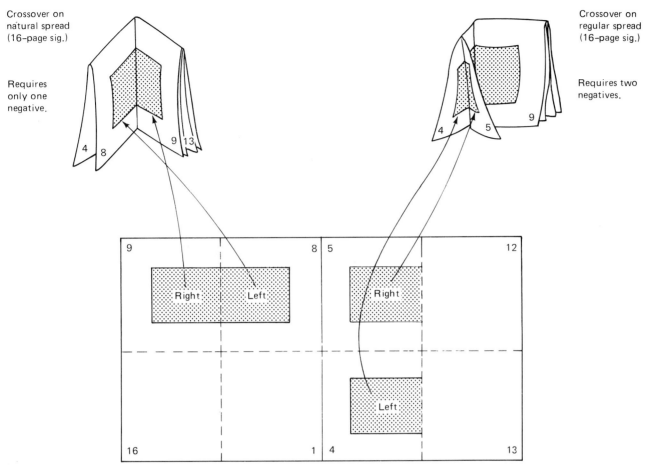

Crossover on natural spread (16-page sig.)

Requires only one negative.

Crossover on regular spread (16-page sig.)

Requires two negatives.

FIG. 14–10
Advantage of using the natural spread in imposition.

Sheet Enhancing

While many nonprinting sheet modifications are performed on press, some may be performed in the bindery. Rotary perforating, scoring, and numbering devices are common. Some plants can perform spot-carbon and laminating services. Die cutting, foil stamping, and embossing presses are sometimes located in or near bindery areas, and many of the new machines that perform such tasks bear little resemblance to printing presses. Chapter 11 discusses sheet enhancement in detail.

Cutting

A common sight in the bindery is the programmable power cutter. Designed for cutting printed sheets apart and trimming off excess edges, this device has been greatly improved in the past decade. Today's cutter contains a microprocessor that allows fully electronic programming of multiple cuts and trims. It has hydraulic clamps to hold large *lifts* (stacks) of paper secure, and it is fitted with an electric eye that stops the blade instantaneously if the hand of its operator should venture too close. Air forced through tiny holes in the working surfaces "floats" the heavy paper stacks like a hydrofoil over water—fifty pounds or more with ease. Some large volume equipment is designed to convey lifts of paper to and from the cutter automatically. Figure 14–11 shows a power cutter and a lift conveyor attachment.

The sequence of a simple cutting program is shown in Figure 14–12. For ease of effort and accuracy, it stresses (a) cutting on the long side first, and (b) rotating the lift only 90 degrees at a time. The program calls for four perimeter (outside) cuts and four inside cuts. Following each cut, the back gauge moves automatically to the next position.

Other operations that involve cutting paper in some fashion are *hole punching* (drilling), *slotting, round cornering,* and *slitting.* These are usually performed on small pedal-operated devices fitted with sharp, rotating or hollow bits or blades. *Die cutting* is discussed in Chapter 11 as an on-press or specialty

FIG. 14–11
A power cutter and lift-conveyor attachment.
POLAR-Mohr

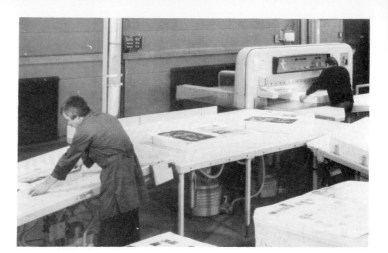

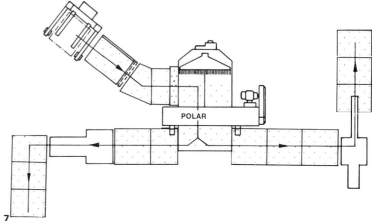

7

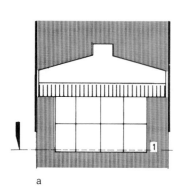

a

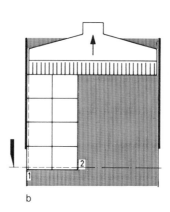

b

FIG. 14–12
A simple programmed cutting sequence.
POLAR-Mohr

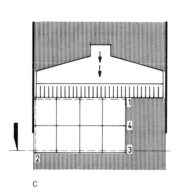

c

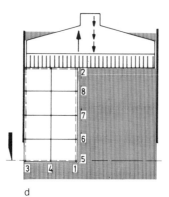

d

finishing process. However, if it is performed with heavy steel dies *(high dies)* that look like huge cookie cutters and must be positioned onto paper manually, it logically may be considered part of binding or finishing. Envelopes are often cut to their open shape with deep dies prior to being glued and formed into their finished shape.

Folding

While web presses are usually fitted with folding mechanisms, sheet-fed presses are not. Large buckle-type folders are used to make the necessary folds after sheets are printed and fully dried. These specialized machines are designed to perform from 4-page to 32-page folds at speeds of more than 12,000 sheets per hour. They are equipped for automatic feeding and collecting of finished pieces.

Many kinds of parallel and right-angle folds are possible. Figure 14–13 shows many popular styles. Figure 14–14 depicts an automatic folder.

Planning a job so that more than one piece can be

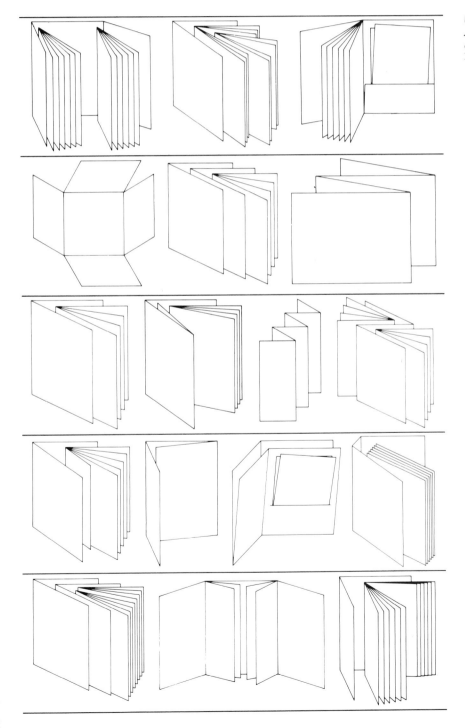

FIG. 14–13
Popular folding patterns, some with added covers.
Hammermill Paper Company

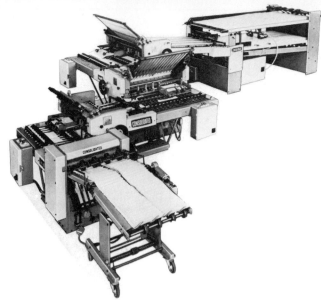

FIG. 14–14
A high-speed, buckle-type automatic folder capable of handling up to 32-page signatures. Perforating, slitting, and scoring attachments may be added.
Consolidated/Bonelli

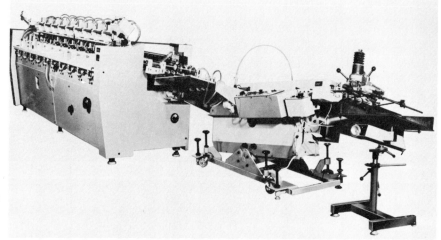

FIG. 14–15
Consolidated Neckar horizontal collator with gluing attachment and jogger. Can be equipped with rollaway stitcher, folder, and front knife trimmer.
Consolidated International Corporation

folded at one time can double, triple, even quadruple folding speed. This *gang folding* is done with folds parallel to one another so the sheet contains a strip of folded pieces ready to be cut apart.

Assembling

Bringing the separate pieces of a job together is usually referred to as *collating* or *gathering*. Collating generally refers to single sheets, whereas gathering denotes signatures; however, automatic "collators" are equipped to handle either.

Automatic collating devices can assemble up to ten sheets or signatures printed on manifold (tissue or onionskin) to heavy cover paper at up to 4,500 sets per hour, and some can stitch, fold, and trim continuously as well. The machine also detects both "misses" or "doubles" and stops, signalling the operator where the problem occurred. Figure 14–15 shows such a device.

Nonadhesive (Mechanical) Binding

Methods of fastening without a glue include *stapling*, *saddle- or side-wire stitching*, *sewing*, and other mechanical means such as *spiral wire* or *plastic comb* (Figure 14–16).

Saddle stitching is a very common method for binding booklets or other publications under 96 pages. A modern automatic saddle stitcher is actually a whole bindery in itself. It is usually connected to or fitted with a collator (or gatherer), a three-knife trimmer, a counter, a stacker, and a bundler, so that all operations are performed in-line (Figure 14–17).

Adhesive Binding

Also known as *perfect binding*, this form of binding has become very popular in the past few years due to improvements in glue formulas and automation of operations. Books up to 1¾-inch thick can be bound on typical equipment. Signatures are stacked flat rather than open as in saddle stitching. The machine clamps the signatures together and transports them over a rotating blade that rips off the folded backbone of the sheets. This edge then passes over a roller which coats it with hot glue. A cover is placed over the book and the cooling glue binds the entire package together tightly. A perfect binding and covering machine is shown in Figure 14–18.

Spot gluing *(tipping)* can be done on- or off-press and is often required when pockets are designed for portfolio-style folders. Many web presses are adapted for high-speed glue binding.

Packaging

The final step in postpress operations is the packaging of finished pieces for delivery. Packaging is determined by type of printing, method and distance of transportation, handling requirements of the customer, and the capabilities of the printer.

Newspapers need only to be tied in counted bundles and thrown into trucks. Higher quality printing may be shrink-wrapped in cellophane or wrapped in paper, stacked on wooden skids, and carefully moved by fork lift. Corrugated cartons packed with Styrofoam chips may be used for high quality goods, especially if they require extra handling or longer storage. Smaller pieces that require individual addressing may be stacked in large trays instead of being wrapped.

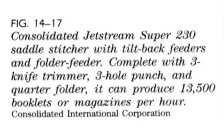
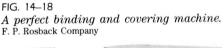

Saddle stitched   Cleat bound   Wire bound   Plastic spiral

FIG. 14–16
Common nonadhesive binding methods.

FIG. 14–17
Consolidated Jetstream Super 230 saddle stitcher with tilt-back feeders and folder-feeder. Complete with 3-knife trimmer, 3-hole punch, and quarter folder, it can produce 13,500 booklets or magazines per hour.
Consolidated International Corporation

FIG. 14–18
A perfect binding and covering machine.
F. P. Rosback Company

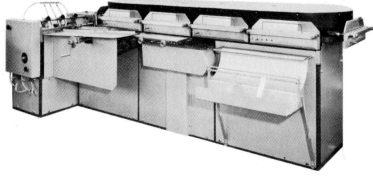

Blade that rips off the backbone before gluing

Perfect bound booklet

The Professional Interface

Controlling Costs and Quality Through Communication

Professionalism is the key to good relations with the graphics and printing specialists with whom we work. Consistent quality requires a commitment to excellence from everyone involved and the good sense not to make someone else's job harder than it has to be. The dividends of such relations are seen regularly by leading editors and designers who receive carefully crafted publications from their printers.

Introduction

Successful people in graphics know the importance of a strong working relationship with printers. On each job, they try to select the appropriate printing firm, prepare all materials carefully and correctly for reproduction, and communicate precisely all information to the printer. Dividends from a professional relationship between designer and printer accrue not only to any single job in process but also to those in the future. Well-conceived designs, prepared by those who have earned professional respect, become a challenge and a source of pride to printers, who relish any chance to demonstrate competence in their own craft.

Establishing a Professional Relationship

Know Your Printer. Begin establishing a working relationship with printing professionals by learning which ones meet your needs. One of the best ways for you to select printers is to talk to their satisfied customers. By joining professional communications or design organizations and attending their meetings, you will be exposed to people who work with printers every day. These people, already involved in graphic

design, company editing, public relations, or advertising, will be glad to give you their recommendations. A list of professional organizations you may wish to join can be found in Appendix C.

If possible, visit the plants of printers you have chosen. If time is limited, invite a representative or salesperson from each firm to visit you and present their samples. Check to see if your company has used any of the firms before, and, if so, ask a colleague whether or not they were proficient and dependable. On-site visits to printing plants are preferable since you can tour the facilities, meet those who will be producing your work, and ask specific questions about production techniques.

Most printers are specialized manufacturers. Ask for samples of work they *regularly* produce which is similar to what you will want. Depending upon your needs, you might ask about web- and sheet-fed press capacity, color printing, and in-house color separation. You may want publications services such as typesetting, art, pasteup, or special bindery and finishing. All of these services vary widely among printers.

Generally, the more in-house capabilities of the plant, the quicker a job can be produced. However, printers

290

located in large metropolitan areas can choose from many "for the trade" suppliers who offer rapid specialty services at competitive rates. Certain materials, such as color separations and embossing dies, are commonly bought from outside sources by even the largest printers.

Many plants require a minimum of ten to twelve *working* days for average jobs, but you will be wise to inquire about special scheduling and production cycles.

Develop a reference file of printers and their capabilities, along with contact persons for each. Be sure to include names of small, medium, and large plants, since you may need any one of them some day.

Talk the Language.   Printers use a vocabulary all their own. Learn as much of their language as you can before discussing your jobs with them. The bibliography at the back of the book provides valuable source materials. Precision in the use of terms is vital, and printing representatives will be of great help in your practical understanding of their special language.

Talk to printers *before* any production begins on your job. Often they can offer time- and money-saving methods of preparation. The more familiar you are with proper printing terminology, the more respect you will command as a professional. Be quick to admit your limitations, however. Insist that the printing salesperson define any language unfamiliar to you. Terms such as *mechanical, font, signature,* and *mask* have several meanings, so clarify their use. Describe your job in the simplest language at first, then allow the salesperson to give you his or her interpretation of what you want. You may be amazed at how many small details must be settled before work begins. Always prepare at least a simple dummy or rough layout in advance.

Before finished materials are presented for printing, be sure all pieces are marked correctly. Label the backs of all art boards, or their cover flaps, with your company name and a brief, descriptive job title. Attach tissue overlays to all pages. Photographs and other material not pasted directly on pages should have an overlay tissue with page and sequence clearly marked to match page overlays. Item "12-A," for example, would indicate page 12, photograph A, and would appear in position A labelled on that page.

Reduction or enlargement percentages and crop marks may be indicated carefully on overlay tissues or, if the printer prefers, on the back or edges of the art or photographs. To avoid misinterpretation, mark only a number—75%, for example. Don't say *75% reduction.*

This would mean that the art or photo would be brought down to 25 percent of its original size. Review the use of the proportion wheel or calculator before you size your material.

On all pasted pages, identify elements such as screen tints, reverses, and surprints by marking instructions on the overlays with a fine, felt-tipped or technical pen. Don't let your markings bleed through to the page below, however. Use a colored marker of the approximate hues desired to indicate color "callouts." Attach any necessary color swatches to each board and identify colors by PANTONE MATCHING SYSTEM® or other number. Finally, be sure *all* materials are ready before you take in the job.

Review each board carefully with the printer. Fifteen minutes of direct communication at this stage can save hours of phone calls and wasted work.

Prepare Materials Correctly.   Too many designers of print cause costly duplication of effort and "makeovers" by assuming they know the most efficient methods of preparing materials for reproduction. A production technique you have used or seen before may not be the best one for today's printing. Review Chapters 6 and 10, which describe preparation of art, photographs, and mechanicals, then *ask the printer* about special requirements.

The following are particularly troublesome:

1.   Problem: *Art containing lines or screen tint patterns too fine to be reduced in size.* Remember that even the smallest images will be reduced in proportion to the reduction of the overall image within which it appears. For example, the dots in a 100-line screen will become 200-line dots when reduced to 50% of original size—perhaps too fine for quality reproduction. *Solution:* Use screen densities coarse enough to withstand the particular reduction—in this case, an 85-line screen. It is often better to cut a mechanical overlay where screen tints are desired and allow the printer to "burn in" the tints at the platemaking stage. Use thicker lines for any art that is to be reduced. Avoid hairlines on original art.

2.   Problem: *Prescreened positive halftones pasted on pages.* A common practice for newspapers, this results in reshooting fine dots and ultimately losing picture detail. *Solution:* Where halftone quality is important, outline in red the exactly sized area of the page where you wish the halftone image to appear. Then submit the photograph, properly marked for sizing, to the printer along with the page. A halftone negative will be made and exposed separately to the plate. This will yield much higher halftone quality.

3.   Problem: *Inaccurately cut mechanical overlays.*

Unfortunately, many persons insist upon preparing all overlays for second colors and screen tints by cutting their own film. All too often, the "ruby" is poorly cut, out of register, or completely unnecessary, and requires extra work by artists or strippers at the plant. *Solution:* Discuss *any* mechanical overlay with the printer before work begins on the original art. In cases where no close register is required, colors and tints can be indicated with marks on a tissue overlay. These marks are referred to as "call-outs." Sometimes colors and tints can be prepared from a single negative. Where exact ("hairline") register is required, the printer will prefer to cut the overlays precisely with special tools or produce effects entirely photomechanically.

4. Problem: *Unnecessary reverses.* Persons who prepare elaborate reverses for the printer by painting or inking in large black areas may not realize that it is a simple matter for the camera operator to photograph any black-on-white type or art as a film positive, then create a reverse at the platemaking stage. *Solution:* Let the printer prepare all but the simplest reverses. Carefully mark the area of the page intended for reversing and let the printer produce the effect.

5. Problem: *Copy density too weak for reproduction.* Few things degrade the quality of printing more than variations in the darkness of original type or art. Rules or artwork done with felt-tipped pens, unevenly inked solids, and faded type strips are far too common. When shooting pages containing such work, the camera operator at the printing plant must attempt to balance the exposure of each negative to avoid losing weak details. This produces uneven quality in these page negatives. *Solution:* Use only solid black drawing inks or border tapes for rules. Be certain all chemicals for the phototypesetter and the camera you use are fresh. Adjust exposure lamp settings on your phototypesetter (but avoid *overexposure*). Calibrate your process camera for correct exposures or have service personnel do this for you.

Follow Through with the Job. To ensure continued good service from any printer, make an extra effort to facilitate the printing work. Make sure you can be reached easily by phone. Read proofs carefully, mark them correctly, and return them ahead of time, if possible. Standard proofreading symbols are shown in Chapter 5. Don't use copyediting symbols on proofs.

If you must view press proofs or wish to be present during some phase of production, be at the plant promptly when called. When printing is completed, pick up original materials immediately if they are not delivered to you with the job.

Protecting Your Printing Budget

Seek Out Good Price Quotations. One of the most important—and least understood—tasks that beginning designers face is obtaining prices for jobs they wish to have printed. For today's tight budgets, it is more important than ever that you make every printing dollar count. An understanding of how printing is priced will help.

Like most other manufacturers, printers base their prices on many factors. These include costs for facilities and equipment, materials, labor, competition, and market conditions. Modern printers, through extensive cost analyses, determine the expense of operating each of their "cost centers" (for example, a camera, press, or folder is one center). After base data ("plant standards") are determined, they know rather precisely how costly each unit of time will be to operate each piece of equipment. Price estimates are based on the estimated time each piece of equipment will be used on a particular job. Variable costs, such as paper and ink, are also accounted for.

Many printers today use specialized computer systems to compute price estimates by factoring in applicable fixed and variable costs from each cost center. Figure 15–1 shows a model of the Super Stewy/MP, a business computer system designed for the smaller printer by Stewart Systems Corporation. Stewart defines a small printer as one with between $400,000 and $2 million in annual sales.

Figure 15–2 shows a typical estimate sheet generated by the fully integrated system. Note that the estimating program (software) plans and estimates the full job, including selection of the best presses, best preparation and binding methods, and best imposition (placement of pages) for the job. Hours (divided into tenths), materials, and outside purchases for each cost center are all indicated.

The estimate shown is for the printing of 4,000 copies of a quarterly magazine for a bank, $8\frac{1}{2} \times 11$ inches, with 32 pages plus cover. We can see that this job will require 17,680 sheets of paper, including spoilage, for four 8-page signatures. Press number 6 has been selected to accommodate the 17×23 inch press sheet for the inside pages; press number 4 will be used for the 25×38 inch cover paper. Also, we can see data generated for each cost center. For example, fotocomp (typesetting) is estimated at 20.2 hours for a total labor cost of $263.

A tentative quote (Figure 15–3) gives a summary of the estimate, and the factors in the markup percentage of 10. This summary is used as a worksheet for determining a final selling price. When the tentative quote is approved by management in the plant, a final quotation, Figure 15–4, is automatically generated by the computer, along with an addressed envelope, and mailed to the customer. The actual dollar amounts shown here are for illustrative purposes only and are *not* necessarily typical of charges by printers.

While you as a customer may never see anything but a final quotation from a system such as the one described here, you are able to take advantage of the computer's power to facilitate your own work. These advantages include:

Multiple Quotations. You are able to receive rapid and accurate quotes on jobs without worrying that you are a nuisance to a printer's estimating personnel.

This allows you to plan printing more easily and to request alternate quotes simultaneously.

Job Status. You are able to call in at any time during the production of a job to determine up-to-the-minute information on the job's progress through the plant and whether or not early delivery is likely. You may also want to know if outside purchases are on schedule, or the extent to which author's alterations, page corrections, and other variable costs are adding excessively to the job price you were originally quoted.

"Finished Goods" Mailing Services. You may have the printer generate mailing lists too small for major mailing houses to handle. Some printers will even serve as your distribution center for printed and packaged materials. After printing, they will store and ship items automatically, as they are needed, to destinations you designate.

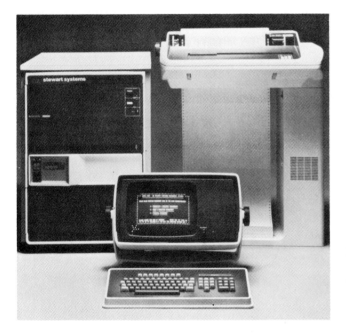

FIG. 15–1
(Left): The Super Stewy/MP computer system for small printers. Pictured are the Micro Nova 200 computer, a line printer, and a video display console, manufactured for Stewart Systems Inc. by Data General.
(Below): The Data General Model 30, part of the new desktop generation of microcomputer systems marketed by Stewart Systems.

```
         STDS. SET      QUANTITY        TOTAL $     $ ADDTL/M
            1           40000.          8585.         173.11




            ORDER SHEET SIZE
       INCHES  8THS  INCHES  8THS          POUNDS        SHEETS          1000 S
         17     6     23     0              759          17680             10




            PRESS SHEET SIZE
       INCHES  8THS  INCHES  8THS       SIGS PAGES/SIG  SHEETS          1000 S
         17     6     23     0            4      8      17680             10

       PRESS   NO. UP  NO. OUT  PLATES      FOLDER  PAGES/SIG   NO. UP   BIND UP
         6     1-SW      1        8           0        0          0        1


       DEPARTMENT      NO.        HRS.    LBR.$   UNITS   MAT.$  PURCH.$  TOT.$

       FOTOCOMP        100        20.2    263.     0.      0.     0.      263.
       PASTE-UP        102        11.3    226.     0.      0.     0.      226.
       ART             103         1.0     20.     0.      0.     0.       20.
       4-CLR SEPS      104         0.0      0.     0.      0.    200.      230.

       CAMERA          106         2.4     54.     1.     33.     0.       87.
       STRIP           107         4.4     89.     1.      0.     0.       89.
       PROOFS          108         1.8     37.     0.      0.     0.       37.
       PLATES          109         2.5     50.     8.     48.     0.       98.
       RDY&WASH          6         1.5    114.     0.      0.     0.      114.
       RDY-BACK          6         1.0     76.     0.      0.     0.       76.

       STOCK           113         0.0      0.  7597.   3494.     0.     3494.
       RUN&DRY           6         4.4    334. 88000.      0.     0.      334.
       INK             117         0.0      0.    70.    175.     0.      175.

            ORDER SHEET SIZE
       INCHES  8THS  INCHES  8THS          POUNDS        SHEETS          1000 S
         26     0     40     0             2820         11750              1




            PRESS SHEET SIZE
       INCHES  8THS  INCHES  8THS       SIGS PAGES/SIG  SHEETS          1000 S
         25     0     38     0            1      4      11750              1

       PRESS   NO. UP  NO. OUT  PLATES      FOLDER  PAGES/SIG   NO. UP   BIND UP
         4     4-SW      4        5           1        4          1        1



       CAMERA          106         0.2      6.     1.      4.     0.       10.
       STRIP           107         2.6     52.     4.      0.     0.       52.
       PROOFS          108         1.1     22.     5.     10.     0.       32.
       PLATES          109         3.8     76.     5.     50.     0.      126.
       RDY&WASH          4         3.7    142.     0.      0.     0.      142.
       RDY-BACK          4         0.4     18.     0.      0.     0.       18.

       STOCK           113         0.0      0.  2820.   1816.     0.     1816.
       CUT-PRESS       114         2.2     45.     0.      0.     0.       45.
       RUN&DRY           4         5.6    246. 23500.      0.     0.      246.
       RUN-BACK          4         2.8    123. 11750.      0.     0.      123.
       INK             117         0.0      0.    15.     58.     0.       58.
       CUT             120         3.3     67.     0.      0.     0.       67.
       FOLD              1         5.2     83.     0.      0.     0.       83.

       SADDLEWIRE      125         9.5    228.     0.      0.     0.      228.
       PACKING         127        10.2    102.   201.     50.     0.      152.
       FREIGHT         129         0.0      0.     0.      0.    115.      132.

            TOTAL       999       101.2   2482.     0.   5741.   315.     8585.


       DEPARTMENT    HRS.    TOT.$

       FOTOCOMP      20.2     263.
       PASTE-UP      11.3     226.
       ART            1.0      20.
       4-CLR SEPS     0.0     230.
       CAMERA         2.7      97.
       STRIP          7.0     141.
       PROOFS         2.9      69.
       PLATES         6.3     225.
       RDY&WASH       4.7     256.
       RDY-BACK       1.4      94.
       STOCK          0.0    5311.
       CUT-PRESS      2.2      45.
       RUN&DRY       10.0     581.
       RUN-BACK       2.8     123.
       INK            0.0     234.
       CUT            3.3      67.
       FOLD           5.2      83.
       SADDLEWIRE     9.5     228.
       PACKING       10.2     152.
       FREIGHT        0.0     132.
```

FIG. 15–2
A typical estimate generated by the
Super Stewy/MP.

294

```
                    TENTATIVE QUOTE

                 DATE:    5/15/85


SHAWMUT BANK                          QUOTE NO:      63
1500 COMMONWEALTH AVE.                SALESMAN:   CROW. W.
BOSTON          MA 02132
ATTN:  GEORGE JONES
TELEPHONE: 617-666-4000

JOB TITLE: QUARTERLY EMPLOYEE MAGAZINE

JOB DESCRIPTION: 32 PAGE 8-1/2 X 11 PLUS COVER
TEXT PRINTS BLACK ON BOTH SIDES
COVER 4 COLORS SIDE 1. BLACK SIDE 2
PAPER: 50# UNCTD OFFSET & 60# C1S COVER
40 PAGES PHOTOCOMP & PASTE-UP. 36 LINE NEGS.
6 TINTS & HALFTONES. 1 REVERSE. 2 COLOR SEPS.
PROOFS REQUIRED. STITCH & CARTON PACK.


TERMS AND CONDITIONS:  NET 30 DAYS


QUANTITY:          20000.      30000.      40000.

STDS SET:             1           1           1

ESTIMATE $:        5298.       6965.       8586.

$ ADDL/M:         181.86      176.80      173.12

% MARKUP:           10.0        10.0        10.0

MARKUP $:         529.81      696.50      858.56

TENTATIVE QUOTE:   5828.       7662.       9444.

$ ADDL/M:         200.04      194.48      190.43
```

FIG. 15–3
The tentative quote: a summary of the estimates.

Obtain the Best Quotes Possible. A wise printing buyer will insist on a firm price quotation on every job. But how many quotes from different printers are enough? Two-thirds of the companies responding to a recent survey prepared for the PICA/Dillard Graphics Management Institute indicated they require at least three quotes from among the four to six printers they use regularly.[1]

More important than the number of quotes you request is the type of quote you receive and from whom. Wallace Stettinius, a prominent printing management consultant, suggests that the most valuable type of quotation is one based upon "product modules."[2] This method of quoting identifies and puts a price on individual components of a job—halftones, pages of composition, proofs, plate burns, even additional per-thousand quantities beyond the basic number of pieces ordered.

```
                    QUOTE
                 DATE:    5/15/85

SHAWMUT BANK                          QUOTE NO:      63
1500 COMMONWEALTH AVE.                SALESMAN:   CROW. W.
BOSTON          MA 02132
ATTN:  GEORGE JONES




JOB TITLE: QUARTERLY EMPLOYEE MAGAZINE

JOB DESCRIPTION: 32 PAGE 8-1/2 X 11 PLUS COVER
TEXT PRINTS BLACK ON BOTH SIDES
COVER 4 COLORS SIDE 1. BLACK SIDE 2
PAPER: 50# UNCTD OFFSET & 60# C1S COVER
40 PAGES PHOTOCOMP & PASTE-UP. 36 LINE NEGS.
6 TINTS & HALFTONES. 1 REVERSE. 2 COLOR SEPS.
PROOFS REQUIRED. STITCH & CARTON PACK.


   QUANTITY         PRICE           $ PER ADD'L M

    20000.      $  5827.93          $    200.04

    30000.      $  7661.55          $    194.48

    40000.      $  9444.16          $    190.43

TERMS AND CONDITIONS:  NET 30 DAYS


                        SINCERELY.
```

FIG. 15–4
The final quotation, in letter form, ready for approval by the customer.

Stettinius counsels against accepting "lump sum" or "process" quotes which either list no details or show only charges for estimated time needed for each production step.

When seeking several quotes, select those printers most likely to give you favorable prices on a particular job. Consult your file of printing companies for those whose specialties match your requirements. Quotes received from printers whose equipment capacity is too large, too small, or too specialized, may not reflect an accurate picture of appropriate prices.

Avoid "Tempered" Quotes.   Quotes that are higher than average, or "tempered," reflect bad experiences printers might have had with a customer. Gerald A. Silver, a respected estimating specialist, suggests that printers often charge more to certain customers with consistently bad habits.[3] These buyers always need rush jobs, ask for too many quotes that are never

FIG. 15–5
*Trade Practices of a Small Printing Company**

PRINTING TRADE CONDITIONS

1. Quotations, Estimates and Taxes   A quotation is subject to review if it is not accepted within 7 days, or work does not commence within 30 days after acceptance, or work does not proceed with continuous production.

Printer has the right to invoice for work in process if production is interrupted by customer for more than 10 working days.

The prices quoted are based on current material and labor costs and are based on straight-time scheduling. Price adjustments may be made at the time an order is placed if material and/or labor costs have increased, or if preferential scheduling is required by the customer to meet customer-desired delivery dates.

All estimates based on visuals, rough dummies, sketches, copies, etc., whether accompanied by verbal or written specifications, shall be considered only tentative. Estimates are subject to revision upon inspection of the finished artwork or manuscript. Customer will be notified of price changes, if any, before the job is put into production.

Local, state, and federal taxes and duties are not included in any price and will be added to all application invoices.

2. Credit and Terms   All estimates and quotations are contingent upon credit approval by the printer. Customers with unsatisfactory credit rating or unverified credit rating must deposit 50% at time of placing order; the balance is due at time of delivery.

Payment shall be net cash on the 10th of the month following the date of invoice (unless otherwise provided in writing) and are delinquent thereafter.

Delinquent accounts will be assessed at ONE AND ONE-HALF PERCENT (1½%) per month service charge. This is an annual rate of 18%. This will become a contractual debt that is due and collectable. Should it be necessary to employ outside services to collect any acount, it is specifically agreed that the customer will pay all such costs, including reasonable attorney's fees and court costs. As security for payment of any sum due or to become due, printer shall have the right, if necessary, to retain possession of and shall have a lien on all customer property in printer's possession including work in process and finished work. The extension of credit or the acceptance of notes, trade acceptances or guarantee of payment shall not affect such security interest and lien.

3. Orders   Orders regularly entered, verbal or written, cannot be cancelled except upon terms that will compensate printer for labor and materials used to date.

4. Experimental Work   Experimental work performed at customer's request, such as sketches, drawings, composition, plates, presswork, and materials will be charged for at current rates and may not be used without consent of the printer.

5. Preparatory Work   Sketches, copy, dummies, and all preparatory work created or furnished by the printer, shall remain his exclusive property and no use of same shall be made, nor any ideas obtained therefrom be used, except upon compensation to be determined by the printer.

6. Conditions of Copy   Estimates for typesetting are based on the receipt of original copy or manuscript clearly typed, double-spaced on 8½" × 11" uncoated stock, on one side only. Condition of copy which deviates from this standard is subject to re-estimating and pricing review by printer at time of submission of copy, unless otherwise specified in estimate.

7. Preparatory Materials   Title to designs, type, artwork, film, plates, dies, tools, and other property created or furnished by printer with respect to the production of goods, shall remain with printer.

8. Alterations   Alterations represent work performed in addition to the original specifications. Such additional work shall be charged at current rates and be supported with documentation upon request.

9. Customer Responsibility   When copy, proofreading, layouts, design, composition, selection of paper stocks, color of ink, etc. are not specified by customer and are left to the discretion of the printer, no further responsibility on the printer's part will be incurred except the use of reasonable care and good judgment. Any customer alterations from the above will be charged for at current rates.

10. Proofs   Proofs shall be submitted with original copy. Corrections are to be made on "master set," returned marked "O.K." or "O.K. with corrections" and signed by customer. If revised proofs are desired, request must be made when proofs are returned. Printer regrets any undetected errors that may occur through production, but cannot be held responsible for errors if the work is printed per customer's O.K. Printer accepts no responsibility for changes communicated verbally. Printer shall not be responsible for errors if the customer has not ordered or has refused to accept proofs or has failed to return proofs with indication of changes, or has instructed printer to proceed without submission of proofs.

11. Press Proofs   Unless specifically provided in printer's quotation, press proofs will be charged for at current rates. An inspection sheet or any form can be submitted for customer approval, at no charge, provided customer is available at the press during the time of makeready. Any changes, corrections, or lost press time due to customer's non-availability, change of mind, indecision, or delay will be charged for at current rates.

12. Color Proofing   Because all differences in equipment, paper, inks, and other conditions between color proofing and production pressroom operations, a reasonable variation in color between color proofs and the completed job shall constitute acceptable delivery. Special inks and proofing stocks will be forwarded to customer's suppliers upon request at current rates.

13. Over Runs or Under Runs   As it is practically impossible to produce exact quantities, it is agreed that the delivery of more or less than the ordered quantities shall be acceptable and invoiced as follows: ordered quantity up to and including 10,000, 10%; 10,001 to 25,000, 5%; 25,001 to 100,000, 3%; more than 100,000, 2%. Over runs or under runs not to exceed 10% on any quantity for business forms shall constitute acceptable delivery. Printer will bill for actual quantity delivered within this tolerance. If customer requires guaranteed "no less than" delivery, percentage tolerance of overage must be doubled.

---

*Courtesy of Premier Printing Corporation, Fullerton, CA.

accepted, and provide poor copy and art. They also deliver material piecemeal, haggle over prices, or constantly change specifications while a job is in process. Protect your own budgets by avoiding such practices.

Know Trade Customs and Practices. According to Silver, printing trade customs were adopted originally by the National Association of Photo-Lithographers at a 1937 annual convention, and were reaffirmed as

recently as 1975.[4] Some form of these customs often accompanies a printing quote as legally binding contract conditions. Figure 15–5 shows a typical example of a small printing company's trade conditions based on standard customs within the trade. While most are self-explanatory, a few are emphasized here:

Item 1. Quotations, Estimates, and Taxes. You should note that estimates based on rough or

FIG. 15–5 (continued)

PRINTING TRADE CONDITIONS

14. Production Schedules Production schedules will be established and adhered to by customer and printer, provided that neither shall incur any liability or penalty for delays due to state of war, riot, civil disorder, fire, strikes, accidents, action of Government or civil authority, and acts of God or other causes beyond the control of customer or printer.

Prices quoted are based on straight-time work. Overtime work caused by the customer's failure to meet deadline, delay in turning in copy, proofs, or other material necessary to complete the work within the time specified, shall be charged for at current overtime rates, over and above the price quoted herein. When agreed deadline is missed due to customer delay, no new delivery date is to be assumed without specific re-scheduling with printer. No time allowance for author's alterations is included in estimating production schedules.

15. Delivery Unless otherwise specified, the price quoted is for a single shipment, without storage, for local customers; and F.O.B. printer's platform for out-of-town customers. Proposals are based on continuous and uninterrupted delivery of complete order, unless specifications distinctly state otherwise. Title for finished work shall pass to the customer upon delivery, to carrier at shipping point, or upon mailing of invoices for finished work, whichever occurs first.

16. Customer-Furnished Material Paper stock, camera copy, film, color separations, and other customer-furnished materials shall be manufactured, packed, and delivered to printer's specifications. Customer-furnished materials shall be handled with frugality and to prevent under spoilage. Paper stock which is running at excessive spoilage will not be processed without client approval. Client must take into consideration and accept printer spoilage allowances when ordering materials for our processing.

Materials delivered from customer or his suppliers are verified with delivery ticket as to cartons, packages, or items shown only. The accuracy of quantities indicated on such tickets cannot be verified and printer cannot accept liability for shortage based on supplier's tickets. All shortages and material defects will be reported immediately to customer for his disposition with his supplier. Wherever possible, printer will assist customer regarding shortages and material defects; however, the customer must bear full responsibility when dealing with his supplier. Printer accepts no responsibility for accuracy, suitability, or usability of customer-furnished materials.

Additional cost due to delays or impaired production caused by specification deficiencies shall be charged to the customer at current rates.

Charges related to delivery from customer to printer are not included in any quotations unless specified. Special priority pickup or delivery service will be provided at current rates upon customer's request.

While printer will provide storage service free of charge for customer-owned artwork, film, and plates as long as space permits, printer can accept no liability for replacing any materials lost or damaged while in printer's storage.

Printer maintains insurance coverage on work produced and all property belonging to customer only during the period of time required for production and delivery. Space permitting, printer will gladly store printed materials for customer under our Storage and Distribution Plan. Ask about complete details.

Unless otherwise agreed in writing, the prices in the proposal do not include a charge for storage of furnished goods, nor for the storage of paper or other mate-

rials customer may furnish. Charges at current rates will be made for handling all paper stock furnished by customer and for storing customer-furnished stock, inserts, covers, or other materials and finished goods. All such property is stored at customer's risk. Printer shall not be liable for any loss or damage thereto, occasioned by any cause, including without limitation, fire, water, leakage, theft, negligence, insects, or breakage.

17. Claims All claims of any kind must be in writing within 10 days after receipt of goods, and in cases of shortages or errors in shipment, must be accompanied by a receiving report showing the number of pieces received and the weight and contents of each piece. Failure to make such claim within the stated period shall constitute irrevocable acceptance and an admission that they fully comply with terms, conditions, and specifications.

18. Sub-contractors Printer may subcontract any or all of its obligations, but shall remain liable to the customer therefor.

19. Reprints and samples Printer has the privilege of displaying samples of the work for advertising purposes and the right to imprint their name on the sample copies unless expressly prohibited by customer.

20. Limitation on Liability Printer's liability shall be limited to stated selling price of any goods, and shall in no event include special or consequential damages, including profits (or profits lost) or claims of any third party with respect to the work.

21. Right of Refusal The customer agrees that the printer may refuse at any time to print any copy, photographs, or illustrations of any kind that in his sole judgment he believes is an invasion of privacy, is degrading, libelous, unlawful, profane, obscene, pornographic, tends to ridicule or embarrass, or is in bad taste. The customer also agrees to defend the printer and hold him harmless in any suit or court action brought against him for alleged damages resulting from his printing any copy, photographs, or illustrations that is felt by others to be degrading, libelous, or harmful to their reputations, images, or standing in the community.

22. Indemnification The customer shall indemnify and hold harmless the printer from any and all loss, cost, expense, and damages on account of any and all manner of claims, demands, actions, and proceedings that may be instituted against the printer on grounds alleging that the said printing violates any copyright or any proprietary right of any person, or that it contains any matter that is libelous or scandalous, or invades any person's right to privacy or other personal rights, except to the extent that the printer has contributed to the matter. The customer agrees to, at the customer's own expense, promptly defend and continue the defense of any such claim, demand, action or proceeding that may be brought against the printer, provided that the printer shall promptly notify the customer with respect thereto, and provided further that the printer shall give to the customer such reasonable time as the exigencies of the situation may permit in which to undertake and continue the defense thereof.

incomplete layouts or materials are just that—estimates. They are not binding quotes. Save time and trouble by presenting your job specifications as accurately as possible from the start, and accept firm quotes promptly to avoid price changes.

Items 4, 5, and 7. Experimental and Preparatory Work and Materials. Although you may not be *charged* by the printer for some of the dummy work or layouts done in your behalf, you must get permission to use the ideas or materials. Printers are especially annoyed by customers who request such work, then take the job to another printer—unless, of course, they specialize in such services for a fee.

Item 8. Alterations. Many buyers of printing are shocked to learn how costly corrections and alterations can be in the final stages of a job's production. Your best protection is to begin with meticulously edited copy and fully approved, complete layouts and materials.

Items 10 and 11. Proofs. Most printers today will not proceed with a job until proofs have been approved. Every hour you delay returning corrected page or galley proofs is an hour your job stands idle. Never hold up a job on the press by being unavailable to view a press sheet. Press time is prohibitively expensive. Note that corrections made by telephone are worthless if you question their accuracy later. Always initial proofs and review them with the salesperson.

Item 13. Over Runs or Under Runs. Note here that printers reserve the right to run somewhat more or fewer pieces than you ordered, and will adjust your bill accordingly. However, you are wise to specify a *minimum* acceptable quantity, especially where important mailing lists must be served. For the sake of customer goodwill, some printers don't actually charge for overage unless the job is unusually complex.

Item 16. Customer Furnished Material. Some printing customers buy and store their own paper to take advantage of quantity discounts. This practice can be troublesome for the printer if the customer delivers paper in poor condition or it arrives too late for proper humidity adjustment in the plant. Sometimes the grain of a customer's paper is running in the wrong direction for presswork and folding, or the dimensions vary too much for proper registration. Such problems can cause expensive delays that wipe out savings gained by furnishing your own paper stock.

Sometimes, however, the advantage to you in furnishing your own paper takes precedence over the general preference of printers to provide it for you. For example, you may wish to order a large quantity of special stock to ensure its continued availability.

You may have plenty of unused floor space for storage, or perhaps you can use some of the paper you purchase in quantity for an in-plant printing operation of your own. Whatever the reason for buying your own paper, *plan* before you purchase. Be sure the company magazine for which you may be buying large quantities of paper in advance won't change format before the paper is used. Also, be certain you have the time and transportation to deliver paper to the printer as it is required. Try to buy sizes and weights that have multiple use. Finally, be reasonably sure the paper will be used before it deteriorates from age.

Your printer will assist you in buying paper. However, you can buy directly from a local distribution firm which represents many paper manufacturers. Review the chapter on paper, and study its varieties and characteristics.

Figure 15–6 shows a typical page from a paper company's catalog. Such information is usually obtained through the firm's graphic arts representative. The paper shown, a popular gloss-coated book stock, is available in many sizes other than the standard 25×38 in both 80- and 100-pound weights.

Paper in less than large volumes is usually sold on the basis of its price per thousand (M) sheets. We can see from the table that per-thousand prices go down as the quantity ordered increases. Say, for example, we wish to buy 16 cartons of 80-pound paper in a 23×35 size for our company publication. We can see that this size is packed 1,100 sheets to the carton. This means we will receive 17,600 sheets. We will pay for the paper at the tabled rate of $140.83 per thousand sheets, or a total of $2,478.61 ($140.83 $\times$ 17.6 thousands). Note that we would have had to pay $173.60 per thousand if we had bought only a single carton.

Two things are worth remembering here, as well: (1) the printer defines paper "grade" as the price per thousand sheets *delivered to the plant*, so freight charges, if any, add to the printer's markup on paper; and, (2) the prices listed in a paper company's catalog often are subject to further discounting, especially on large quantities.

Another item commonly furnished to printers by their customers is color separation films for four-color work. Many printers do not produce in-house color separations. They should be consulted, though, before you order such materials from outside suppliers. They will be happy to suggest trade color separators who consistently produce materials that meet the highest standards. Color density values and lines per inch

CENTURA GLOSS OFFSET ENAMEL *Consolidated*

For Letterpress or Offset Printing
Coated 2 Sides

				PRICE PER CWT			
			BKN CTN	1 CTN	4 CTNS	16 CTNS	5M LBS	10M LBS
White, Gloss			186.70	127.65	112.30	103.55	97.75	93.20
White, Skids			. . . .	. . . .	. . . .	102.10	96.35	91.90

BASIS	SIZE	M WT	SHEETS PER CTN	PRICE PER 1000 SHEETS					
WHITE									
80	17½ x 22½	66	2400	123.22	84.25	74.12	68.34	64.52	61.51
	19 x 25	80	2000	149.36	102.12	89.84	82.84	78.20	74.56
	23 x 35	136	1100	253.91	173.60	152.73	140.83	132.94	126.75
	24 x 36	146	1000	272.58	186.37	163.96	151.18	142.72	136.07
	25 x 38	160	1000	298.72	204.24	179.68	165.68	156.40	149.12
100	17½ x 22½	83	1800	154.96	105.95	93.21	85.95	81.13	77.36
	19 x 25	100	1600	186.70	127.65	112.30	103.55	97.75	93.20
	23 x 35	169	900	315.52	215.73	189.79	175.00	165.20	157.51
	24 x 36	182	800	339.79	232.32	204.39	188.46	177.91	169.62
	25 x 38	200	800	373.40	255.30	224.60	207.10	195.50	186.40
WHITE, SKIDS									
80	23 x 35	136		. . . .	. . . .	. . . .	138.86	131.04	124.98
	24 x 36	146		. . . .	. . . .	. . . .	149.07	140.67	134.17
	25 x 38	160		. . . .	. . . .	. . . .	163.36	154.16	147.04
100	24 x 36	182		. . . .	. . . .	. . . .	185.82	175.36	167.26
	25 x 38	200		. . . .	. . . .	. . . .	204.20	192.70	183.80

APPROX. CALIPER: Gloss Bs. 80—.00375, Bs. 100—.00475
500 Sheets Per Package
*Must Be Converted

FIG. 15–6

Page from a paper supplier's price catalog.
Butler Paper, a company of Great Northern Nekosa Corp.

should match the paper and production methods planned for the job, and Color Key or Chromalin proofs are often suggested. Review the chapter on color printing before you proceed with purchasing.

Conclusion

The tips provided here for working with printers are but a beginning. Your own experiences will soon add to them greatly. Your knowledge of customs and standards, as well as your attention to professionalism in all dealings with printers will mean better production of your work and a smoother, more satisfying transition into the world of communication graphics. Printers are the last to see your jobs before they are final. Keep them on *your* side.

Notes

[1]"Dependability and Quality Rate Tops in Printer Selection," *Graphic Arts Monthly*, March 1978, pp. 46–48.
[2]Wallace Stettinius, *Management Planning and Control: A Printer's Path to Profitability* (Arlington, Va.: Graphic Communications Center, 1976), pp. 200–211.
[3]Gerald A. Silver, *Professional Printing Estimating* (Philadelphia: North American Publishing Company, 1975), p. 70.
[4]Gerald A. Silver, "Trade Customs in Printing—Part I," *Graphic Arts Monthly*, October 1978, p. 68.

Appendix A

Selected Display Alphabets

abcdefghijklmnopqrstuvwxyz
ABCDEFGHIJKLMNOPQRSTUVWXYZ
1234567890&$$¢£%
AMAÆÆ̃AHAKÆ̃PARAASSTT
ÇØÆ̃ṎŒ̊ẞÇ̂ǿæ̃œ̊fi
(:;,.!?·-˝”́—/#@**)[†‡§«»1234567890]

abcdefghijklmnopqrstuvwxyz
ABCDEFGHIJKLMNOPQRSTUVWXYZ
1234567890&$$¢£%
AMAÆÆ̃AHAKÆ̃PARAASSTT
ÇØÆ̃ṎŒ̊ẞÇ̂ǿæ̃œ̊fi
(:;,.!?·-˝”́—/#@**)[†‡§«»1234567890]

ITC BENGUIAT MEDIUM

abcdefghijklmnopqrstuvwxyz
ABCDEFGHIJKLMNOPQRSTUVWXYZ
1234567890(&,.:,!?'""--$¢%/)

abcdefghijklmnopqrstuvwxyz
ABCDEFGHIJKLMNOPQRSTUVWXYZ
1234567890(&,.:,!?'""--$¢%/)

AVANT GARDE GOTHIC BOOK

abccdeefghijjklmnopqrsstuvwxyz
ABCDEFGHIJKLMNOPQRSTUUVWXYZ
1234567890(&.,:;!?"—."'*$¢%)

abccdeefghijjklmnopqrsstuvwxyz
ABCDEFGHIJKLMNOPQRSTUUVWXYZ
1234567890(&.,:;!?"—."'*$¢%)

KORINNA BOLD

abcdefgghijklmnopqrstuvwxyz
ABCDEFGHIJKLMNOPQRSTUVWXYZ
1234567890(&.,:;!?""''"-*$¢%)

abcdefgghijklmnopqrstuvwxyz
ABCDEFGHIJKLMNOPQRSTUVWXYZ
1234567890(&.,:;!?""''"-*$¢%)

FRIZ QUADRATA

abcdeefghijkklmnopqrstuvwxyz
ABCDEFGHIJJKKLMNOPQRRSTUVWXYZ
1234567890&&$$¢£%
ÇØÆŒßçøäẽæôåœfl
(:;,.!?.¨¨¯——""'""**/#«»)[]

abcdeefghijkklmnopqrstuvwxyz
ABCDEFGHIJJKKLMNOPQRRSTUVWXYZ
1234567890&&$$¢£%
ÇØÆŒßçøäẽæôåœfl
(:;,.!?.¨¨¯——""'""**/#«»)[]

ITC TIFFANY DEMI

abcdefghijklmnopqrstuvwxyz
ABCDEFGHIJKLMNOPQRSTUVWXYZ
1234567890 (&.,:;!?""'–-*$¢/)

abcdefghijklmnopqrstuvwxyz
ABCDEFGHIJKLMNOPQRSTUVWXYZ
1234567890 (&.,:;!?""'–-*$¢/)

ITC GARAMOND BOOK

abcdeffghijklmnopqrstuvwxyz
ABCDEFGHIJKLMNOPQRSTUVWXYZ
12345678890&$$$¢ƒ£%@
ÇĘŁØÆŒßçęłøäãôêfi˛¸ˇ˘˜˙˚
(:;,.!?."""'/#**)[†‡§«»12345678890]aeilmorst

abcdeffghijklmnopqrstuvwxyz
ABCDEFGHIJKLMNOPQRSTUVWXYZ
12345678890&$$$¢ƒ£%@
ÇĘŁØÆŒßçęłøäãôêfi˛¸ˇ˘˜˙˚
(:;,.!?."""'/#**)[†‡§«»12345678890]aeilmorst

ITC MODERN NO. 216 BOLD

abcdefghijklmnopqrstuvwxyz
ABCDEFGHIJKLMNOPQRSTUVWXYZ
1234567890 (&.,:;!?'""-[]*$¢%/)

abcdefghijklmnopqrstuvwxyz
ABCDEFGHIJKLMNOPQRSTUVWXYZ
1234567890 (&.,:;!?'""-[]*$¢%/)

SOUVENIR BOLD

aabcdeeffghijkklmnopqrrssttuvwxyzz
ABCDEEFGHIJKLLMNOPQRSTUVWXYZ
1234567890(&.,:;!?""""-*$¢%)()

aabcdeeffghijkklmnopqrrssttuvwxyzz
ABCDEEFGHIJKLLMNOPQRSTUVWXYZ
1234567890(&.,:;!?""""-*$¢%)()

ITC SERIF GOTHIC BOLD

abcdefghijklmnopqrstuvwxyz
ABCDEFGHIJKLMNOPQRSTUVWXYZ
1234567890(&.,:;!?"""'·*$¢%)

abcdefghijklmnopqrstuvwxyz
ABCDEFGHIJKLMNOPQRSTUVWXYZ
1234567890(&.,:;!?"""'·*$¢%)

ITALIA BOLD

abcdeefghijklmnopqrstuvwwwxyyz
ABCDEFGGHIJKLMNOPQRSTUVWXYZ
1234567890&$¢£%
AACACEAFAGAHTKAIAIAMMNTRRASSTHUTVVWW
ÇØÆŒßčøœêêffffifffiflffl
(.;,.!i?¿""-*/#«») ( )

abcdeefghijklmnopqrstuvvwwwxyyz
ABCDEFGGHIJKLMNOPQRSTUVWXYZ
1234567890&$¢£%
AACACEAFAGAHTKAIAIAMMNTRRASSTHUTVVWW
ÇØÆŒßčøœêêffffifffiflffl
(.;,.!i?¿""-*/#«») ( )

ITC LUBALIN GRAPH MEDIUM

abcdeefghijklmnopqrstuvwxyz
ABCDEFGHIJKLMNOPQRSTUVWXYZ
1234567890(&&.,:;!?""''-—*$$¢£%)[ ]

abcdeefghijklmnopqrstuvwxyz
ABCDEFGHIJKLMNOPQRSTUVWXYZ
1234567890(&&.,:;!?""''-—*$$¢£%)[ ]

ITC AMERICAN TYPEWRITER BOLD

ABCDEFGHIJ
JKLMNOPQR
STUVWWXYZ?
&⊙&*ÆŒØÇ
$1234567890!
[old style figures]1234567890
abcdefghijkl
mnopqrstuvw
xyzøœæçThro
tkhhhpqnme
ß$$¢¢(£#";")
AABBCDE
FGHIJKLM
NOPQRRSTU
VVVWWXYZ

ABCDEFGHIJ
JKLMNOPQR
STUVWWXYZ?
&⊙&*ÆŒØÇ
$1234567890!
[ ]1234567890
abcdefghijkl
mnopqrstuvw
xyzøœæçThro
tkhhhpqnme
ß$$¢¢(£#";")
AABBCDE
FGHIJKLM
NOPQRRSTU
VVVWWXYZ

ITC BOOKMAN MEDIUM
ITALIC WITH SWASH

abcdefghijklmno
pqrstuvwxyzABC
DEFGHIJKLMN
OPQRSTUVWX
YZ1234567890 12
34567890&$¢£%
AÇÐEŁØÄÃÊÇÉßą
çędłøæœ!i?¿:;,.--
‹(')»"/#*()[]†‡§«»ſ

START LETTERS: ſ f v w ſt th of

VARIATIONS: g v w x y y z

SWASH CAPS: A B C D E F G H
I J K L M N O P Q
R S T U V W X Y Z

ALTERNATES: A E I L ?

FINALS: d d L e e k r t y ⁓ ⁓ ⁓

abcdefghijklmno
pqrstuvwxyzABC
DEFGHIJKLMN
OPQRSTUVWX
YZ1234567890 12
34567890&$¢£%
AÇÐEŁØÄÃÊÇÉßą
çędłøæœ!i?¿:;,.--
‹(')»"/#*()[]†‡§«»ſ

START LETTERS: ſ f v w ſt th of

VARIATIONS: g v w x y y z

SWASH CAPS: A B C D E F G H
I J K L M N O P Q
R S T U V W X Y Z

ALTERNATES: A E I L ?

FINALS: d d L e e k r t y ⁓ ⁓ ⁓

ITC ZAPF CHANCERY DEMI

Appendix B

Per-Pica Character Counts
for Fifty Popular Typefaces*

TYPEFACE	6-pt.	8-pt.	9-pt	10-pt.	11-pt.	12-pt.
ITC American Typewriter® Medium	3.80	3.00	2.63	2.40	2.23	2.05
Americana	3.63	2.83	2.56	2.33	2.16	1.96
ITC Avant Garde Gothic® Medium	3.63	2.83	2.56	2.33	2.16	1.96
ITC Benguiat® Book	3.90	3.10	2.73	2.46	2.30	2.13
Bookman	4.20	3.30	2.96	2.66	2.43	2.26
ITC Caslon No. 224® Book	4.11	3.21	2.90	2.56	2.40	2.23
Century Old Style	4.46	3.43	3.16	2.86	2.58	2.40
Century Schoolbook	3.96	3.11	2.76	2.50	2.33	2.16
ITC Cheltenham® Book	4.20	3.30	2.96	2.66	2.43	2.26
ITC Clearface® Regular	4.41	3.41	3.11	2.80	2.56	2.38
ITC Eras® Book	4.36	3.40	3.06	2.76	2.50	2.36
Excelsior®	3.56	2.78	2.46	2.28	2.08	1.93
ITC Fenice® Regular	4.31	3.33	3.03	2.73	2.48	2.33
Folio Light	4.31	3.33	3.03	2.73	2.48	2.33
ITC Friz Quadrata	4.11	3.21	2.90	2.56	2.40	2.23
Futura Light	4.11	3.21	2.90	2.56	2.40	2.23
ITC Galliard™	4.31	3.33	3.03	2.73	2.48	2.33
ITC Garamond™ Book	4.20	3.30	2.96	2.66	2.43	2.26
Gill Sans	4.56	3.53	3.23	2.93	2.66	2.46
Goudy Old Style	4.51	3.50	3.16	2.86	2.63	2.43
Helvetica® Light	4.06	3.16	2.83	2.56	2.36	2.20
Helvetica®	3.86	3.03	2.70	2.46	2.28	2.10
ITC Italia Book	4.26	3.31	3.01	2.66	2.46	2.30
Janson®	4.20	3.30	2.96	2.66	2.43	2.26
ITC Kabel® Book	4.41	3.41	3.11	2.80	2.56	2.38
Kennerly	4.31	3.33	3.03	2.73	2.48	2.33
ITC Korinna®	4.36	3.40	3.06	2.76	2.50	2.36
ITC Lubalin Graph® Book	3.80	3.00	2.63	2.40	2.23	2.05
Melior®	3.90	3.10	2.73	2.46	2.30	2.13
Memphis® Medium	4.20	3.30	2.96	2.66	2.43	2.26
ITC Modern No. 216™ Medium	3.76	2.93	2.60	2.38	2.20	2.01
ITC Newtext® Light	3.76	2.93	2.60	2.38	2.20	2.01

*These counts are approximate, calculated from alphabet length conversion tables published by Allied Linotype and International Typeface Corporation. They should be accurate enough for most typefitting applications. If absolute precision is required, type manufacturers can provide exact counts for every size and style.

® indicates a registered trademark of Allied Corporation or its licensees.

™ indicates a trademark of Allied Corporation or its licensees.

The initials ITC when accompanied by ® or ™ specify either a registered trademark or a trademark of the International Typeface Corporation.

TYPEFACE	6-pt.	8-pt.	9-pt	10-pt.	11-pt.	12-pt.
Optima®	4.20	3.30	2.96	2.66	2.43	2.26
Palatino®	3.96	3.11	2.76	2.50	2.33	2.16
Plantin	4.26	3.31	3.01	2.66	2.46	2.30
ITC Quorum® Book	4.51	3.50	3.16	2.86	2.63	2.43
ITC Serif Gothic® Light	4.31	3.33	3.03	2.73	2.48	2.33
ITC Serif Gothic®	4.11	3.21	2.90	2.56	2.40	2.23
ITC Souvenir® Light	4.36	3.40	3.06	2.76	2.50	2.36
ITC Souvenir® Medium	4.11	3.21	2.90	2.56	2.40	2.23
Spartan® Light	4.76	3.71	3.40	3.11	2.83	2.58
Stymie Light	4.31	3.33	3.03	2.73	2.48	2.33
ITC Tiffany Medium	3.81	3.01	2.66	2.43	2.26	2.08
Times Roman®	4.06	3.16	2.83	2.56	2.36	2.20
Trade Gothic®	4.31	3.33	3.03	2.73	2.48	2.33
Univers® 55	3.86	3.03	2.70	2.46	2.28	2.10
Venus Light	4.66	3.63	3.26	3.01	2.73	2.50
Video™ Medium	4.16	3.23	2.93	2.63	2.41	2.26
ITC Zapf Book® Light	4.11	3.21	2.90	2.56	2.40	2.23
ITC Zapf International® Medium	4.31	3.33	3.03	2.73	2.48	2.33

® indicates a registered trademark of Allied Corporation or its licensees.

™ indicates a trademark of Allied Corporation or its licensees.

The initials ITC when accompanied by ® or ™ specify either a registered trademark or a trademark of the International Typeface Corporation.

Appendix C

Selected Sources of Information on Communication Graphics and Publications Design

American Association of Advertising Agencies (AAAA)
666 Third Avenue
New York, New York 10017

American Advertising Federation (AAF)
1225 Connecticut Avenue, NW
Washington, D.C. 20036

American Newspaper Publishers Association (ANPA)
The Newspaper Center
Box 17407
Dulles International Airport
Washington, D.C. 20041

American Society of Magazine Editors (ASME)
575 Lexington Avenue
New York, New York 10022

American Society of Newspaper Editors
Box 17004
Washington, D.C. 20041

American Paper Institute (API)
260 Madison Avenue
New York, New York 10016

Direct Marketing Association (DMA)
6 E. 43rd Street
New York, New York 10017

Education Council of the Graphic Arts Industry, Inc.
4615 Forbes Avenue
Pittsburgh, Pennsylvania 15213

Graphic Arts Technical Foundation (GATF)
4615 Forbes Avenue
Pittsburgh, Pennsylvania 15213

International Association of Business Communicators (IABC)
1730 Pennsylvania Avenue, NW
Washington, D.C. 20006

Magazine Publishers Association (MPA)
575 Lexington Avenue
New York, New York 10022

National Association of Printing Ink Manufacturers (NAPIM)
550 Mamoroneck Avenue
Harrison, New York 10528

National Association of Printers and Lithographers (NAPL)
780 Palisade Avenue
Teaneck, New Jersey 07666

National Paper Trade Association (NPTA)
111 Great Neck Road
Great Neck, New York 11021

National Press Photographers Association (NPPA)
P.O. Box 1146
Durham, North Carolina 27702

Printing Industries of America (PIA)
1730 N. Lynn Street
Arlington, Virginia 22209

Public Relations Society of America, Inc.
845 Third Avenue
New York, New York 10022

Society of Professional Journalists, Sigma Delta Chi (SPJ/SDX)
840 North Lake Shore Drive
Suite 801 W.
Chicago, Illinois 60611

Society of Newspaper Design (SND)
The Newspaper Center
Box 17290
Dulles International Airport
Washington, D.C. 20041

Technical Association of the Graphic Arts (TAGA)
P.O. Box 3064, Federal Station
Rochester, New York 14614

Typographers International Association (TIA)
2266 Hall Place, N.W.
Washington, D.C. 20007

Women in Communications, Inc.
P.O. Box 9561
Austin, Texas 78766

Glossary

abstract mark   A symbol which depicts an organization's overall essence, philosophy, or sense of purpose.

abstraction   The kinesthetic quality of a visual event reduced to its basic components.

additive color primaries   The light primaries red, green, and blue, which form color effects when merged.

alley   The space between columns on a page.

alphaglyph   A symbol formed completely or partially by the initials of an organization.

alphanumerics   Letters and numbers in an alphabet.

archiving   Recording and managing information generated by an organization.

area composition   On-screen composing of a single page or facing-page spread, either through interactive or passive operator input and previewing.

ascender   The portion of a lower-case letter which extends above the mean (average) height of such letters.

author's alterations (AAs)   Corrections or additions made by the author after type has already been set.

base line   The imaginary horizontal axis on which a line of type rests.

basic size   The size in inches adopted as standard for a grade of paper.

basic (basis) weight   The weight in pounds of a ream (500 sheets) of the basic size of a given paper.

bleed   Any image which extends to the edge of a paper page.

blockout film   A red, self-adhesive film used to create a window patch on a page pasteup.

body   Formerly referring to the full thickness of a piece of metal type, the term is now used to indicate the true point size of type, which must account for the presence of both ascenders and descenders.

calendering   The smoothing of paper during its manufacture, through the use of heated metal rollers.

calligraphic   A member of the Hand-formed Group in the Four-Group Classification System for type.

calligraphy   "Beautiful writing" created with broad tipped writing instruments usually held at a constant angle.

combination flat   A masking sheet onto which both line

and halftone negatives have been taped in preparation for burning a plate.

caption   A small line or block of type which identifies what is in a picture. Also called a cutline, a term which dates back to the time when pictures were printed from zinc "cuts" mounted on a wood or metal base.

cast coating   Pressing freshly coated paper against a hot, polished surface to produce a mirrorlike finish.

CVI   Center of Visual Impact, the focal point on a page.

certification mark   A symbol used for quality assurance.

collective mark   A symbol used by groups such as trade associations.

color separation   Breaking down a color original (slide or print) into the four records of yellow, magenta, cyan, and black, necessary to reproduce its images in full-color printing. Separation may involve more than the normal four colors.

comprehensive (COMP)   A fairly detailed version of a design for an advertisement or other printed piece. Type is simulated carefully and images are often done in marker. A finished comp may utilize actual color photoreproductions, typeset or transfer headlines, and simulated or typeset body copy.

continuous tone original   An image with smooth gradations from light to dark; usually a photograph or shaded art from which a halftone is made.

contoured wrap   Type set to fit around a curved or shaped area.

crash printing   High impact letterpress printing, whereby the relief plate or type presses hard enough against a carbonless form to transfer its image down through all parts of the set.

CRT   Cathode Ray Tube. The display screen of a television, computer terminal, or other electronic device. Very high resolution screens are required where typeset images must be generated.

cursive   A member of the Hand-formed Group in the Four-Group Classification System for type.

cyclical design path   A decision making process in design which does not flow in a linear fashion, but allows feedback and evaluation at any stage.

cylinder press   A press which prints from a flat form of type or plates by bringing paper wrapped around a rotating cylinder into contact with the inked images.

data base sharing   Interacting with systems containing large amounts of general or specialized information.

decorative type   Designs which contain shading or coloring, three-dimensional effects, contoured outlines, or are otherwise ornate. A member of the Specialized Group in the Four-Group Classification System.

demand printing   A term for customized printing done "on demand" through electronic methods, allowing rapid assembly and reproduction of books, manuals, and other materials where highest commercial quality is not required.

descender   The portion of a lower-case letter which projects below the average (mean) line of such letters.

die-cutting   The punching out of shapes through the use of sharp steel rules formed to create the desired image.

diffusion transfer   A system for making line or halftone images by exposing them to a paper negative, then transfering the image to a paper receiving sheet through the use of a single-bath processor. Both black-and-white and color transfers are possible.

digital typesetter   A device capable of storing type fonts as magnetic signals, generating letters made up of thousands of tiny, overlapping lines instead of the whole-letter images produced from photo-optic typesetters.

disk drive   The record/playback unit of a computer, into which magnetic diskettes may be inserted.

display type   Point sizes of 14 and above. Type fonts intended for display usually require special design so that letter shapes and spacing are pleasing in large sizes.

doctor blade   A tempered steel blade which scrapes ink from the top surface of a gravure cylinder during printing.

dropout (highlight) halftone   A halftone which has been exposed (or retouched) so that the tiny dots in the lightest highlight areas have been eliminated. This allows the paper color, usually white, to show through in these areas and creates improved image contrast.

dry offset   A modification of regular offset printing. A relief plate is used to print against a rubber blanket, eliminating the need for water to contact the plate.

dummy   A set of folded sheets showing the placement of pages within a printed piece. A dummy may contain images as well.

duotones   Two-color images made from black-and-white continuous tone originals.

early roman type   Roman type—including both old style and transitional variations—having bracketed serifs, light to moderate contrast in stroke, thickness, and some backward stress in rounded letter strokes. A member of the Serif Group in the Four-Group Classification System.

electronic color scanner   A device which separates and records on film the color elements necessary to reproduce images in full color. It optically examines the original image bit by bit, then electronically modifies and reproduces the image as positive or negative records for printing.

electronic prepress system   A fully integrated electronic system for interactive image creation and composition of pages. It allows a high degree of image manipulation, correction, text/graphics merging, and color separation.

electrotype   A relief plate made from a molded image that has been metal coated in an electrolytic plating tank.

em   A designation of type spacing equal to the square of the point size. Also known as a quad, it is the standard indention for a paragraph of body type.

embossing   Creation of an inkless raised image by compressing the paper or other substrate between a die and a counter die.

en   One-half of an em. Also used to designate the space required for typeset numbers in tabular composition.

engraving   Reproduction of images etched or inscribed into smooth metal. When the image is inked the surface is wiped clean, allowing only the ink remaining in the depressions to be transferred to the printed page.

exception word dictionary   A list of words and their proper hyphenation, consulted by the computer prior to its applying logic rules during hyphenation and justification. User-defined dictionaries may also be consulted.

extension strokes   The following parts of a letter: ascender, descender, or swash.

family   Variations of a parent typeface, all bearing the name of that face but differing in weight and/or style.

flat color   Images printed from a single nonprocess or process ink.

folio   The name, date, and/or page number printed in small type on each spread of a publication.

font   All capital and lower case letters, numbers, ligatures, and special characters required to reproduce a particular typeface in a single size and style. Also used to describe a master film strip, disk, or other storage medium capable of producing all sizes of one style of type.

format (for a publication)   The overall configuration of the page, including paper page size, type page size, number of columns, and grid plan, if any.

formatting for typesetting   Creation of single-key or software activated commands which allow rapid setup of type size, style, line width, leading, indents, tabular columns, and other information needed by the typesetter.

formation   How evenly the fibers are spread to form a sheet of paper.

Fourdrinier   A large machine used to make paper from wet pulp by forming, pressing, drying, and winding it onto huge rolls.

front-end system   An integrated, multiterminal system

designed to drive one or more typesetters or image setters through sophisticated software programs. Logic may be concentrated in one main central processing unit and shared with low cost terminals, or distributed among intelligent terminals.

full color   Usually called four-color process, this type of color requires the overprinting of dots of yellow, magenta, cyan, and black to recreate a color image. Often full color contains more than four color inks, however.

galley   A strip of photographic paper containing typeset images. A galley proof is usually a photocopy of the original typeset galley.

ghost   An unwanted image caused either by chemical problems related to ink drying and absorption, or by mechanical problems related to the manner in which pages are imposed on a plate for printing.

global search and replace   An editing function which allows one to seek out and replace any word or string of words which appears anywhere within a set of electronic files.

glossy   A photograph with a shiny finish.

glyph   A stylized symbol that suggests function, product, service, or raw material.

gothic type   Type having no serifs and some variation in stroke thickness. A member of the Sans Serif Group in the Four-Group Classification System.

grain   The alignment of cellulose fibers in paper.

grid   A design pattern of intersecting rectangles and the spaces between them.

group   A major classification of type based upon very broad design differences.

H&J   Hyphenation and justification of type lines, usually assumed to be computer controlled.

halftone   A paper positive or film negative record of the light and dark areas of a continuous tone image captured as thousands of tiny dots of various shapes and sizes.

hand-formed group   One of the major divisions in the Four-Group Classification System, it contains those types which resemble handwriting or hand lettering.

hard copy proof   Any record of electronically generated information that is output to paper.

heatset web   Web offset printing using specialized inks which, when heated and rapidly cooled, set the images so they will not smear at the delivery end of the press.

high-key photograph   One which contains mostly light images or a bright background, with few prominent dark areas.

hue   The common name of a color.

icon   A visual code or symbol used to convey information efficiently.

illustration   A term often used to describe both pictures and art in publications. The terms *image* and *graphic* are also used frequently to describe non-typographic elements.

image assembly   The procedure for preparing films for plate exposure. Also commonly referred to as *stripping*.

image setter   A device capable of setting not only type but also line art, halftones, tint patterns, borders, and any other images which might appear on a page.

imposition   The correct placement of pages on a plate so that the printed, folded sheet will be in proper page sequence.

indexing   Highlighting content categories through the use of differing typographic treatments.

intaglio   A major principle of printing whereby ink is squeezed from a depression in the image carrier onto the paper or other substrate.

intensity of a color   Its relative saturation or purity.

interfacing   Enabling one electronic system to share information with another. In typesetting this usually means the establishment of an electronic link between a typesetter or image setter and a personal computer through the use of a translation device or program.

italics   Specially designed, forward-slanting versions of roman type designs. Non-roman types which slant are often called *oblique* rather than roman, since they are not derived from roman designs.

kerning   Adjusting space between certain letter combinations to produce a more pleasing fit. Usually assumed to be automatically controlled by computer programs.

lands   The tops of the cell walls in gravure, which form a smooth surface over which the doctor blade travels.

laser   Light Amplification by Stimulated Emission of Radiation.

layout integration   A term describing the use of Gestaltist principles of attraction, continuity, and similarity, to maximize the cohesiveness of elements in a layout.

leading   Space between lines of type, measured in points or fractions of points.

legibility   A term usually used to describe how well display type can be read and comprehended.

ligature   A combination of letters such as *ff* or *ffi*, designed and typeset as one image for aesthetic reasons.

line conversion   A high contrast image made by exposing a continuous tone original to film or paper without using a halftone screen.

line original   Any original art, type, or other image which is made up of solid tones. This includes line art to which shading films have been added to give the appearance of gray areas or tints.

linecaster   A machine which casts a full line of type as a single piece from hot metal.

line gauge A metal ruler, sometimes called a pica pole or pica ruler, used to measure type and other graphic elements.

linear design path The process by which decisions affecting a design are made in a linear fashion from initial concept to final evaluation.

lithography A major printing method whereby a flat stone or porous metal plate containing an ink receptive image may be dampened, then inked, and the image transferred directly to paper. Its modern application is in offset printing, where the image is transferred to a rubber blanket, then to paper.

low-key photograph One that is made up of predominantly dark tones, with few prominent highlights.

main strokes The following parts of a letter: stem, hairline, curve, serif, bar, and cross.

manufactured color Color formed during printing as dots and solids of the four process inks overlap.

masking (or stripping) film A two-layer, peel-apart film used to block out sections of photographs or to produce areas for printing in tints or color.

matrix A small brass mold (mat), circulated in a linecasting machine, from which individual letters are cast in hot metal.

matte coating A layer of coating pigment, usually applied by a blade, and often thin in relation to other coatings.

maximum rag parameter The shortest length permitted for a ragged line, as defined by a hyphenation and justification program in typesetting.

mean line The average height of lower-case letters for a particular typeface.

microprocessor The central processing unit (CPU) containing the microprocessing chip or electronic "brain" of a computer.

mnemonic coding Simplified typesetting commands based upon easy to remember letter combinations. Example: ps for point size. Mnemosyne was the Greek goddess of memory.

mode The overall configuration of a section or column of type. Common modes include justified, ragged left or right, centered, staggered, indented, wrapped, contoured, and molded.

modern roman Roman type having great contrast in stroke thickness, flat serifs with few or no brackets, and no stress in round letter strokes. A member of the Serif Group in the Four-Group Classification for type.

monoline A serifless type of even thickness throughout. A member of the Sans Serif Group in the Four-Group Classification System for type.

monogram Distinctive initials presented without a border or background.

monoseal A seal-like symbol utilizing initial letters in a stylized typeface.

mortise A hole cut into the interior of an image, usually so that type may appear in that area.

multisector kerning Automatic letter fitting in typesetting. Based on the computer program's ability to evaluate the aesthetic space needed between individual letter pairs by assigning numerical values to each point or sector of that space.

nameplate The title of a publication, usually appearing in large type at the top of the front cover.

negative (minus) leading Reduction of white space between typeset lines beyond that obtained with solid setting.

negative space Space surrounding or framing the elements of a design; sometimes called white space.

network An integrated electronic system capable of sharing information easily between its subgroups or clusters.

nonimpact printing Any imaging method which requires no plate or striking mechanism.

notched halftone A picture with a section cut from one edge to allow type or other images to overlap into its area.

novelty Any type of unique or novel design. A member of the Specialized Group in the Four-Group Classification System for type.

OCR Optical Character Recognition.

offset printing The most common method of printing today. Often still called lithography, it utilizes the planographic principle of printing from a flat plate which receives both ink and water. Unlike true lithography, however, the image is transferred first to a blanket, then to paper.

old style roman Roman type which has minimal contrast in stroke thickness, bracketed serifs, and pronounced backward stress in round letter strokes.

on-line Devices, typefaces, or software directly accessible to a system.

pagination Electronic generation of multiple pages of a publication or document through the use of highly sophisticated software programs.

paper page area The finished size of one page after it has been printed and trimmed.

papyrus A writing substrate developed in ancient Egypt, made from overlapping strips of the papyrus plant.

perception The process by which we become aware of information through the senses.

perfect binding A type of adhesive binding which requires that the pages of signatures first be separated, then clamped to receive the adhesive. Usually, a cover is added.

perfecting Printing on both sides of a web or sheet with one pass through the press.

peripheral An external device connected to a computer.

photoengraving A photographic method of creating a relief printing plate by exposing images onto a flat metal surface containing a light sensitive coating, then etching the surface with acid.

photographic quality variables In the context of this book, the technical characteristics of grain, gradation, and sharpness of image.

photo-optic typesetter A device which images letters by passing a beam of light through a master film strip or disk.

pica A 12-point increment of space usually used to designate the width of columns or the overall dimensions of elements on a page. A pica is one-sixth of an inch.

pigment The coloring matter used in printing inks and paints.

pixel A tiny electronic image signal or picture element on a CRT screen.

planography The principle of printing from a flat surface through chemical means.

platen press Based on early press designs, this form of letterpress employs a flat bed for holding type and a flat platen on which the paper is carried to meet the type. In one press design both bed and platen move together vertically during printing. In the other, only the platen moves.

point Smallest increment of the point system, used to measure the height of type and the thickness of leading and other elements. A point is .0138″.

posterization The conversion of a continuous tone image into three or more basic tones or colors for dramatic effect.

prepress proof Any color proof, either overlays or laminated sheets, which shows the finished color images prior to printing.

primary colors The three pigment colors—red, yellow, and blue—from which all other colors can be produced, at least in theory.

process camera A camera designed to reproduce line and halftone images on film or paper, while at the same time reducing or enlarging the images as needed.

product module quotation A printing quotation which places a price on each component of the job so that a customer can easily evaluate how changes would affect cost.

progressive proofs Printed color proofs showing each individual color and its successive combination with other colors.

proportion wheel A two-piece plastic device used to scale and crop photographs and other images.

pulp A mixture of cellulose fibers, water, and other filler ingredients used in the manufacture of paper.

RAM Random Access Memory, the non-permanent computer memory accessible to a user for programs and files.

raster image processor A device which converts electronic digital data into signals needed by raster-based output devices such as intelligent copiers and laser typesetters.

readability A term used to describe how well body type can be read and comprehended.

relief printing Any printing method which utilizes a raised surface. Letterpress is but one of several relief printing methods.

representation What we see and recognize from our environment and experience.

reverse A white image (usually type) formed within a darker image area.

reverse leading The setting of multicolumn images through the actual backing up of the photosensitive paper in a typesetter or image setter or the simulation of such through software.

ROM Read Only Memory, the inaccessible permanent memory which contains the machine language and operating/assembly system of the computer.

rough A final idea for an advertisement or other design, usually done full size in pencil or marker.

round serif Non-roman types with rounded finishing strokes. A member of the Serif Group in the Four-Group Classification System for type.

roman type Type derived from the latin alphabet, employing both thick and thin strokes and serifs in its design. The term is also applied in proofreading to distinguish a straight standing type from a leaning (italic) one.

rotary press A type of letterpress which prints by impressing paper between curved plates and curved impression cylinders at high speed. Most such presses are web-fed, printing from large rolls of paper.

rotogravure Gravure printing through the use of rotary presses as opposed to sheet-fed presses.

run-around Type set to fit around a rectangular image area.

sans serif group A major division of the Four-Group Classification System. It includes those types which do not have finishing strokes known as serifs.

scoring The creasing of heavy paperboards or covers to allow them to fold without cracking.

screen printing One of the major printing processes, it uses the stencil principle whereby ink is passed through the surface of the image carrier to the substrate to be printed below.

screen ruling A designation of lines or dots per inch present in a given halftone image.

script type Type resembling handwriting, with flowing, connected strokes that lean to the right. A member of the Hand-Formed Group in the Four-Group Classification System.

seal A geometrically-shaped symbol, utilizing full words in the design, not initials.

secondary color Colors which appear between the

primary colors on the color wheel: orange, green, and violet.

segmentation   The extraction and use of several small sections of an image, often to provide continuity between several pages of a related article.

serif group   A major division of the Four-Group Classification System, encompassing all type designs with finishing strokes called serifs.

service mark   A name, symbol, or other visual device used to identify a service-related entity.

setoff   The accidental transfer of an image from the top surface of a printed sheet to the bottom surface of the sheet above.

shallow relief   Printing from photopolymer plastic plates by letterpress.

sharp serif type   Non-roman type with small, sharp finishing strokes. A member of the Serif Group in the Four-Group Classification System.

sheet enhancement   Any operation such as embossing, foiling, or die-cutting, which adds to the appearance and impact of printing.

soft copy previewer   A display device which allows the operator to see an electronic proof of a page, usually in exact size and type style, prior to its being set.

specialized group   A major division of the Four-Group Classification System. It includes types which are novel or decorative in design.

square serif type   Non-roman type with squared, blunted, or slab finishing strokes. A member of the Serif Group of the Four-Group Classification System.

standing head   A repeated element such as a section heading, index, or personal column title in a publication.

stereotype   A letterpress printing plate made by impressing type or another relief plate into dampened cardboard-like material, then pouring hot metal into the resulting mold.

subgroups   Important subdivisions of type groups, containing members with common design characteristics not found in other members of the main group.

subtractive primary   A combination of additive light primaries.

surprint   A dark image (usually type) printed over a lighter image area.

symbolism   A system of symbols representing designated objects or meanings.

text type   Type resembling early hand lettering forms made from wide-tipped writing instruments which create broad main strokes and thin connecting lines. A

member of the Hand-Formed Group of the Four-Group Classification System. The term is also used to describe body type or reading matter of under 14-point size.

texture   The apparent density or visual pattern produced by a mass of type.

thermography   Printing with ink that contains rosin, then heating the image to create a raised effect.

thumbnail   A miniature sketch of a layout.

tracking   A term used to describe the overall adjustment of space unit values between all letter combinations in typesetting.

trademark   A name, symbol, or other visual device used to identify a product or distinguish it from other products.

transitional roman type   Roman type designs having moderate contrast in stroke thickness, bracket serifs, and moderate backward stress in round letter strokes. A member of the Serif Group in the Four-Group Classification System.

transparency   A positive color image on a clear film base (a color slide) used in making color-separation films for full-color printing.

type page area   The area taken up by images on the page. It excludes the margin areas on all sides.

typography   The study and design of typographic images and messages that are both functional and attractive.

unit system   A method of designating the width of individual letters and the space between them by assigning each a value based on fractions of the em.

value   The relationship of a color to black or white.

vehicle   The substance in printing ink or paint in which the coloring pigment and other additives are suspended.

velox   A positive print—either line or halftone—made by the contact exposure of a film negative to photographic paper. Development and fixing is similar to that of orthochromatic films.

vignette   An image which fades out gradually at the edges.

visual syntax   The purposive sequencing of important design elements on a page or facing pages to enhance understanding of the subject matter.

window   A small area on a page spread, usually set off by rules, where a quotation or other important part of a story is highlighted.

xerography   Reproduction of images through the use of dry ink (toner) fused to paper from an electrically charged selenium drum.

Selected Bibliography

ADAMS, J. MICHAEL and DAVID D. FAUX, *Printing Technology: A Medium of Visual Communications.* North Scituate, Mass.: Duxbury Press, 1977.

ARNOLD, EDMUND C., *Modern Newspaper Design.* New York: Harper & Row, Publishers, 1969.

ARNOLD, EDMUND C., *Ink on Paper 2: A Handbook of the Graphic Arts.* New York: Harper & Row, Publishers, 1972.

BERRYMAN, GREGG, *Notes on Graphic Design and Visual Communication.* Los Altos, Calif.: William Kaufman, Inc., 1979.

BUREAU, WILLIAM H., *What the Printer Should Know about Paper.* Pittsburgh: Graphic Arts Technical Foundation, 1982.

CLICK, J. W. and RUSSELL N. BAIRD, *Magazine Editing and Production*, 3rd ed. Dubuque, Iowa: Wm. C. Brown Company Publishers, 1983.

CRAIG, JAMES, *Designing with Type.* New York: Watson-Guptill, 1971.

CRAIG, JAMES, *Production for the Graphic Designer.* New York: Watson-Guptill, 1974.

DONDIS, DONIS A., *A Primer of Visual Literacy.* Cambridge, Mass.: The MIT Press, 1973.

DURBIN, HAROLD C., *Printing and Computer Terminology.* Easton, Pa.: Durbin Associates, 1980.

GARCIA, MARIO R., *Contemporary Newspaper Design: A Structural Approach.* Englewood Cliffs, N.J.: Prentice-Hall, Inc., 1981.

GOTTSCHALL, EDWARD M., *Graphic Communication '80s.* Englewood Cliffs, N.J.: Prentice-Hall, Inc., 1981.

GRAHAM, WALTER B., *Complete Guide to Pasteup.* Philadelphia: North American Publishing Company, 1979.

HURLBURT, ALLEN, *The Grid: A Modular System for the Design and Production of Newspapers, Magazines, and Books.* New York: Van Nostrand Reinhold Company, 1978.

LIEBERMAN, J. BEN, *Types of Typefaces and How to Recognize Them.* New York: Sterling Publishing Co., Inc., 1967.

MINTZ, PATRICIA BARNES, *Dictionary of Graphic Arts Terms.* New York: Van Nostrand Reinhold Company, 1981.

NELSON, ROY PAUL, *Publication Design, 3rd ed.* Dubuque, Iowa: Wm. C. Brown Company Publishers, 1983.

Pocket Pal: A Graphic Arts Production Handbook, 13th ed. New York: International Paper Company, 1984.

Printing Ink Handbook, 4th ed. Harrison, N. Y.: National Association of Printing Ink Manufacturers, Inc., 1980.

SALTMAN, DAVID, *Paper Basics: Forestry, Manufacture, Selection, Purchasing, Mathematics and Metrics, Recycling.* New York: Van Nostrand Reinhold Company, 1978.

SCHLEMMER, RICHARD M., *Handbook of Advertising Art Production*, 3rd ed. Englewood Cliffs, N.J.: Prentice-Hall, Inc., 1984.

SCHRAMM, WILBUR and DONALD F. ROBERTS, eds., *The Process and Effects of Mass Communication.* Urbana: University of Illinois Press, 1971.

SELAME, ELINOR and JOE SELAME, *Developing a Corporate Identity: How to Stand Out in the Crowd.* New York: Chain Store Age Books, 1975.

SILVER, GERALD A., *Professional Printing Estimating.* Philadelphia: North American Publishing Company, 1975.

STETTINIUS, WALLACE, *Management Planning and Control: The Printer's Path to Profitability.* Arlington, Va.: Graphic Communications Center, 1976.

STEVENSON, GEORGE A., *Graphic Arts Encyclopedia*, 2nd ed. New York: McGraw-Hill Book Company, 1979.

STRAUSS, VICTOR, *The Printing Industry: An Introduction to its Many Branches, Processes and Products.* New York: R. R. Bowker Company, 1967.

TOUSSAINT, J. A., ed., *Highlights of Printing History.* Washington, D.C.: U.S. Government Printing Office, 1941.

TURNBULL, ARTHUR T. and RUSSELL N. BAIRD, *The Graphics of Communication*, 4th ed. New York: Holt, Reinhart and Winston, 1980.

Index